Environmental Science
Fundamentals, Ethics and Laws

Environmental Science
Fundamentals, Ethics and Laws

Ashish Shukla
Senior Researcher
Dept. of Civil Engg., Architecture Building
Coventry University, UK

Renu Singh
Scientist
Centre for Environment Science and Climate Resilient Agriculture
Indian Agricultural Research Institute, New Delhi

Anil Kumar
Assistant Professor
Department of Energy
Maulana Azad National Institute of Technology
Bhopal, India

I.K. International Publishing House Pvt. Ltd.

NEW DELHI • BANGALORE

Published by

I.K. International Publishing House Pvt. Ltd.
S-25, Green Park Extension
Uphaar Cinema Market
New Delhi–110 016 (India)
E-mail:info@ikinternational.com

ISBN 978-93-82332-84-8

Published by Krishan Makhijani for I.K. International Publishing House Pvt. Ltd., S-25, Green Park Extension, Uphaar Cinema Market, New Delhi–110 016 and Printed by Rekha Printers Pvt. Ltd., Okhla Industrial Area, Phase II, New Delhi–110 020.

**Dedicated to
Teachers, Parents
and Family**

Perface

Environment is the sum total of all surroundings or conditions in which an organism lives including abiotic (soil, water, air) and biotic components (plants, animals, microorganisms). It is the complex set of physical, geographical, biological, social, cultural and political conditions that surrounds an individual and that ultimately affects the organism. All the living organisms depend on each other and on environment for their needs. The food we eat, the air we breathe, the water we drink, the clothes we wear and the house in which we live, all come from the environment. Environment fulfils all the necessities of the human beings, but it cannot fulfil the greed of the human beings. Human beings are deteriorating the environment day by day through rapid industrialization and urbanization. Large-scale deforestation, enormous increase in number of industries, factories, transport vehicles, huge amount of plastics, garbage, electronics and other wastes, reckless application of fertilizers, insecticides lead to the air, water and soil pollution. Today, the world is facing the major environmental problem that is climate change which leads to the melting of glaciers, rise in sea level and change in weather pattern. All these conditions bring out the increase in number of cyclones, floods and droughts. Nobody can survive without the environment; the time has come to think about the sustainable development and to conserve our environment.

The present book has been framed with the intention of providing consolidated information on each and every aspect of the environment. The book comprises 12 chapters opening with the introduction and importance of environmental sciences, then physics of natural resources, structure and functions of ecosystem, biodiversity and conservation, environmental pollution, environmental impact assessment, human population and environment, waste pollution and management, environmental laws, social issues, global warming and ends with the carbon credits. The objective of this compendium is to create awareness among the scientists, faculties, researchers and students about each and every component of environment, its importance and to educate them about the environmental laws and other social issues. The book provides in-depth information on types of natural resources and associated problems, structure and functions of ecosystem, biogeographical classification of India, sources and effects of different kinds of pollution, methodology of environmental impact assessment, environmental and human health, hazardous waste management and treatment technologies, carbon sequestration, global warming and difference between carbon credits and carbon taxes.

We have consulted various numbers of books, journals, and websites during the preparing the text and also a large number of works were discussed. Hence, we can not declare very much distinctiveness. Many sorts of errors must have crept in. As authors we bear responsibility for all interpretations, opinions and errors in this work. We would thankfully welcome further suggestions and comments for the improvement of the book. We hope this compendium will be a landmark for the people who are concerned about the environment and meet the need scope of the readers.

It is a pleasure for us to convey our sincere gratitude to, Chairman Prof. G. K. Mehta and Director Dr. Appu Kuttan K.K., Maulana Azad National Institute of Technology, Bhopal; Dr. Tim Davis, Head, Civil Engineering, Architecture and Building, Coventry University, Coventry (UK); and Dr. Shiv Dhar Singh, Head, Centre for Environment Science and Climate Resilient Agriculture, Indian Agricultural Research Institute, New Delhi, for their constant and consistent encouragement.

Authors are highly grateful to Padmashree Prof. M.S. Sodha (FNA), Prof. G. N. Tiwari, CES, IIT Delhi, Prof. Avinash Chandra, CES, IIT Delhi, Dr. Shuli Liu, Reader, Civil Engineering, Architecture and Building, Coventry University Coventry UK, Prof. S. N. Varma, Ex. Vice Chancellor, RGPV, Bhopal, Prof. R. K. Khitoliya, HOD, Dept. of Civil Engineering, Punjab Engineering College, Chandigarh, Prof. Samsher, Dept. of Mech. Engg., Delhi Technological University, Delhi, and Dr. Om Prakash Chaursia, Professor, NIT Patna for their significance and utmost moral support. We are also highly indebted to our colleagues and students.

We deeply acknowledge the patience, co-operation and support of Ms. Poonam Shukla (w/o wife of Dr. Ashish Shukla), Director, World Energy and Environment Technology Ltd (WEENTech) UK.

My (Dr. Anil Kumar) wife Ms. Abhilasha, son Tijil Kumar and daughter Idika Kumar have encouraged me throughout this long and time-consuming project which took many hours away from them. Without their continuing support it would never been completed. I will always grateful to them for their patience.

Last but not the least, I express my deep gratitude to our respected parents for their blessings.

Authors

About Authors

Dr. Ashish Shukla is a senior researcher in the Department Civil engineering, Architecture and Building, Coventry University. He has more than seven years of research experience in the field of building engineering physics, energy management, energy generation within building envelopes, environmental physics and solar thermal engineering. Dr. Shukla has worked on various projects funded by the EPSRC, TSB, Welsh Assembly Government, Low Carbon Research Institute and Eon. He has expertise in thermal/mathematical modelling of energy systems, product design and development and computer simulations. He is involved in design and development of renewable energy applications and environmental problems. He has published research papers in journals and at international conferences of world repute. He is also a reviewer for several journals in the field of sustainable energy.

Dr. Renu Singh is a scientist, Centre for Environment Science and Climate Resilient Agriculture, Indian Agricultural Research Institute, New Delhi. She holds master's degree in Environmental Sciences and M.Tech. in Environmental Sciences and Engineering and Ph.D. degree in Environmental Impact Assessment from Indian Institute of Technology Delhi. She has more than ten years of research experience in the field of Environmental Impact Assessment, Biofuels, Ecology and Environmental Microbiology. Dr. Singh has worked on various projects related to Climate change and biofuels funded by different agencies. She also has the exposure to remote sensing and dispersion modeling. She has published research papers in journals and at conferences in the field of environmental Sciences.

Dr. Anil Kumar is Assistant Professor in Department of Energy, Maulana Azad National Institute of Technology (MANIT), Bhopal, India. Before joining MANIT, Bhopal, he was Lecturer in the Department of Mechanical Engineering in the University Institute of Technology, Rajiv Gandhi Proudyogiki Vishwavidyalaya, Bhopal (MP State Technological University). He has more than ten years of research and teaching experience in higher technical education.

He holds bachelor's degree in Mechanical Engineering followed by M.Tech. in Energy Technology and Ph.D. degree in solar thermal technologies from Indian Institute of Technology Delhi. Dr Kumar's main area of interest is solar thermal technology, distribution of energy generation, clean energy technologies and application in buildings. He has published several books like, 'Energy Environment Ecology and Society', 'Fundamentals of Mechanical Engineering' and 'Advance Internal Combustion Engine'. Dr. Kumar has published more than 45 research articles in journals and at international conferences. Presently, Dr. Kumar is reviewer to seven national and international journals. In his tenure at MANIT, he has supervised ten M.Tech. students and is supervising six Ph.D. students. Dr. Kumar has carried various experimental projects in the area of solar water heating, solar drying for different crops, active and passive greenhouse dryer, solar collector assisted greenhouse dryers, and ground heat exchangers.

Contents

Perface *vii*
About Authors *ix*

1. Introduction **1**
1.1 Environment 1
1.2 Scope and Importance 2
1.3 Importance of Environmental Study 2
1.4 Need for Public Awareness 3

2. Physics of Natural Resources **5**
2.1 Introduction 5
2.2 Forest Resources 6
2.3 Over-exploitation of Forests 7
2.4 Deforestation 7
2.5 Water Resources 8
2.6 Floods 13
2.7 Drought 13
2.8 Mineral Resources 14
2.9 Nitrogen Cycle 14
2.10 The Importance of the Carbon Cycle 15
2.11 Food Resources 20
2.12 Energy Resources 26

3. Structure and Functions of Ecosystem **36**
3.1 Introduction 36
3.2 Classification of Ecosystem 36
3.3 Approach to Ecosystem 37
3.4 Food Chain 38
3.5 Food Web 41
3.6 Ecological Pyramids 41
3.7 Forest Ecosystem 42
3.8 Grassland Ecosystem 45

3.9 Aquatic Ecosystem 46
3.10 The Synergistic Threats to Stream Ecosystems 52

4. Ecosystem Biodiversity and Conservation **56**
4.1 Introduction 56
4.2 Level of Biodiversity 56
4.3 Biogeographical Classification of India 60
4.4 Value of Biodiversity 60
4.5 Biodiversity Around the World and India 63
4.6 Classification of Species 66
4.7 Threats to Biodiversity 67
4.8 Conservation of Biodiversity 68
4.9 Case Study 71

5. Environmental Pollution **73**
5.1 Introduction 73
5.2 Water Pollution 73
5.3 Sources of Water Pollution 74
5.4 Impact of Domestic Waste on Water Pollution 75
5.5 Impact of Industrial Activities on Water Pollution 76
5.6 Impact of Agriculture on Water Pollution 76
5.7 Types of Water Pollution (Fresh Water Mainly) 76
5.8 Measuring Oxygen Content 77
5.9 Plant Nutrients and Eutrophication 78
5.10 Other Sources of Water Pollution 78
5.11 Land Pollution 81
5.12 Pollution of Land and Water 84
5.13 Pollution in Soil 84
5.14 Control of Soil Pollution 90
5.15 Natural Land Pollution 91
5.16 Marine Pollution 91
5.17 Noise Pollution 97
5.18 Framework Standards for Noise Measurement at Machines 101

6. Environment Impact Assessment **104**
6.1 Introduction 104
6.2 Basic Principles of EIA 108
6.3 Methodologies for Predicting Impact in an EIA 113
6.4 Environment Impact Evaluation Methods 114
6.5 Benefits of EIA 118

7. Human Population and Environment **120**
7.1 Population Growth, Variation Among Nations 120
7.2 Population Explosion – Family Welfare Programme 121
7.3 Environment and Human Health 126
7.4 Human Rights 137
7.5 Value Education 140
7.6 HIV/AIDS 147
7.7 Woman and Child Welfare 148
7.8 Role of Information Technology in Environment and Human Health 151

8. Waste Pollution and Management **152**
8.1 Introduction 152
8.2 Solid Waste Management: causes and Control Measures 152
8.3 Effects and Control Measures of Urban and Industrial Wastes 154
8.4 Integrated Waste Management 155
8.5 Technologies 160
8.6 Waste Management Concepts 160
8.7 Energy from Waste 162
8.8 Introduction to the Resource Conservation and Recovery Act (RCRA) 165
8.9 Guidance on the Use of Section 7003 of RCRA 171
8.10 Composting 171
8.11 Hazardous Waste Treatment 173
8.12 Conclusions and Role of Individuals in Prevention of Pollution 183

9. Environment Laws **185**
9.1 Introduction 185
9.2 Development Phases and Trends in Environmental Law 187
9.3 Environment (Protection) Act, 1986 194
9.4 The Water (Prevention & Control of Pollution) Rules, 1975 202
9.5 Ministry of Environment and Forests 211
9.6 Ministry of Environment and Forests the Municipal Solid Waste (Management and Handling) Rules, 2000 218

10. Unsustainable to Sustainable Development: Social Issues **231**
10.1 Introduction 231
10.2 Urban Problems Related to Energy 232
10.3 Water Conservation, Rainwater Harvesting and Watershed Management 233
10.4 Resettlement and Rehabilitation of People: Its Problems and Concerns 236
10.5 Environmental Ethics: Issues and Possible Solutions 237
10.6 The Ethical Basis of Environment Education and Awareness 241
10.7 Climate Change, Global Warming, Acid Rain, Ozone Layer Depletion, Nuclear Accidents and Holocaust 243

10.8 Wasteland Reclamation 248
10.9 Acts and Legislations 253
10.10 Issues Involved in Enforcement of Environmental Legislation 260
10.11 Role of Non-Governmental Organizations (NGOs) 265

11. Global Warming and Carbon Sequestration **267**
11.1 Introduction 267
11.2 Contributors of Greenhouse Gas 269
11.3 Impact of Greenhouse Gas 270
11.4 Global Warming 272
11.5 Climate Change 275
11.6 Carbon Sequestration 290

12. Financial Promotion for Sustainable Development: Carbon Credits **300**
12.1 Introduction 300
12.2 Compliance v Voluntary Markets 302
12.3 Carbon Tax 305

Appendix: The Batteries (Management and Handling) Rules, 2001* **308**

Index **315**

Chapter 1

Introduction

1.1 ENVIRONMENT

The physical, chemical and biological presence of living and non-living things outside an individual species is called its *environment*. Each and every living organism (whether it belongs to the higher or the lowest life form) has its own environment.

According to ISO 14001, *environment* can be defined as, "surroundings in which an organisation operates, including air, water, land, natural resources, flora, fauna, human and interrelations".

The environment serves as

(i) Source of all raw materials (like oxygen, carbon, hydrogen, phosphorus, calcium, potassium, iron, etc.).

(ii) Source of all energy.

(iii) Receiving and reprocessing plant for all wastes.

(iv) Buffer from harsh conditions.

The main categories of environment are biotic and abiotic environments. The abiotic environment can be further classified into atmosphere (air), lithosphere (soil), and hydrosphere (water). The biotic environment is called biosphere.

The biotic environment includes assembly of plants, animals, and microbes. The abiotic environment/component includes components such as physical factors and physiochemical substances (light, heat, wind, moisture, tide, solar energy, pH, soil, topography minerals, etc.). The biotic and abiotic components constantly affect each other and cannot be isolated from each other.

1.1.1 Components of Environment

The life supporting environment of the earth (called *biosphere*) is composed of the following three media:

(1) Air (atmosphere)

(2) Land (lithosphere)

(3) Water (hydrosphere)

Environmental Science, an interdisciplinary subject, comprises Earth Science, Life Science, and Social Science. In essence, environment science begins by understanding how ecosystem works.

1.2 SCOPE AND IMPORTANCE

Environment science discipline has multiple and multilevel benefits. This study is important and necessary not only for children but also for everyone. The benefits are summarized as follows:

(1) This study creates awareness among the people to know about renewable and non-renewable resources of the region. The endowment and potential, patterns of utilisation and balance of various resources available for the future use in the state or a country are analyzed in the study.

(2) It provides the knowledge about ecological systems, causes, effects and relationships between the components.

(3) It provides necessary information about biodiversity richness and the potential dangers to the species of the plants, animals and microorganisms in the environment.

(4) This study enables one to understand the causes and consequences of natural and man-induced disasters, pollution and measures to minimise their effects.

(5) It enables one to evaluate alternative responses to environmental issues before deciding an alternative course of action.

(6) This study enables environmentally literate citizens (By knowing the environmental acts, rights, legislations, etc.) to make appropriate judgement and decisions for the protection and improvement of the earth.

(7) This study exposes the problems of over pollution, health, hygiene, etc. and the role of arts, science, and technology in eliminating/ minimising the evils from the society.

(8) This study tries to identify and develop appropriate and indigenous eco-friendly skills and technologies for various environmental issues.

(9) It teaches the citizens about the need for sustainable utilization of resources, as these resources are inherited from our ancestors to the younger generation without deteriorating their quality.

(10) This study enables theoretical knowledge into practice and the multiple uses of environment.

1.3 IMPORTANCE OF ENVIRONMENTAL SCIENCES

Environmental science is based upon a comprehensive view of various environmental systems. It aims to make the citizens competent to do scientific work and to find their practical solutions to meet the current environmental problems. The citizen acquires the ability to analyse the environmental parameters like the aquatic, terrestrial and atmospheric systems and their interactions with the biosphere and atmosphere. Some of the imperative aspects of environmental sciences are summarised below:

(1) World population is increasing at an alarming rate especially in developing countries.

(2) The natural resource endowment on the earth is limited.

(3) The methods and techniques of exploiting natural resources are advanced.

(4) The resources are over-exploited and there is no foresight of leaving the resources to the future generation.

(5) The unplanned exploitation of natural resources leads to pollution of all types and at all levels.

(6) The pollution and degraded environment seriously affect the health of all living things on the earth, including man.
(7) The people should take a combined responsibility for the deteriorating environment and begin to take appropriate actions to save the earth.
(8) Education and training are needed to save the biodiversity and species extinction.
(9) The urban areas coupled with industries, are the major source of pollution.
(10) The number and area of the protected area should be increased so as to protect the wildlife.
(11) This study enables the people to understand the complexities of the environment and need for the people to adopt appropriate activities and pursue sustainable development, which are harmonic with the environment.
(12) This study motivates the students to get involved in the community action and to participate in various environmental and management projects.
(13) It is high time to reorient educational systems and curriculum towards these needs.
(14) Environmental studies take a multidisciplinary approach to the study of human interactions with the natural environment. It integrates different approaches of the humanities, social sciences, biological sciences and physical sciences and applies these approaches to investigate environmental concerns.
(15) Environmental study is a key instrument for bringing about the changes in the knowledge, values, behaviour and lifestyles required to achieve sustainability and stability within and among countries.
(16) Environmental studies describe the interrelationship among organisms, the environment and all factors, which influence life on the earth, including atmospheric conditions, food chains, the water cycle, etc. It is a basic science which is important for one and all.

1.4 NEED FOR PUBLIC AWARENESS

Besides natural and manmade hazards, the earth's surface is polluted by a lagre number of sources. The land is polluted by millions of tonnes of solid wastes disposed of from mines and urban centres while industries and cities discharge enormous quantity of liquid wastes. In the same way, air is polluted by burning of fossil fuels and wastes. As a consequence of all these disasters and pollutions, the environment is overburdened. The level of burden varies between countries and time to time.

Agriculture, the life line of Indian economy, is threatened by the application of enormous quantity of chemical fertilizers and pesticides to crops. This has not only polluted the land and waste resources but also the food grains, vegetables and other food products. The long term effects to all living organisms could threaten the life on the earth in the future. The natural and man-made forest fires, deforestation, pollution, etc., lead to extinction of sensitive species and loss of biodiversity in different environments. In industrial sector the pollution generated in several industries is irreparable in nature. These lead to global warming, acid rain, stratospheric ozone layer depletion, snow melt, rise in sea level and the consequent impacts to various components of the environment. The low lying countries like Maldives (average elevation above sea level is one metre), the Netherlands (one third of land area is already below the sea level) and a number of other countries are worried and concerned about their future. The individuals and industries, responsible for depleting and deteriorating the resources of environment, should realise and take appropriate measures to minimise the sources of pollution initially and subsequently eliminate them for the sustenance of life on the earth. The urgent need of more environmental awareness is utmost important for well-being of our ecosystems and its inhabitants as well. There can be many ways to address the issue using diverse approaches for improving public awareness towards environmental issues:

1. Individuals of schools, colleges, industries, service centres, village, urban centres, etc., should realise the importance of day-to-day environmental issues.
2. The individuals should practise environmental conservation principles and create awareness among family members.
3. The individuals could expose the problems by writing in the newspapers/ discuss about the environmental evils in forums and make the people aware of the same.
4. There is a need to meet the people and discuss again and again, so that the problem is alive till it is eliminated.
5. The governments (central and state) have instituted awards for individuals and organisations for their services to environmental management. The villages, panchayat unions, taluks, districts and other units can be made aware of such units.
6. The non-governmental organisations, in India and abroad, are doing tremendous efforts in conserving the environment.
7. Several universities and colleges have started environmental sciences/environmental studies departments. They conduct courses and research works related to local and regional environmental issues.
8. Several government/aided institutions conduct and support environmental researches, organise seminars, conferences, etc.

Learning Outcomes

Major learning contribution from this chapter is about Definition, Scope & Importance, Environment, Biosphere, Ecosystem, Need for public awareness, Effects of human activities on environment, Basics of environmental impact assessment, Sustainable Development.

Exercise

1. Write a 100 word essay about your views on growing importance of environmental sciences.

Chapter 2

Physics of Natural Resources

2.1 INTRODUCTION

The word resources means "a source of supply or support generally held in reserve". By resource, one can refer to any physical entity which has limited availability. A natural resource is the stock that can be drawn from nature, i.e., air, water, minerals, land, vegetation, animals, solar energy and raw materials, for supporting life. For man, resources are those materials which are needed for his survival and prosperity. The nature of resources varies from society to society, depending upon culture, level of development and the nature of work of that particular society.

Natural resources can be classified as:

(1) **Renewable resources:** The resources that have the inherent capacity to reappear, or replenish themselves by quick recycling, reproduction and replacement within a reasonable time and maintain themselves, e.g., soil, water and living organisms.

(2) **Non-renewable resources:** The resources that lack the ability for recycling and replacement, e.g., fossil fuels like coal, petroleum and minerals.

In general, people are not aware that everything we use in our day-to-day life is derived from natural resources. For example, fat which comes from vegetation and animal source, salt and spices which comes from minerals and herb plants, milk and cheese from animals and vegetation, bicycle made from iron ore and chemical by-products, etc. Also by human intervention one natural resource can be used in several forms, e.g., wood coming from tree can be used to make house, or can be used for cooking, paper can also be made from wood. One can say man utilises these resources in various ways by processing them to meet his needs. Human interaction with natural re sources made environmental science a very imperative subject for the present generation. While utilising the resources we are affecting the nature some way or other in order to maximise the benefits for the living being.

The table below shows an example in what way natural resources have been utilised:

Table 2.1 Natural and man-made resources.

Natural resources	*Products made by humans*
Water reservoirs, rivers	Electric power
Solar energy	Electric and thermal power
Vegetables	Chips, crisps
Forest	Furniture
Wind	Electrical power

Natural resources are very important for the living beings on the earth. They not only make a country powerful and prosperous but also come up with responsibility to use and manage them sustainably.

2.2 FOREST RESOURCES

Forest is one of the most important renewable natural resources. Owing to its richness in a variety of resources, forests have been rampantly exploited to feed the needs of the growing population. Such unsustainable ruthless exploitation of the forests has led to a variety of environmental problems — many of which potentially threaten the existence of several living species including human beings. However, in recent years there has been growing concern among policy makers and scientific community on the need to save these forests from over-exploitation. Figure 2.1 shows vegetation on the earth and in the following section a detailed account of the importance of forests, deforestation, its causes and consequences with a few important case studies will be discussed.

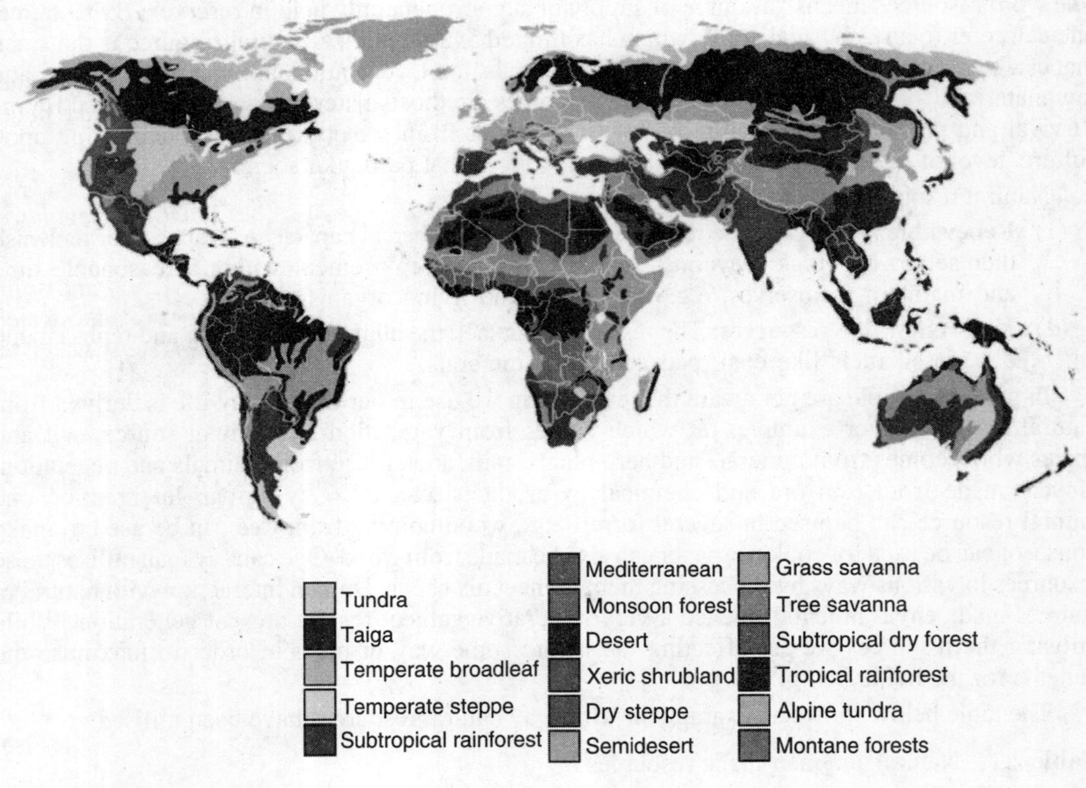

Fig. 2.1 Forest cover in the world.
Source: (http://en.wikipedia.org/wiki/Vegetation).

2.2.1 Uses

The chief economic product of forests is *timber*, which is primarily used as building materials such as lumber, plywood, hardboard, etc.

Provides raw material for a variety of industries: provides pulp for paper industry, ingredients for pharmaceutical industries, etc.

Provides fuelwood: fuelwood is the major source of household energy in the developing world. Worldwide about half the timber cut each year is used as fuel for heating and cooking, especially in less developed countries.

Provides employment: to millions of people, the world trade in forest produce is estimated to be 100 billion U.S dollars annually.

Serves as habitat: for a variety of species, which live on the edge of the forests and in the adjoining ecosystems, deep and continuous forests are important to other species, which require large habitats without human interference.

Serves as an important biodiversity: reserve and as a gene reserve of a variety of wild species. This biodiversity is a very vital source for evolving new economic varieties in agriculture, horticulture, veterinary and medical fields.

Moderates greenhouse effects: by absorbing atmospheric CO_2 they act as big sinks for CO_2 and moderate greenhouse effect.

Regulates stream flow: vast expansion of forests and other vegetation slows the runoff of water and allows water to percolate into the soil thus helping and regulating the stream flow. Further the plant and leaf debris on the forest floor slows water as it runs along the ground. This reduces erosion by allowing water to soak into the soil, rather than runoff.

Regulates earth's temperature regimes and increases rainfall occurrences, balances/regulates CO_2*,* O_2*, nutrient and water cycles.*

Increases the water holding capacity of the soil: because of the thick layer, loose soil and soil-retaining capacity of the tree roots, forests are vitally important for preserving adequate water supplies. Tree roots penetrate compacted soils and increase soil porosity. This allows water to percolate into the soil and increase the water holding capacity of the soil.

Checks soil erosion/silting/landslides/floods.

Also have aesthetic and touristic values.

2.3 OVER-EXPLOITATION OF FORESTS

Forests being the storehouse of a variety of resources, which directly benefit the human society, are being plundered for short-term gains. Though the exploitation of forest resources has been happening since earliest times, the scale of exploitation was very much limited to local areas and so the environmental problems were not so widespread. But in the last few decades, exploitation at any cost attitude which was prevalent in most of the countries has led to the over-exploitation of the forests. Deforestation problem is not only prevailing in the developing countries but also increasing due to increase in population. The consequence of such over-exploitation has resulted in a variety of environmental and social problems— a few of which are global in nature, which threaten the living environment.

2.4 DEFORESTATION

Deforestation refers to the removal of plants and trees in the forests. According to the Food and Agriculture Organisation (FAO, 1997) of the United Nations, the annual rates of deforestation in the developing countries is about 15.5 million hectares for the period 1980-1990 and 13.7 million hectares for 1990-1995. This clearly shows the increasing trend of the deforestation. It has also been estimated that the total forest area lost during the 15-year period was approximately 200 million hectares. To put this figure inperspective, 200 million hectares is more than the total area of Mexico or Indonesia.

2.4.1 Causes of Deforestation

1. Slash and burn farming
2. Commercial agriculture
3. Cattle ranching and livestock grazing
4. Mining and petroleum exploitation
5. Infrastructure development
6. Fuelwood collection
7. Charcoal kilns

2.4.2 Consequences of Deforestation

1. Economic loss
2. Loss of biodiversity
3. Destructs the habitats of various species
4. Reduction in stream flow
5. Increases the rate of global warming
6. Disruption of weather patterns
7. Degradation of soil and acceleration of the rate of soil erosion
8. Induces and accelerates mass movement/landslides
9. Increases flood frequency, magnitude/severity
10. Breaks the water cycle
11. Breaks the nutrients cycle
12. Loss of forests puts additional pressure on the pristine forests.
13. Increase in draught-hunger cycle.

2.5 WATER RESOURCES

Water is the most important natural source for the living beings on the earth. Specially, over increasing population is putting lots of pressure on the availability of potable water around the world especially in African continent and India. Around the world, human activities and natural forces are reducing available water resources. Only a tiny fraction of the planet's abundant water is available to us as fresh water. About 97% is found in the oceans and is too salty for drinking, irrigation and industry.

The remaining 3% is fresh water. About 2.997% of it is locked in the ice caps or glaciers or buried so deep that it costs too much to extract. Only about 0.0035% of earth's total volume of water is easily available to us as soil moisture, exploitable groundwater, water vapour, and lakes and streams.

Although awareness of the need to better manage and protect water has grown over the last decade, economic criteria and political considerations still tend to drive water policy at all levels. Science and best practices are rarely given adequate consideration. Pressure on water resources is increasing mainly as a result of human activities, namely, urbanisation, population growth, increased living standards, growing competition for water and increasing power demand from hydroelectricity. These are aggravated by climate change and variations in natural conditions. Few methods being developed by scientists around the world, e.g., rain water harvesting but still, in order to achieve water conservation and effective utilisation a long way is ahead. More and more, policy makers, social scientists, climate change experts and government officials are evaluating water quantity and quality together, and also coordinating water management efforts across borders.

2.5.1 Forms of Water

Water exists naturally in different forms and locations: in the air, on the surface, below the ground, and in the oceans. Fresh water accounts for only 2.5% of the earth's water, and most of it is frozen in glaciers and icecaps. The remaining unfrozen fresh water is mainly found as groundwater, with only a small fraction present above ground or in the air. Looking at how water moves through the earth's water cycle will help us to understand how water interacts with the environment and how much is available for human use.

Precipitation – rain, snow, dew, etc. – plays the key role in renewing water resources and in defining microclimate conditions and biodiversity. Depending on the local conditions precipitation may feed rivers and lakes, replenish groundwater or can return to the air by evaporation.

Glaciers store water as snow and ice, releasing varying amounts of water into local streams depending on the season. But many are shrinking as a result of climate change.

River basins are a useful "natural unit" for the management of water resources and many of them are shared by more than one country. The largest river basins include the Amazon and Congo Zaire basins. River flows can vary greatly from one season to the next and from one climatic region to another. Because lakes store large amounts of water, they can reduce seasonal differences in how much water flows in rivers and streams.

Wetlands – including swamps, bogs, marshes, and lagoons – cover 6% of the world's land surface and play a key role in local ecosystems and water resources. Many of them have been destroyed, but the remaining wetlands can still play an important role in preventing floods and promoting river flows.

Almost all of the fresh water is found below the surface of earth as groundwater. Generally of high quality, groundwater is withdrawn to supply drinking water and support farming in dry climates. The resource is considered renewable as long as groundwater is not withdrawn faster than nature can replenish it, but in many dry regions the groundwater does not renew itself or only very slowly. Few countries measure the quality of groundwater or the rate at which it is being exploited.

2.5.2 Indian Water Resources

In India, in an area of 3290 lakh hectares, 4000 billion cubic metres(bcm) of rainfall occurs annually. Out of the total, 41% is lost by evaporation, 40% is lost by runoff, 10% is retained by soil-moisture and remaining 9% seeps in for recharging groundwater. Of the 40% stream flow water or runoff water, 8% is used for irrigation, 2% for domesticuse, 4% for industry and 12% for electric generation. In available total water resource of 1869 bcm, the usable, water resources is

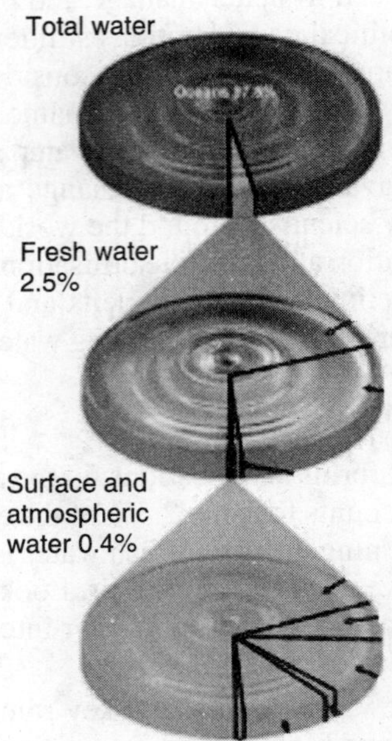

Fig. 2.2 Global distribution of the world`s water.

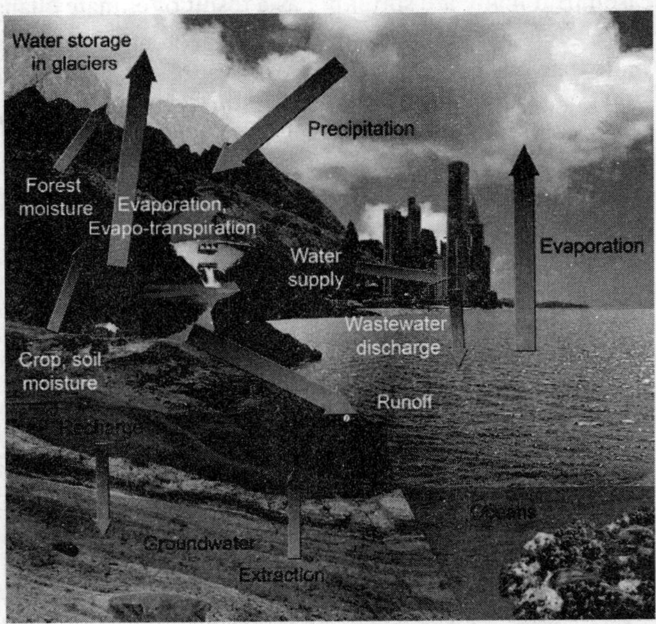

Fig. 2.3 Schematic of the hydrologic cycle components in the present-day setting.

only 1122 bcm, which consists surface water 690 bcm and ground water 432 bcm. If one considers the present per capita availability of water resources is 1122 cm, which makes India water stress country and if the water sources are used with the same rate by 2050 it is likely to be reduced to only 748 cm.

When the countries per capita water availability is less than 1700 cm it is considered as water stress country.

2.5.3 Classification of Water Related Diseases

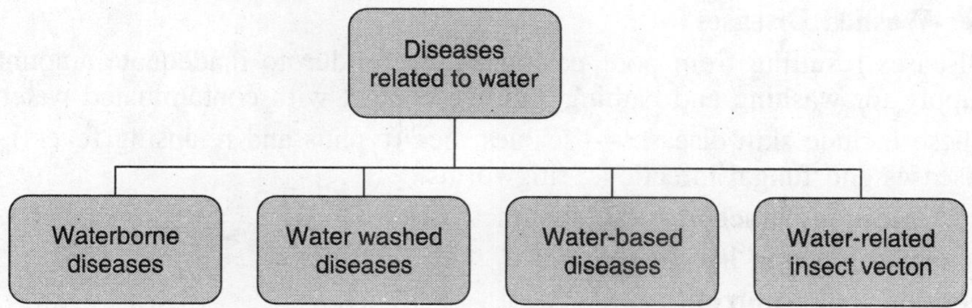

Fig. 2.4 Classification of water related diseases.

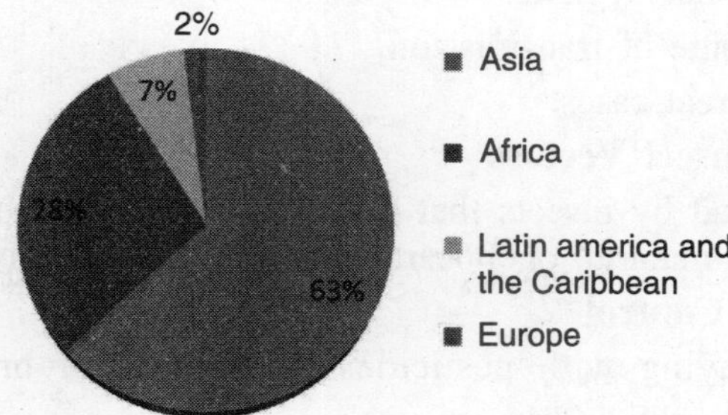

Fig. 2.5 Worldwide access to safe water.

2.5.4 Waterborne Diseases

Diseases caused by ingestion of water contaminated by human or animal excrement, which contain pathogenic microorganisms.

Includes cholera, typhoid, amoebic and bacillary dysentery, hepatitis A & E and other diarrhoeal diseases.

In addition, waterborne diseases can be caused by the pollution of water with chemicals that have an adverse effect on health.

(i) Other Diarrhoeal Diseases and their source:

- Giardiasis (protozoan)
- Cryptosporidiosis (bacteria)
- Camphylobacteriosis (bacteria)
- Viral Gastroenteritis (virus)
- Cyclosporiasis (parasite)

Prevention and Control

- Generally, through improvement of microbiological water quality, either through water treatment or source protection.
- Maintain safe water chain.
- Improvement in sanitation facilities.
- Promotion of hygiene practices.
- Identification and treatment of cases (disease surveillance).
- Treatment of wastewater/sewage.

(ii) Water-Washed Diseases

- Diseases resulting from poor personal hygiene due to inadequate amounts of water supply for washing and bathing and eye contact with contaminated water.
- These include skin diseases— scabies, lice (typhus and relapsing fevers), tick-borne diseases and fungal infections, ringworms.
- Eye infections: trachoma and conjunctivitis.
- Parasitic infection like jiggers.

Prevention and Control

- Adequate quantity of water supply.
- Closer, easier access to water
- Health education to improve personal and domestic hygiene
- Improve personal hygiene.
- Disrupt the route of transmission.
- Identify and treat cases.

(iii) Water-related Insect Vectors

- Diseases spread by insects that breed or feed near contaminated water or bite near water such as malaria, onchocerciasis, trypanasomiasis and yellow fever.

Prevention and Control

- Through spraying with pesticides destruction of breeding grounds (improving environmental sanitation).
- Construction of piped water supplies
- Reducing man-insect contact.
- Destroy breeding sites.
- Improving on housing conditions.
- Decrease the need for visiting infected areas.
- Identify and treat cases.

(iv) Water-based Diseases

- Are transmitted by hosts which live in water or require water as part of their life cycle, e.g., schistosomiasis and guinea worm.
- Transmission is more likely due to human activities like fishing, swimming, farming-rice.

Prevention and Control

- Prevent water contamination with excreta (water protection and safe disposal of excreta).

- Control direct contact of human beings with water sources.
- Prevent human contact with infected or suspicious water bodies.
- Control intermediate vector population.
- Provide protective wear.
- Identification and treatment of cases.

2.6 FLOODS

Floods occur most commonly when water from heavy rainfall, melting ice, or snow or a combination of these, exceeds the carry capacity of the receiving river system. During the floods the river carries fertile sediment and deposits it on the level land along its lower course. Such areas are called floodplains which are very fertile. Some rivers with extensive floodplains are: the Nile, Ganga, Brahmaputra, Yellow River, Mekong, and Godavari. Often, the floodplains become areas for human settlement. People are lured to them by the productive soils, reliable water supply, inexpensive means of transport and fishing potential. To reduce the flooding of the floodplains, flood banks are built along the river. Another way of controlling floods is through impounding river water by building upstream dams.

In India floods bring much havoc causing loss of life and property each year. Due to floods, the plains have become silted with mud and sand, thus affecting the cultivable land areas. Extinction of civilization in some coastal areas is mainly due to such natural calamities as flood. The national commission on floods has calculated that the land area prone to floods has doubled from 20 million hectares in 1971 to 40 million hectares in 1980. Flood damages cost the country INR 21 crore in 1951 which increased to INR 1130 crore in 1977. It was INR 128 crore per year during the decade 1960-1970, increasing to INR 739 crore per year during 1970-78. Assam, Bihar, Orissa, U.P, and West Bengal are worst suffering states in India. Through modern technology and scientific knowledge, there is need of a proper understanding of the ecosystem so that changes could be forecast well in time. Thus, management of rainfall and resultant runoff is very important. Such management can be best based on a natural unit called watershed. A watershed is an area bounded by the divide line of water flow. Thus it may be drainage basin or stream. The Himalayas are one of the most critical watersheds in the world. Our water regimes in the mountain ranges are threatened resulting in the depletion of water resources. The damage to reservoirs and irrigation systems and misuse of Himalayan slopes are increasing every year; which leads to increase in the costs for control measures during the flood season each year. The vast hydroelectric power potential can be harnessed from the Himalayan watershed only when proper control measures are taken. These include soil and land use survey, soil conservation in catchments of river valley projects and flood-prone rivers, afforestation/social forestry programmes, drought prone area development programmes and desert development and control of shifting cultivation.

2.7 DROUGHT

Drought is a devastating phenomenon. In comparison to fast onset disasters, drought destroys an area slowly, taking hold and tightening its grip with time. In severe cases, drought can last for many years, and can have devastating effects on agriculture, water supplies, local vegetation and microclimate.

In general, drought is defined as an extended period-a-season, a year, several years of deficient rainfall relative to the statistical multiyear average for the region. The lack of rainfall leads to inadequate water required by plants, animals, and human beings. A drought leads to other disasters, namely food insecurity, famine, malnutrition, epidemics and displacement of population from one area to another.

Rural communities can sometimes cope with one or two successive rain failures and crop or cattle losses: the situation becomes a crucial emergency when they have exhausted all their purchasing resources, food stocks and usual coping mechanisms.

These natural phenomena of drought, flood, etc., affect the different cycles of natural microclimate, e.g., nitrogen cycle, carbon cycle; which further make problem worse and havoc for civilisation. These different cycle are discussed in the next section.

2.8 MINERAL RESOURCES

A mineral is an element of an inorganic compound that occurs naturally and is solid. Mineral resources of the earth include metals and materials extracted from the soil. It usually has a crystalline internal structure made up of an orderly, three-dimensional arrangements of atoms or ions. All of the earth's crust, except the rather small portions composed of organic materials, is made of minerals. Mineral resources are limited and thus non-renewable.

The most frequently expressed concern is that high levels of consumption of minerals will lead to their depletion, thereby retarding industrial growth. However, estimates have suggested that the mineral reserves are adequate to support human development for the next few decades. Moreover, advanced scientific technology has helped in the discovery of newer reserves from time to time.

2.9 NITROGEN CYCLE

Nitrogen is essential to life on the earth. Nitrogen gas makes up 78 per cent of the air we breathe.

The nitrogen cycle is the continuous series of natural processes by which nitrogen passes from the air to the soil, to plants, and ultimately to sustain all animal life, and then returns back to the air or soil through decay or denitrification (the loss of nitrogen).

Nitrogen is a basic element of all living cells. Through nitrogen fixation, nitrogen gas is converted into inorganic nitrogen compounds to be used by plants, animals, and people. This conversion is accomplished through the nitrogen cycle, in which nitrogen as a gas is carried to the earth's surface in precipitation. Nitrogen is then used by plants and incorporated in their tissues as plant protein. Nitrogen then passes through the food chain to animals and humans.

Human intrusion in the nitrogen cycle can result in more or less nitrogen being cycled as part of the natural system. For example, cultivation of croplands, harvesting of crops, and cutting of forests has caused a decline in the natural occurring nitrogen in the soil. Soil nitrogen can be replenished with manufactured nitrogen fertilizer, animal manure and legumes. Nitrogen from over-fertilization of plant life, industrial discharges, and human and animal waste discharges can add too much nitrogen to the natural system and may have an impact on soil, water, and air quality.

2.9.1 Forms of Nitrogen

Nitrogen is present in the environment in many forms. The predominant form is nitrogen gas. Nitrate and ammonia are the two forms used by plants. Organic nitrogen is mineralized by microorganisms to create these two forms that are found in our soils and water. Microorganisms to create these two forms that are found in our soils and water.

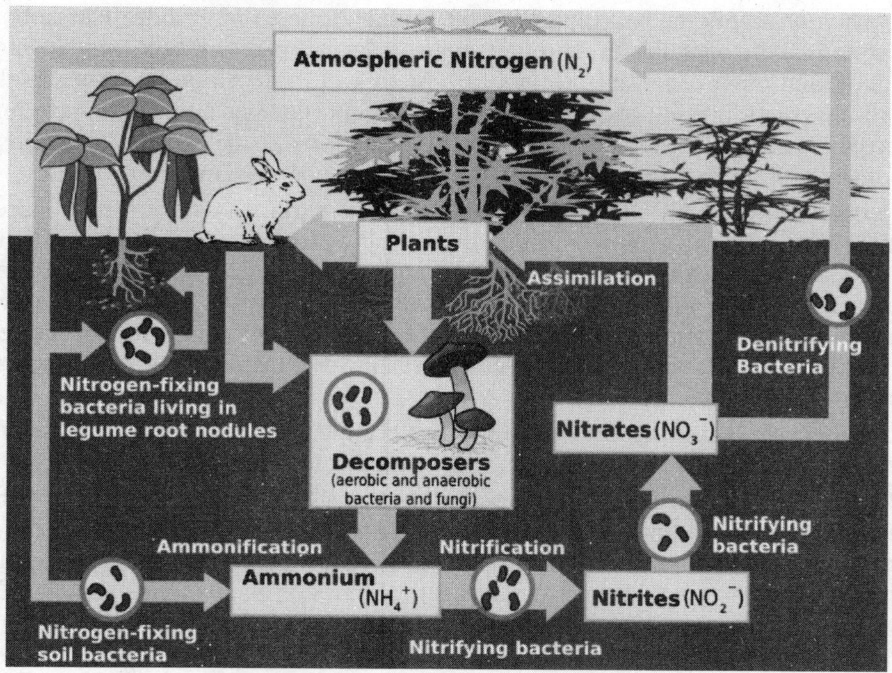

Fig. 2.6 Atmospheric nitrogen cycle.

Lifestyles and behaviours are the most important factors to quality of life (diet, exercise, etc). The environment is another key factor affecting quality of life. We need to make decisions and develop policies that balance economics, resource utilization, and quality of life.

We are exposed to nitrogen in the food we eat, the water we drink, and the air we breathe. Drinking water sources are susceptible to watershed run-off. Groundwater contamination from over-fertilization of plant life, human and animal waste discharges can occur. Excessive amounts of nitrogen as nitrate in drinking water may cause health problems. In infants under 6 months of age, nitrogen as nitrate may transform into nitrite and cause Blue Baby Syndrome (BBS). Requirements of the Safe Drinking Water Act set a maximum contaminant level for nitrate nitrogen in community drinking water supplies at 10 parts per million.

2.10 THE IMPORTANCE OF THE CARBON CYCLE

Carbon is an element found in all living substances as well as in many inorganic materials. Both diamond and coal are nearly pure carbon, but with different structures. Carbon is a key element for life, composing almost half of the dry mass of the earth's plants (that is, the mass when all water is removed). The carbon cycle is the exchange of carbon among three *reservoirs* or storage places: the land, the oceans, and the atmosphere. The amount of carbon in these reservoirs is so large that it is expressed in gigatons (Gt): 109 metric tons (1 metric ton equals 1000 kilograms or 2200 pounds).

The atmosphere is the smallest pool of actively cycling carbon, in this reservoir carbon stays less than a thousand years or so. The land, its plants and animals, which scientists call the terrestrial biosphere, is the next largest reservoir of carbon. The oceans are the earth's largest active carbon reservoir by so far.

The *carbon budget* is the balance of carbon among the three reservoirs. The carbon cycle is vitally important to life on the earth. Through photosynthesis and respiration, it is the way the earth produces food and other renewable resources. Through decomposition, it serves as the earth's waste disposal system. In addition, the carbon cycle is important because carbon-containing gases in the atmosphere affect the earth's climate. Increased carbon dioxide (CO_2) in the atmosphere has been responsible for more than half of the global warming observed in recent decades.

The processes by which carbon moves through the earth's reservoirs take place on very different time scales. The *short-term carbon cycle* includes processes that transfer carbon from one reservoir to another in a matter of years, including photosynthesis, plant and animal respiration, and the movement of CO_2 across the air-sea interface. Other processes, such as the transformation of carbon into limestone and its subsequent release as the rock is weathered, occur very slowly (thousands to millions of years). Scientists call these slower transfers of carbon among reservoirs the *long-term carbon cycle*.

The carbon cycle is inextricably linked to other chemical cycles, including those of nitrogen, phosphorus, and sulfur, as well as to the global hydrological cycle.

Actively cycling carbon in its three reservoirs affects human life every day. Carbon in the atmosphere serves as food for plants (in the presence of sunlight); carbon in the soil serves as energy for the growth of miniature animals called microbes; carbon in plants feeds humans and other animals; and carbon in the oceans makes up the homes of many marine animals. Obviously, carbon is critically important in all the reservoirs where it is found.

2.10.1 The Atmosphere

Carbon composes much less than 1% of the atmosphere. Even this small percentage adds up to 750 Gt of carbon, which is a lot (although, as noted earlier, much less than in the other two reservoirs). Carbon in the atmosphere occurs almost entirely as CO_2; small amounts of methane (CH_4), carbon monoxide (CO), and chlorofluorocarbons (CFCs) are also present. These gases are all radiatively active, and can trap heat near the surface of the earth. That's why they are also considered greenhouse gases.

2.10.2 Photosynthesis and Respiration

In the short-term carbon cycle, photosynthesis and respiration are the primary processes that involve the atmosphere. Photosynthesis is the process by which green plants make their food. In this process, plants combine CO_2 and water using light energy to make carbon-containing compounds (sugars and starches, called carbohydrates). Oxygen is produced during the reaction and released to the atmosphere. On land, plants use CO_2 from the atmosphere. In the oceans, phytoplankton use CO_2 dissolved in seawater; much of this dissolved CO_2 also originally comes from the atmosphere. Photosynthesis also releases oxygen to the atmosphere. Thus, thanks to the activity of photosynthetic organisms, all other forms of life on the earth have oxygen to breathe and food to eat—even carnivores feed on animals that eat plants.

Respiration is the chemical process by which carbon-containing compounds are broken down within cells. It is essentially the opposite of photosynthesis: oxygen and carbohydrates react to produce CO_2 and water, releasing energy during the process. Living organisms use the energy released by respiration to power everything they do. For example, reading this page requires energy, and that energy is supplied by respiration.

2.10.3 The Oceans

The oceans absorb and store about 39,000 Gt of carbon—more than 60 times as much as the atmosphere (see Fig. 2.7). Over 95% of oceanic carbon is in the form of dissolved inorganic carbon (particularly dissolved CO_2, bicarbonate and carbonate ions); the remainder comprises various forms of organic carbon (living organic matter, particulate and dissolved organic carbon). Scientists make a distinction between the inorganic and organic forms of carbon because the two forms go through different chemical reactions and also are affected in different ways by physical ocean processes.

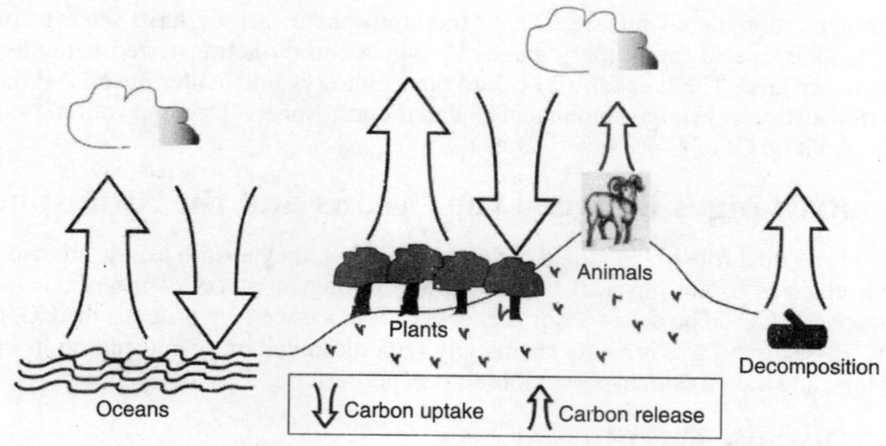

Fig. 2.7 Cycle of photosynthesis and respiration.

2.10.4 The Terrestrial Biosphere

Within this reservoir, the largest pool of carbon (roughly 65.5 million Gt) is in sedimentary rocks—mostly limestone, carbonate rocks, and also organic sediments containing fossil fuels. However, aside from human use of these fuels (which will be discussed in the next chapter), rocks do not play a role in the short-term carbon cycle. Carbon storage in plants and soils is more important for carbon cycle. Plants and soils store and release carbon on time scales ranging from seasons to thousands of years, and they contain about 2200 Gt carbon (C)—almost three times as much total carbon as the atmosphere and more organic (but not total) carbon than in the oceans.

Not only is the amount of carbon in the terrestrial biosphere big, but it is also readily modified by human activities. Therefore, this reservoir may be thought of as the most active part of the global carbon cycle. However, it is difficult, to pin down the role of the terrestrial biosphere in the global carbon cycle because of the complex biology underlying carbon storage, the great heterogeneity of vegetation and soils, and the somewhat unpredictable nature of human land use and land management.

2.10.5 Release of Fossil Fuels to the Atmosphere

When carbon in the short-term carbon cycle changes form or flows to a different reservoir (for example, moving from the land to the atmosphere when organic matter is burnt), the total amount of carbon in the cycle remains the same. The short-term carbon cycle is roughly balanced, and the amount of carbon-containing greenhouse gases is steady.

Fossil fuels, on the other hand, are extracted from underground, so they are not a natural part of the short-term cycle. Thus, when fossil fuels are burnt, carbon from long-term storage is added to the short-term cycle. When this happens, not only does the greenhouse effect increase because there are more carbon-containing greenhouse gases in the atmosphere, but the balance of the cycle is disrupted. Under current conditions, this means that either the terrestrial biosphere, or the oceans, or both may respond to the additional carbon in ways that we cannot yet predict. They may even store less carbon in future than they do today.

2.10.6 Other Emissions from the Terrestrial Biosphere

Plants, soils, and animals all release CO_2 to the atmosphere. As we have seen, respiration by terrestrial plants is one of the primary means by which carbon is transferred naturally from the atmosphere to the land. The breakdown of dead plants into organic matter in soil and their further decomposition also release large amounts of CO_2 to the atmosphere. Decomposition itself releases, on average, about 60 Gt of C per year.

2.10.7 Exchanges Between the Oceans and the Atmosphere

The oceans play a dual role in the global carbon cycle. First, they absorb excess atmospheric CO_2, and second, changes in the physical, chemical, and biological states of the oceans may affect future atmospheric CO_2. The ocean absorbs excess CO_2 in a three-step process: the CO_2 gas passes through the air-sea interface, it reacts chemically with dissolved inorganic carbon in the ocean's surface waters, and then it is transported into the deeper waters.

2.10.8 Oceanic Circulation

Oceanic circulation is important in the carbon cycle because it affects carbon storage in the ocean. The ocean waters circulate mainly because of three factors: winds, which create surface currents; differences in saltiness throughout the oceans (caused by differences between evaporation and precipitation as well as by freezing and thawing); and differences in water temperature.

The flow pattern is sometimes referred to as a *conveyor belt* because the water follows a continuous looping path, taking thousands of years to make a full circuit.

Difference in temperature (thermal forcing) pushes the tropical surface waters towards the pole. CO_2 is more soluble in cold water than in warm water, so surface water from the tropics continues to absorb additional CO_2 as it moves toward higher latitudes and colder temperatures. In the rainy North Atlantic, the addition of fresh water—which weighs less than saltwater—pushes the more saline water (and the extra CO_2) towards the ocean floor. Only the movement of carbon from shallow to deeper waters in the conveyor- belt circulation allows large amounts of CO_2 to be stored in the oceans. The surface water has a limited capacity to absorb CO_2. It will even emit CO_2 to the atmosphere if the local atmospheric concentration is lower than the concentration in the surface water.

(i) ***Chemical Processes:*** Chemical processes in the ocean also affect carbon storage. When CO_2 enters the ocean, it reacts chemically with the water to form carbonate and bicarbonate ions. (Ions are chemical compounds of atoms that have gained or lost electrons, leaving them with a positive or negative charge.) As more of these ions are formed, more atmospheric CO_2 can enter the ocean. This reaction depends on the temperature, the amount of dissolved inorganic carbon already in the surface water, and, to a lesser extent, the water's salinity and alkalinity (opposite of acidity).

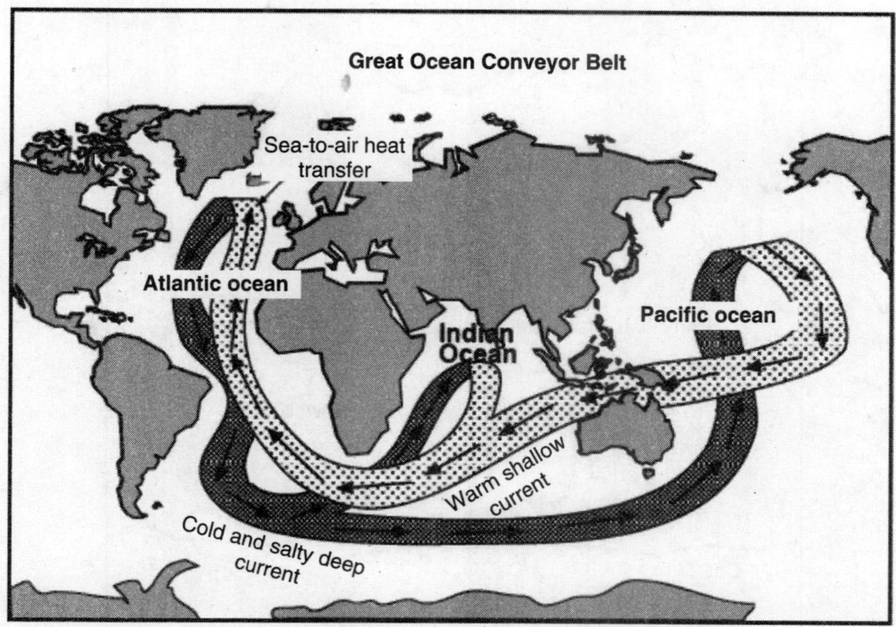

Fig. 2.8 Oceanic circulation.

(ii) Biological Processes: Biological processes in the oceans are driven by the marine *biota* (oceanic plants and animals). The marine biota are believed to play a minor role, if any at all, in the oceans' ability to absorb the CO_2 added to the carbon cycle by humans.

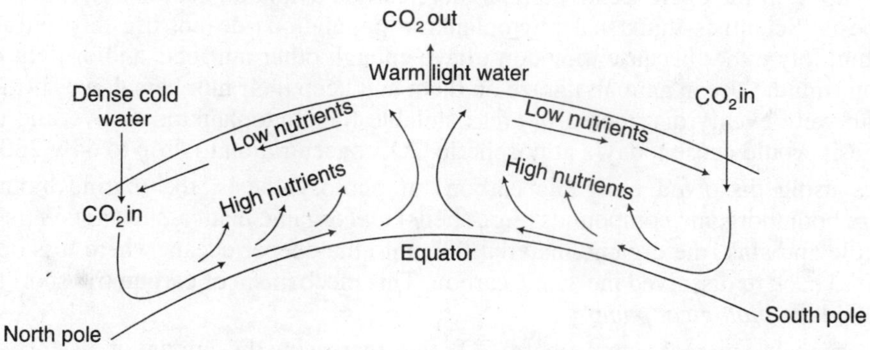

Fig. 2.9(a) Biological processes.

In fact, marine biota contribute only about 3 Gt of C to the world's total carbon balance.

However, the marine biota play a crucial role, in keeping the level of atmospheric CO_2 steady. Phytoplankton (microscopic ocean plants) take up dissolved inorganic carbon through photosynthesis, just as terrestrial plants take up atmospheric CO_2 during photosynthesis. If that dissolved inorganic carbon remained in the ocean waters, the oceans could not absorb as much CO_2

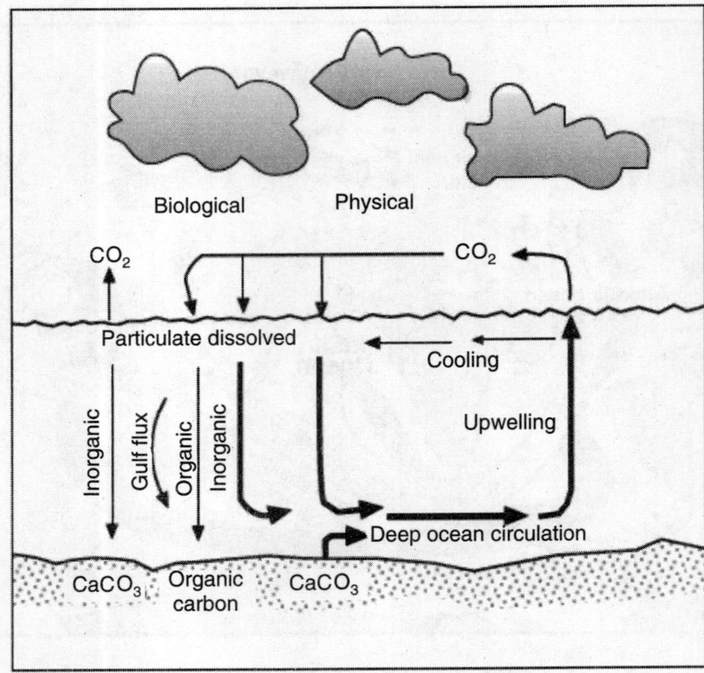

Fig. 2.9(b) Biological processes.

from the atmosphere as they do. In fact, if there were no oceanic photosynthesis, the atmospheric concentration of CO_2 would be approximately 520 ppm—much higher than the mid-1990s level of 350–360 ppm. In the open ocean, there is more carbon available than the marine biota use for photosynthesis. Scientists think that phytoplankton populations do not use this extra carbon to grow and multiply either because they don't have enough other nutrients and/or light or because zooplankton (minute ocean animals) graze on them and keep their numbers down. If the nutrients in the ocean were evenly distributed in areas suitable for phytoplankton to live, the increase in photosynthesis would cause today's atmospheric CO_2 concentration to drop to only 230 ppm.

Besides using dissolved inorganic carbon for photosynthesis, the marine biota use it to manufacture both inorganic compounds (e.g., shells) and organic matter, such as cell tissue. As the organisms die and sink, the organic material falls into the deeper ocean, where it is decomposed and converted back to dissolved inorganic carbon. This mechanism of carbon transport to the deep ocean is called the *biological pump.*

If the ocean takes up more atmospheric CO_2, thus increasing the amount of dissolved inorganic carbon in the shallower waters, this increase won't be passed on to the deeper waters via the biological pump. Photosynthesis is the first step in the biological pump, and, as we just noted, there is already more carbon in the ocean that the marine biota can use for photosynthesis.

2.11 FOOD RESOURCES

Food is any substance consumed to provide nutritional value to support the body. Food is usually of a plant or animal origin, and contains nutrients, e.g., fats, proteins, carbohydrates, minerals and vitamins. From the start of nature people secured food through mainly two methods either by hunting (or gathering) and agricultural. As civilisations grew, food supply chain developed in most countries which provide different types of natural and processed food. Due to the growing number of living beings on the planet earth pressure on food resources is increasing day by day.

2.11.1 World Food Problems

Hunger, which is usually preceded by food shortages, is caused by a complex set of events and circumstances (climatic, social, economical, and political) that differ from place to place and from time to time. Though hunger has been a part of human experience from centuries and a dominant feature of life in many low-income countries, the causes of hunger and starvation are not very well understood. Our awareness of the main causes of hunger and starvation has been hampered by myths and misperceptions about the interplay between hunger and population growth, land use, farm size, technology, trade, environment, and other factors.

2.11.2 Population Growth and Land Use

There is a myth that the hunger and malnutrition afflicting many of the populations of poor nations are the direct result of rapid population growth and the persistence of traditional agricultural technologies. This perception implies that if population increases were to be curbed and modern technologies widely adopted, the problem would disappear. However, the population explosion is not the main cause of hunger and malnutrition, even though it exacerbates the situation. The growth of global food and agricultural production has been faster than the unprecedented population growth of the past 50 years. Hungry part of the world suffers from the lack of food, caused mainly by poverty. The poverty remains deep-rooted in 65% of the world's population.

Poverty and rapid population growth are positively correlated. Where per capita income increases, population growth rates decline and vice versa. In other words, the higher the incidence of poverty, the higher is the population growth, and consequently hunger and malnutrition afflict relatively more people. That means poverty, rather than population growth, is the leading cause of hunger and malnutrition. It is also evident that most of the people afflicted by hunger and malnutrition live in the poorest parts of the world (particularly South Asian and Sub-Saharan African countries) where unemployment is high, income distribution is skewed, and the standard of living is low—reinforcing the obvious connection between hunger and poverty, not between hunger and population growth.

2.11.3 Scarcity of Arable Land

Scarcity of agricultural land is not the primary cause of food shortages, although it does exacerbate the problem. There is adequate arable land for cultivation and food production in the world. For instance, Bangladesh, where starvation and malnutrition are prevalent, has just half the people per cultivated acre compared to Taiwan. Yet Taiwan has no starvation and malnutrition. Ethiopia, Somalia, Sudan, Mozambique, and Bangladesh have adequate arable land for food and agricultural production, but they have been thought of as the countries hardest hit by hunger and starvation for many years. China has twice as many people per cultivated acre as India. Yet people in China do not suffer from hunger to the extent experienced in India. China, with the world's largest population (around 1.2 billion), is self-sufficient in food and agricultural products, and produces grain and meat in large quantities for the domestic and world markets.

Hunger is a more complicated phenomenon than merely a problem that can be rectified by simply expanding agricultural production, although in most instances, expanding agricultural output is a necessary condition to feed the growing number of people. This is because of issues surrounding hunger go to the heart of the political economy of nations. The key obstacle in alleviating hunger is that the rural poor populations in most developing countries primarily depend and live on local agricultural production, exercise little control over the prices they receive and the productive resources they need to produce efficiently. When the control of resources is in the hands of the actual farmers and tenants rather than in the hands of absentee landlords, the farmers are likely to make efficient use of their land.

When farmers own land and work for themselves, they have the motivation to work hard to make the land more productive.

2.11.4 Technology in Food Production

It is often assumed that world food shortages can be eliminated by increasing food and agricultural production through the application of modern technology. It is also argued that supplying modern inputs—such as large-scale irrigation, chemical fertilizers, farm machinery, and pesticides—can improve the productive capacity of the land. However, when a new agricultural technology enters a system characterized by unequal power relationships, it brings greater profits only to those who already have some combination of land, financial resources, creditworthiness, and political influence.

For example, a research study completed by the International Labor Organization (ILO) showed that in the South Asian countries (Pakistan, India, Thailand, Malaysia, the Philippines, and Indonesia), where the focus was on increased agricultural production and where the gross national product (GNP) has risen, the majority of the rural population was worse off than ever before.

2.11.5 Farm Size and Credit Services

The disparities in the availability of institutional support systems in the distribution of services between large and small farmers work to the disadvantage of the latter. Since large farmers are conventionally considered to be more productive than small farmers, the big landholders are provided with public subsidies and credit facilities. The small farm operators are usually disqualified from farm credit loans because of their disadvantaged economic condition and the generally conservative lending practices of financial institutions. Small farmers have low equity positions and can offer little security, which implies high cost for lenders. The lending institutions often limit access of small farm operators to the capital market by imposing rigid rules on lending in order to fully protect the loan capital. Small farmers are often excluded from the modern marketing process, and have high input costs relative to large farmers because they lack bargaining power and do not buy farm inputs in bulk. In addition, tradition plays a large role in the system of production in developing countries. Agricultural and extension service systems consistently fail to serve the majority of small farmers effectively—especially in developing countries. This is partly because small farmers lack the formal channels to communicate their needs and ideas to the public sector. Also the new technologies are quite slow in replacing old techniques. The small farm systems are highly diverse and location specific, make adoption of new technologies very difficult. Public institutions cannot afford to accurately adapt appropriate technologies to each local set of circumstances.

2.11.6 Import-Export Balance

Another misconception about the causes of world food shortage is that a developing country may not be fully exploiting its natural endowments to produce and export crops for acquiring foreign exchange to import food and industrial goods. However, in many developing countries, structural adjustment and economic development programs have tried to promote agricultural exports at the expense of basic food needs. The people who exploit the natural resources and produce agricultural exports are not the poor farmers who suffer from hunger and malnutrition. Those who need food are not the same people who benefit from foreign exchange earned by agricultural exports. Even when a portion of the foreign earnings is used to import food, the imported goods are not basic staples; rather, they consist of items geared to the eating habits of the more well-off urban dwellers, comparable to food preferences of those in more affluent countries.

Attempting to achieve the Euro-American standards of living is impossible for developing countries, based both on projections of their future resource availability and on population growth. According to research studies each person in the United States consumes about 23 times more goods and services than the average person in developing countries. The high levels of prosperity and quality of life in the United States and Europe were achieved through high consumption of natural resources. However, these basic resources are finite and are not unlimited in supply in any country of the world.

Developing countries have not instituted economic incentives to produce enough food for domestic consumption, as well as to extend agricultural production for export, because they lack the necessary means of production. They have failed to promote food security through necessary measures to raise productivity, yield, and storage. For example, in many Sub-Sahara African countries, food self-sufficiency declined steadily during the 1980s. Producers of food crops were hit hard by cheap imports. Many governments find it more expedient and economically advantageous to import food to cover deficits of staple food than to make the necessary investments to increase domestic production and to redistribute real incomes in favor of the poor. And because of embedded structural dependency relationships, these government decisions are reinforced by a confluence of interests of the local elite with those of foreign investors and international financial institutions.

2.11.7 Environmental Issues

Poverty and environmental degradation are closely linked, often in a self-perpetuating negative spiral in which poverty accelerates environmental degradation, and degradation results in or exacerbates poverty while simultaneously reducing the productivity of the resource, and hence food production. While poverty is not the only cause of environmental degradation, it does pose the most serious environmental threat in many low-income countries. This problem is further heightened because many millions of people, living near the subsistence minimum, have exploited natural resources with inappropriate technologies in order to survive. The very success of export-oriented agriculture undermines both the position of the poor as well as the environment. When export commodity prices rise, small self-sufficient farmers are pushed off onto marginal land by cash crop producers seeking to profit from the higher commodity prices.

(i) **Salinity:** Dryland salinity is a groundwater problem – salinity is only a symptom. Salinity occurs when there is an imbalance between inputs of water into groundwater within a catchment and the amount of groundwater leaving the catchment. This imbalance may occur, for example, when native trees are cleared for cropping.

When more water goes into a catchment than comes out, the water-table rises, bringing salt to the surface. Salts are carried in water moving through soils and the landscape. Salts can then accumulate at the soil surface when groundwater evaporates from shallow water-tables.

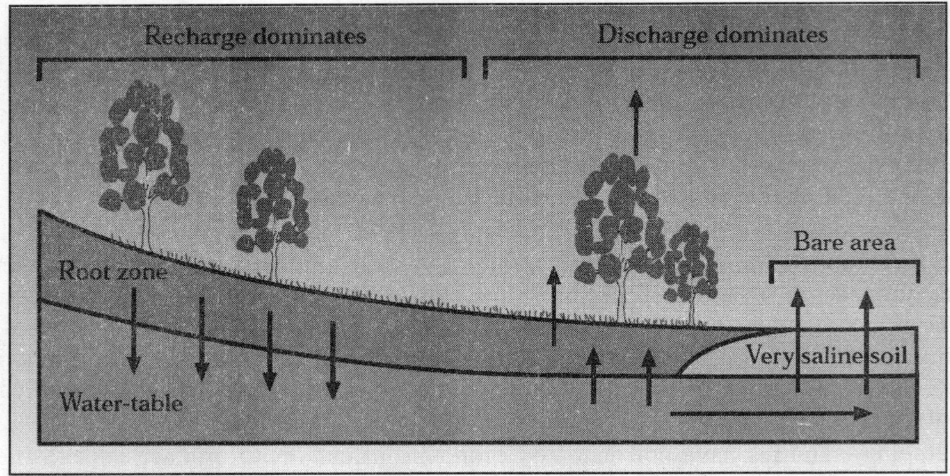

Fig. 2.10 Salinity imbalances between inputs of water into groundwater.

(ii) Inputs to groundwater are known as recharge: Recharge can come either from the small amount of rainfall that percolates below the root zone of plants, or from the water seeping into the groundwater from streams, rivers, lakes and dams. The amount of water that percolates below the root zone of crops and pastures can be 10–100 times that percolating below trees. While all sources of recharge can contribute to dryland salinity, increased groundwater recharge under crops and pastures is the major cause of dryland salinity.

(iii) Loss of water from groundwater are known as discharge: Discharge can occur by subsurface lateral flow when the groundwater flows directly into a stream or river or when it evaporates from soils or transpires from plants. As the groundwater level comes closer to the soil surface, discharge into streams and from soils and plants increases, resulting in increased stream and soil salinity (particularly if the groundwater is saline). If water-table levels are regularly less than about 2 m deep, salts can build up at the soil surface, eventually killing plants and leaving the soil bare and salt-crusted.

However, it should be remembered that groundwater also discharges through transpiration from vegetation on the perimeter of the bare areas, where water-tables are deeper (e.g. within 4 m of the surface) and surface soil salinity is less obvious. Groundwaters are often saline in discharge areas, with an electrical conductivity ranging from 6 to over 60 dS/m.

Dryland salinity will be controlled by restoring the balance in a catchment (Figure 2.11). This can be achieved in two ways:

- reducing groundwater recharge; or

• increasing groundwater discharge.

Planting trees can help in both these processes.

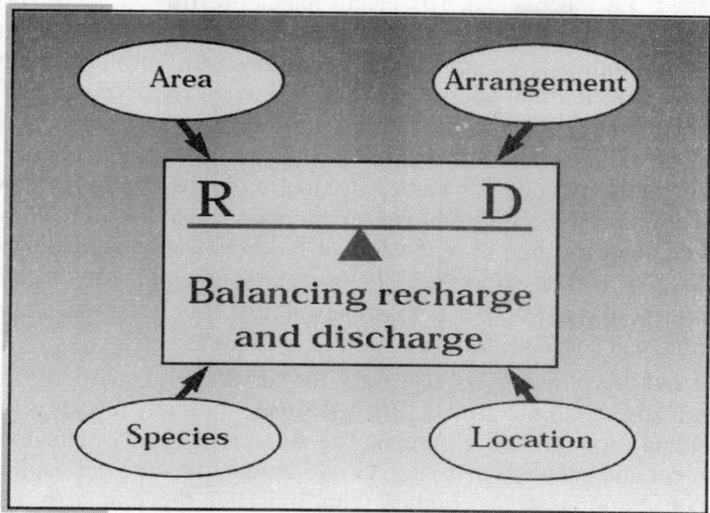

Fig. 2.11 Restoring the balance between groundwater recharge and discharge for dryland salinity control.

There are four factors that can be manipulated when designing agroforestry systems for controlling dryland salinity mentioned below:

- area planted;
- the arrangement of the trees;
- their location within a catchment; and
- the tree species selected.

All of these can impact on recharge and/or discharge of groundwater and all need to be considered when undertaking tree planting for dryland salinity control. The principles associated with these factors are described below in general. The optimum design will always be dictated by specific landscape and climatic conditions, as well as by the overall land management objectives and restrictions relevant to the site. Also, the design may be varied to capture some of the other positive benefits of agroforestry (see table at the end of this chapter). Thus, the final design will most likely be a compromise between salinity control and other factors.

(iv) Waterlogging: Waterlogging is caused by the same processes as dryland salinity. The difference is that salts do not accumulate at the soil surface, either because the groundwater is of very low salinity (less than 6 dS/m) or it flows out of the soil (i.e. from a small spring), flushing the salts away. Waterlogging problems can also be ephemeral, such as when a perched water table develops on impermeable subsoil in a wet season. Most of the information given in this chapter is also directly relevant to management of waterlogging, except the references to salt accumulation and its effects. Thus, waterlogging will not be discussed separately.

2.12 ENERGY RESOURCES

Energy is the capacity of matter to do work in relation to forces acting on it. Five forms of energy are often distinguished: (i) mechanical, (ii) electromagnetic, (iii) thermal, (iv) chemical, and (v) nuclear. Energy conversion takes place among and within these forms in the process of human production and consumption, as well as in nature; for example burning fuels to produce heat converts chemical energy to thermal energy (i.e. burning wood releases stored chemical energy, but all is not converted to heat; if heat is the desired form of conversion, conversion to other forms are losses). The efficiency of conversion from one form to another depends on the extent to which unwanted dissipation can be avoided. Unwanted dissipation in the process of conversion and use is energy lost. The most common energy sources are biomass, solar, fossil, hydropower, wind, and waves. The sources can be classified as renewable or non-renewable depending on whether they can supply continuously or will be exhausted. The primary source of energy available to us comes from the sun. The concept of renewable and non-renewable energy is a relative one—it is basically a distinction about time and space scale.

The availability and use of energy is necessary for survival and prosperity. It is basic to any society's development and economic growth. Energy is used for many purposes in daily life and in the process of social and economic development The amount of energy consumed is an indicator of level of development and standard of living. Gross national income per capita is higher where per capita energy consumption is also higher. This indicates the link between prosperity and energy use in the present world—rich countries use more energy. Expanding economic activity and population growth are the two basic factors behind increases in energy consumption. In any country, where economic growth is necessary and population growth is high, energy demand will continue to rise in the years to come. Energy consumption patterns and the rise in demand, their sources, and ways in which they are harnessed and utilized have implications for the environment and natural resources, which ultimately affect overall development.

2.12.1 Traditional (Biomass)

This includes fuelwood, agricultural residues, and animal waste. Biomass is the traditional source of energy, which is predominantly used in rural and agricultural sector. People in rural areas do not have to pay for biomass—it is either available free from nature (fuelwood collected from a nearby forest) or is a by-product of agricultural activities (agricultural residues and animal dung). In the past, there was an abundant supply of biomass from the forest and agricultural activities. The situation has been changing in recent times due to increase in population, long-term decline in forest areas (despite recent increases), and rising demand for energy. Biomass is becoming scarce and biomass sources are under increasing pressure, with potentially adverse consequences on agriculture and the environment. However, rural people continue to use biomass as an energy source because they have no other options and they do not have to pay for it.

Commercial: Commercial energy, also known as conventional energy, is traded in the market. This comprises coal, electricity supplied through the national grid, and petroleum products.

Alternative energy: Biogas, micro-hydro, solar, and wind energy come under this heading. Biogas is a more efficient way of utilizing some biomass (animal dung) as energy. Hydropower (including micro-hydro), biogas, solar, and wind energy are renewable; within their regeneration capacity, these can supply energy indefinitely.

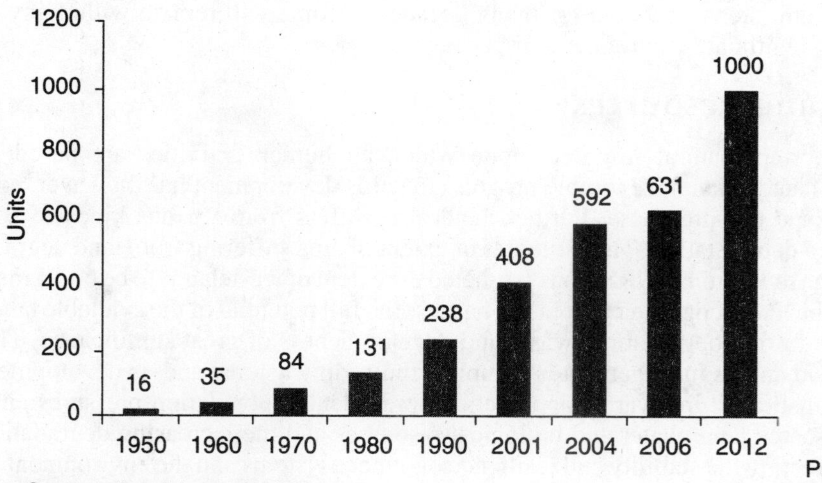

Fig. 2.12 Growth in India`s per capita consumption.

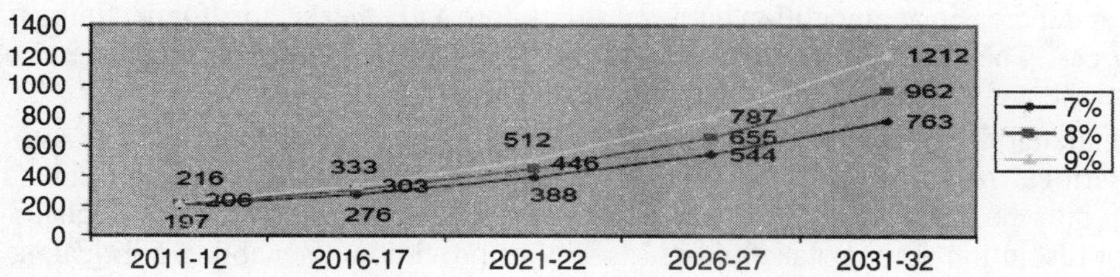

Fig. 2.13 Capacity requirements (in GW).

Growth Projections (at 9% GDP)

Particulars	2008	2012	2017	2022	2027
Generation capacity (GW)	124	216	233	612	780
Per capita consumption (kWh)	631	1000	1300	1800	2800

Fig. 2.14 Growth projections for generation and consumption.

Currently, India's per capita GHG emissions are 24% of global average, 5% of the US, 12% of EU, 11% of Japan.

In cumulative terms, India has 2% of global GHG emissions. According to World Energy Outlook's projections, China's share in cumulative emissions between 1900 to 2030 would rise to 16% approaching that of the US (25%) and the EU (18%). India's cumulative emissions are projected to reach 4% which would be comparable to those of Japan.

Our GHG emissions now, and for many decades to come will remain within any reasonable measure of our legitimate environmental space.

2.12.2 Land Resources

The most important natural resource, upon which all human activities are based since time immemorial, is land. Man's inexorable progress towards development has, however, considerably damaged our land resource base. Further, land also suffers from various kinds of soil erosion, degradation and deforestation. The estimates of extent of area suffering from land degradation vary from 38.40 m.ha to 187 m.ha. NRSA has estimated the extent of wastelands to be 63.85 m.ha which is about 20% of the total geographical area. To harness the full potential of the available land resources and prevent its further degradation, wasteland development is of great significance. The problem of degraded land and its management is complex, multi-dimensional and its development requires a scientific, holistic and innovative approach. Unprecedented population pressures and demands of society on scarce land, water and biological resources and the increasing degradation of these resources is affecting the stability and resilience of our ecosystems and the environment as a whole. Globally, the expansion of human settlements and infrastructure, intensification of agriculture, and expansion of agriculture into marginal areas and fragile ecosystems emphasizes the need for integrated planning and management of land resources.

These trends are also exacerbating conflicts over access and rights to land, water and biological resources, and increasing competition between agriculture and other sectors for declining per capita land resources. They affect food security in many developing countries, global environmental balance and the well-being of present and future generations.

The challenge is to develop and promote sustainable and productive land use systems and to protect critical resources and ecosystems through balancing land, water and other resource uses. It can only be achieved by providing a basis for negotiation, participatory decision-making and conflict resolution among stakeholders, as well as providing an enabling political, social and economic environment.

To accelerate the pace of development of wastelands/degraded lands and to have focused attention in this regard, the Government had set up the National Wastelands Development Board in 1985 under the Ministry of Environment & Forests.

Later a separate Department of Wastelands Development in the Ministry of Rural Development and Poverty Alleviation was created in 1992 and the National Wastelands Development Board was transferred under the Department. In April 1999, the Department of Land Resources was created by changing the nomenclature of the Department of Wastelands Development to act as a Nodal Agency in the field of Land Resource Management. All land-based Programmes/Schemes, which were earlier being implemented by different departments in the Ministry of Rural Development, were brought within the purview of the new Department.

The Department of Land Resources comprises two Divisions, namely the Wastelands Development Division and the Land Reforms Division. Secretary, Ministry of Rural Development heads the Department. He is assisted by one Addl. Secretary, one Joint Secretary, three Directors, two Dy. Inspector General's of Forests and other officers.

The Wastelands Development Division has been implementing various programmes for the development of wastelands/degraded lands such as the Integrated Wastelands Development Programme, the Drought Prone Areas Programme, and the Desert Development Programme.

These Programmes were started in different years. DPAP was started in 1973-74 and DDP in 1977 whereas IWDP was started in 1988-89. Till 1995-96, the programmes were being implemented on sector basis and after 1995-96; they are being implemented on watershed basis.

The Land Reforms Division has been monitoring the progress of implementation of various land reforms measures such as abolition of Zamindari System, distribution of ceiling surplus lands, consolidation of land holdings, implementation of tenancy laws, etc. In addition, it has been administering the Land Acquisition Act, 1894. A draft Bill on Resettlement and Rehabilitation of Project Affected Persons/Families is also under formulation in this Division.

The plan allocation for the Department of Land Resources has been increased from Rs. 324 crore in 1999-2000 to Rs. 1000 crore for 2002-2003. Year-wise approved outlays of the Department since 1999-2000 is given in Annexure-XXVII

Presently 972 blocks of 182 districts in 16 states are covered under Drought Prone Areas Programme (DPAP). Similarly, 235 blocks of 40 districts in 7 states are covered under Desert Development Programme (DDP). The coverage under Integrated Wastelands Development Programme (IWDP) extends generally to blocks not covered in the above programmes.

Projects covering an area of 139.72 lakh hectares have been taken up from 1.4.1995 to 31.3.2002 under the three programmes, namely, DPAP, DDP and IWDP. In addition, an area of 63.50 lakh hectares had been taken up for development prior to 31.3.99 under Employment Assurance Scheme (EAS).

Besides the above schemes, the Wastelands Development Division also implements the Technology Development, Extension & Training (TDET) Scheme and the Investment Promotional Scheme (IPS).

A statement showing state-wise release of funds under these programmes, as on 31.1.2003, is in Annexure –XXVIII

In the area of land reforms, the task of abolition of intermediary tenures has been completed all over the country. Besides, an area of 5.39 million acres of ceiling surplus land has been distributed to 5.65 million rural poor, 50% of which constitute SC/ST beneficiaries. An area of 14.74 million acres of Government wastelands and 2.18 million acres of Bhoodan land has been distributed among the eligible rural poor. Similarly, 0.43 million acres of alienated land has been restored to Scheduled Tribes. Consolidation of landholdings has taken place in an area of 161.53 million acres. 582 districts have been brought fewer than 100% Centrally Sponsored Scheme of Computerization of Land Records and its operationalisation has been extended to 2970 talukas/ tehsils/blocks.

For the purpose of strengthening of survey and settlement operations for updating the land records and other allied matters, a 50:50 Centrally Sponsored Scheme of Strengthening of Revenue Administration and Updating of Land Records is under implementation in collaboration with the State Governments. Besides, amendment of Land Acquisition Act, 1894 and a Bill on Resettlement and Rehabilitation of Project Affected Persons/ Families is under formulation. With the formation of a separate Department of Land Resources, all the Watershed Development Programmes of the Ministry of Rural Development have already been brought within its purview.

However, the programmes relating to conservation, development and management of land resources remain scattered in different Ministries and Departments. At the Joint Session of Parliament in February, 2000, the President of India had made the following announcement:

"There is an imperative need to put in place an integrated mechanism capable of responding effectively to the challenges of managing our scarce land resources, especially those arising from globalization, liberalization and privatization. The Government will, therefore, bring all the programmes and schemes as well as the institutional infrastructure relating to land in rural areas, under the control of the newly created Department of Land Resources in the Ministry of Rural Development."

The Department of Land Resources has already taken further action for implementation of the above announcement. It is hoped that a final decision in the matter will be taken very shortly.

2.12.3 Soil

The soil covering the surface of the earth has taken millions of years to form and we must learn to respect it. Soil is formed at a rate of only 1 cm every 100 to 400 years and it takes 3 000 to 12 000 years to build enough soil to form productive land. This means that soil is a non-renewable resource and once destroyed it is gone forever.

If we disregard this, a time will come when there would not be enough soil left to sustain life on the earth, because the soil is a necessary growth medium for plants, a home for certain insects and animals, as well as a medium from which we get minerals. It is important therefore to treat soil, especially topsoil, as a living entity.

2.12.3.1 *Soil erosion*

- Soil erosion occurs when soil is removed through the action of wind and water at a greater rate than it is formed.
- When a raindrop hits soil that is not protected by a cover of vegetation and where there are no roots to bind the soil, it has the impact of a bullet.
- Soil particles are loosened, washed down the slope of the land and either end up in the valley or are washed away out to sea by streams and rivers.
- Erosion removes the top soilfirst. Once this nutrient-rich layer is gone, few plants will grow in the soil again.
- Without soil and plants the land becomes desert-like and is unable to support life.

2.12.3.2 *Causes of soil erosion*

Erosion occurs when farming practices are not compatible with the fact that soil can be washed away or blown away. These practices are:

- Overstocking and overgrazing
- Inappropriate farming techniques such as deep ploughing land 2 or 3 times a year to produce annual crops
- Lack of crop rotation
- Planting crops down the contour instead of along it.

2.12.3.3 *Water erosion*

Water erosion causes two sets of problems:

- A non-site loss of agricultural potential
- An off-site effect of downstream movement of sediment, causing flooding and the silting up of reservoirs.

2.12.3.4 Sheet erosion

- Soil erosion is characterised by the down slope removal of soil particles within a thin sheet of water.
- Sheet erosion occurs when the entire surface of a field is gradually eroded in more or less uniform way.
- It is a gradual process and it is not immediately obvious that soil is being lost.

2.12.3.5 Gully erosion (dongas)

Dongas usually occur near the bottom of slopes and are caused by the removal of soil and soft rock as a result of concentrated run-off that forms a deep channel or gully.

On steep land, there is often the danger of gullies forming. Water running downhill cuts a channel deep into the soil and where there is a sudden fall, a gully head forms at the lower end of the channel and gradually works its way back uphill. As it does so, it deepens and widens the scar that the gully makes over the hillside. Gully erosion is related to stream bank erosion, in which fast-flowing rivers and streams increasingly cut down their own banks.

2.12.3.6 Rill erosion (channel erosion)

Channel erosion can occur on steep land or on land that slopes more gently. Because there are always irregularities in a field, water finds hollows in which to settle and low-lying channels through which to run.

As the soil from these channels is washed away, channels or miniature dongas are formed in the field.

2.12.3.7 Wind erosion

Wind erosion occurs when the land surface is left bare in regions that are arid enough, as a result of low rainfall, to allow the soil to dry out, and flat enough to allow the wind to carry the soil away over several consecutive days. Land may become susceptible to wind erosion through excessive grazing by animals, as grazing removes the protective plant cover, and animals' hooves break up the soil, especially round watering points. Arable land that has been left bare is also a major problem.

2.12.4 Factors Determining Soil Erosion

There are various factors determining soil erodibility of which the following are the most important.

2.12.4.1 Slope

If the landscape has the steeper slope there will be more erosion, because of increase in the velocity (swiftness) of water-flow. The length of the slope is very important, because the greater the size of the sloping area, the greater the concentration of the flooding water.

2.12.4.2 Soil texture

Soil texture is the size distribution of soil particles. The size of particles never changes. A sandy soil, therefore, remains sandy and a clayey soil remains clayey. The three main particles are sand, silt and clay. The more sandy a soil the easier it could erode.

2.12.4.3 Soil structure

The term soil structure means the grouping or arrangement of soil particles. Over-cultivation and compaction cause the soil to lose its structure and cohesion (ability to stick together) and it erodes more easily.

2.12.4.4 Terrain unit

The crest (top of slope) is usually well drained as soil moisture moves downhill, leaving air in the pore spaces most of thetime. Over time, the fine (clay) particles are carried down slope leaving the soil sandy. Plant roots can penetrate easily to deep levels and withdraw enough soil water from there. These soils have a lower erosion potential and are normally more stable.

In the mid slope soil moisture moving from the crest starts to dam up as a result of the clay-rich soil just downhill. The soils are moderately well drained with a higher erosion potential.

In the foot slope the soil has been waterlogged (saturated with water) as a result of the long-term accumulation of clay which does not allow water to infiltrate. Plants that grow on these soils are limited to those that can adapt their root systems to grow laterally above the hard clayey layer. These imperfectly drained soils have a high erosion potential.

2.12.4.5 Organic material

Organic material is the "glue" that binds the soil particles together and plays an important part in preventing soil erosion. Organic matter is the main source of energy for soil organisms, both plant and animal. It also influences the infiltration capacity of the soil, therefore reducing water run-off.

2.12.4.6 Vegetation cover

The loss of protective vegetation through overgrazing, ploughing and fire makes soil vulnerable to being swept away by wind and water. Plants provide protective cover on the land and prevent soil erosion for the following reasons:

Plants slow down water as it flows over the land and this allows much of the rain to soak into the ground.

Plant roots hold the soil in position and prevent it from being blown or washed away.

Plants break the impact of a raindrop before it hits the soil, reducing the soil's ability to erode.

Plants in wetlands and on the banks of rivers are important as they slow down the flow of the water and their roots bind the soil, preventing erosion.

2.12.4.7 Land use

Grass is the best natural soil protector against soil erosion because of its relatively dense cover. Small grains, such as wheat, offer considerable obstruction to surface wash. Row crops such as maize and potatoes offer little cover during the early growth stages and thereby encourage erosion. Fallowed areas, where no crop is grown and the entire residue has been incorporated into the soil, are most subject to erosion.

2.12.5 Preventing Soil Erosion

Some of the following measures can be implemented to prevent soil erosion:

- The use of contour ploughing and windbreaks
- Leave unploughed grass strips between ploughed lands (strip cropping)

- Make sure that there are always plants growing on the soil, and that the soil is rich in humus
- Avoid overgrazing
- Allow indigenous plants to grow along riverbanks
- Conserve wetlands
- Cultivate land, using a crop rotation system
- Minimum or no tillage
- Encourage water infiltration and reduce water run-off.

2.12.5.1 *Desertification*

Desertification is persistent degradation of dry land ecosystems by variations in climate and human activities. Home to a third of the human population in 2000, dry lands occupy nearly half of the earth's land area. Across the world, desertification affects the livelihood of millions of people who rely on the benefits that dry land ecosystems can provide. In dry lands, water scarcity limits the production of crops, forage, wood, and other services ecosystems provide to humans. Dry lands are therefore highly vulnerable to increases in human pressures and climatic variability, especially sub-Saharan and Central Asian dry lands.

Atleast 10 to 20% of dry lands are already degraded, and ongoing desertification threatens the world's poorest populations and the prospects of poverty reduction. Therefore, desertification is one of the greatest environmental challenges and a major barrier to meeting basic human needs in dry lands.

2.12.5.2 *Linkage between desertification and human well-being*

In dry lands, more people depend on ecosystem services for their basic needs than in any other ecosystem. Indeed, many of their resources, such as crops, livestock, fuelwood, and construction materials, depend on the growth of plants. Vegetation in any place around the world depends on water availability and local climatic conditions. Fluctuations in the services supplied by ecosystems are normal, especially in dry lands, where water supply is irregular and scarce.

However, when a dry land ecosystem is no longer capable to recover from previous pressures, a downward spiral of desertification may follow, though it is not inevitable.

Desertification affects a wide range of services provided by ecosystems to humans: products such as food and water, natural processes such as climate regulation, but also non-material services such as recreation, and supporting services such as soil conservation. Changes can be quantified and methods are available to prevent, reduce, or reverse them. When faced with desertification, people often respond by making use of land that is even less productive, transforming pieces of rangeland into cultivated land, or moving towards cities or even to other countries. This can lead to unsustainable agricultural practices, further land degradation, exacerbated urban sprawl, and socio-political problems. Desertification takes place in dry lands all over the world. Some 10 to 20% of all dry lands may already be degraded, but the precise extent of desertification is difficult to estimate, because only few comprehensive assessments have been done so far.

A large majority of dry land populations live in developing countries. Compared to the rest of the world, these populations lag far behind in terms of human well-being, per capita income, and infant mortality. The situation is worst in the dry lands of Asia and Africa. Dry land populations are often marginalized and unable to play a role in decision-making processes that affect their well-being and making them even more vulnerable.

Desertification has environmental impacts that go beyond the areas directly affected. For instance, loss of vegetation can increase the formation of large dust clouds that can cause health problems in more densely populated areas, thousands of kilometres away. Moreover, the social and political impacts of desertification also reach non-dry land areas. For example, human migrations from dry lands to cities and other countries can harm political and economic stability.

2.12.5.3 Major causes of desertification

Desertification is caused by a combination of social, political, economic, and natural factors which vary from region to region.

Policies that can lead to an unsustainable use of resources and lack of infrastructures are major contributors to land degradation. Agriculture can play either a positive or a negative role, depending on how it is managed. If policies favouring sedentary farming over nomadic herding in regions more suited to grazing can contribute to desertification.

The process of globalization both contributes to desertification and helps prevent it. Studies have shown that, in some cases, trade liberalization, economic reforms, and export-oriented production in drylands can promote desertification. In other cases, enlarged markets outside of the drylands also contribute to successful agricultural improvements.

Historically, dryland livelihoods have been based on a mixture of hunting, gathering, farming, and herding. This mixture varied with time, place, and culture, since the harsh conditions forced people to be flexible in land use. Population growth has led to the extension of cultivated lands and the irrigation of these lands has brought desertification, as well as other environmental problems.

2.12.5.4 Prevention or reverse desertification

Effective prevention of desertification requires management and policies to govern and promote the sustainable resource use. As rehabilitation is difficult and costly, prevention should be given preference over rehabilitation.

In order to prevent, stop or reverse desertification, both at local and global levels it becomes imperative to adopt major policy interventions and changes in management approaches. Prevention is a lot more cost-effective than rehabilitation, and this should be taken into account in policy decisions.

Addressing desertification is critical and essential for meeting the Millennium Development Goals which aim to eradicate extreme poverty and ensure environmental sustainability amongst other objectives. The creation of a "culture of prevention" that promotes alternative livelihoods and conservation strategies can go a long way towards protecting dry lands both when desertification is just beginning and when it is ongoing. It requires a change in governments' and peoples' attitudes. Building on long-term experience and active innovation, dry land populations can prevent desertification by improving agricultural and grazing practices in a sustainable way. Even once land has been degraded, rehabilitation and restoration measures can help restore lost ecosystem services. The success of rehabilitation practices depends on the availability of human resources, funds, infrastructures, and government policies. It requires a combination of policies and technologies and more important the close involvement of local communities.

Learning Outcomes

The main learning outcomes from this chapter should be fundamental understanding of: Types of Natural Resources and associated problems;

- Forest resources ;
- Water resources;
- Mineral resources;
- Food resources;
- Energy resources; and
- Land resources.

Readers also advised to carry out following exercise:

Exercise

1. Write a 1000 words essay on "Recent trends in energy sources in India".
2. Summarise learning outcomes in 200 words about what you have learnt from this chapter.
3. Give a brief description of the present and past trends about availability of natural resources.

Chapter 3

Structure and Functions of Ecosystem

3.1 INTRODUCTION

Earlier we defined ecology and ecosystem; in order to enhance our understanding of ecology and ecosystem, we shall have a small review of these terms.

Ecology comes from the Greek words *logos* (study of) and *oikos* (house or place where one lives). It can be surmised as the Study of the "House" in which We Live. This term was first introduced by Ernst Haeckel in 1869. Ecology can be defined more specifically as the study of the interactions between organisms and the non living components of their environment. This also involves collecting information about organisms and their environment, looking for patterns, and seeking to explain reasons behind these patterns.

The biosphere is composed of smaller units called ecosystems. An ecosystem includes all the organisms and the nonliving environment that are found in a particular place. Ecosystems can be as large or as small as we decide. Any place consisting of living/non-living organism, you decide to study can be considered an ecosystem, e.g., you may choose to study your back yard, a multi storeyed apartment, a mountain range or a forest/zoo. Any of these would be considered an Ecosystem which has various levels of organisation.

The simplest level of organization in ecosystem is that of the organism. An organism refers to a particular organism in an ecosystem, say cat, dog, etc. A population includes all the members of the same organism that live in one place at one time. All the different populations that live in a particular area make up a community. The physical location of a community is called the habitat. Ecosystem is in turn a level of organisation and has one higher level of organisation called biosphere (Fig. 3.1).

The diversity of an ecosystem is a measure of the number of different species there, and how common each species is. Ecosystems are very complex and can contain hundreds or even thousands of interacting species. Each organism or species in the community has a role or profession in that community and in ecology; this is the organism's niche.

3.2 CLASSIFICATION OF ECOSYSTEM

An ecosystem can be classified as below:

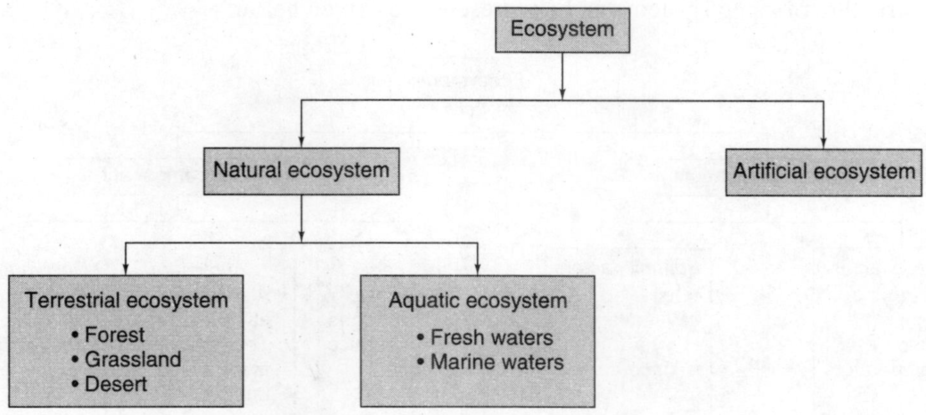

Fig. 3.1 Classification for ecosystem.

There is further classification in the above chart, but for a beginner, it is enough to concentrate on these areas only. Moreover, the study of artificial ecosystem is not the scope of an environmental scientist. The environmentalists deal with natural creations and management only. Moreover the system in artificial ecosystem does not offer much to study. Therefore, we are more interested in natural ecosystem and don't consider artificial ecosystem.

3.3 APPROACH TO ECOSYSTEM

With an ecosystem comprising large number of species, it is impractical to study the interaction of each organism with another. It is impossible to approach an ecosystem by studying the individual organism–environment relationship. Therefore, we study an ecosystem following a wholesome approach comprising two aspects (attributes) of an ecosystem given below;

(1) Structure or architectural process

(2) Function or working process

3.3.1 Structure of an Ecosystem

By architecture or structure of an ecosystem, we mean

- The composition of biological community including species, numbers, biomass, life history and distribution in space, etc.
- The quantity and distribution of non-living materials like nutrients, water, etc.
- The conditions of existence such as temperature, light, etc.

An ecosystem possesses both living components and biotic factors and non-living or abiotic factors.

The non-living factors, called abiotic factors, are physical and chemical characteristics of the environment. They include solar energy (amount of sunlight), oxygen, CO_2, water, temperature, humidity, pH, and availability of nitrogen.

The living components of the environment are called biotic factors. They include all the Living Things that affect an organism. Biotic components are often categorized as producers, consumers, and decomposers.

The structure of an ecosystem can be represented as given below.

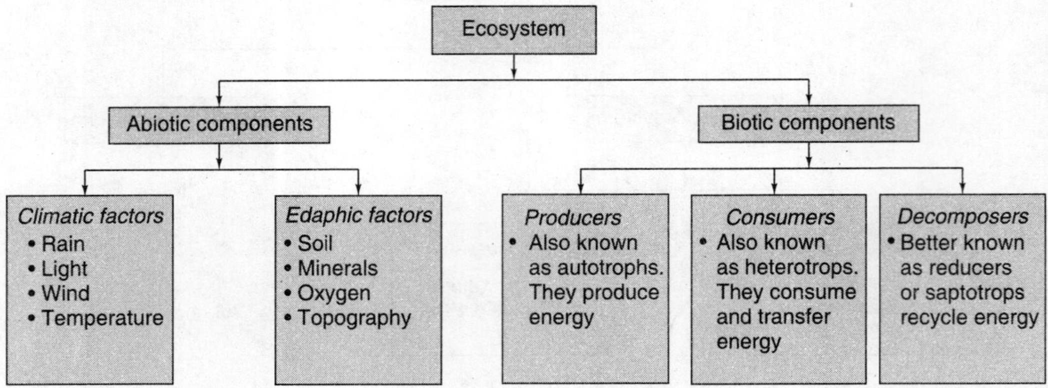

Fig. 3.2 Structure of an ecosystem.

3.3.2 Function of an Ecosystem

The function of an ecosystem is a broad, vast and often confused topic. The function of an ecosystem can be best studied by understanding the history of ecological studies. The function of an ecosystem can be studied under the three heads.

1. Trophic level interaction
2. Ecological succession
3. Biogeochemistry

Trophic level interaction deals with how the members of an ecosystem are connected to each other, based upon their nutritional needs. Changes in the features/members of an ecosystem over a period of time are dealt by ecological succession; whereas biogeochemistry focusses upon the cycling of essential materials in an ecosystem.

As we would be discussing about ecological succession and biogeochemistry in the coming chapters, we shall confine to trophic level interaction alone in this chapter. For examination purposes, the student may also stop with explaining the constituents of trophic level interaction.

Trophic level interaction was developed by zoologist Charles Elton. It deals with who eats who and is eaten by whom in an ecosystem. The study of trophic level interaction in an ecosystem gives us an idea about the energy flow through the ecosystem.

The trophic level interaction involves three concepts, namely:

1. Food chain
2. Food web
3. Ecological pyramids.

3.4 FOOD CHAIN

In an ecosystem one can observe the transfer or flow of energy from one trophic level to other in succession. A trophic level can be defined as the number of links by which it is separated from the producer, or as the position of the organism in the food chain. The patterns of eating and being eaten forms a linear chain called food chain which can always be traced back to the producers. Thus, primary producers trap radiant energy of the sun and transfer it to chemical or potential energy of organic compounds such as carbohydrates, proteins and fats.

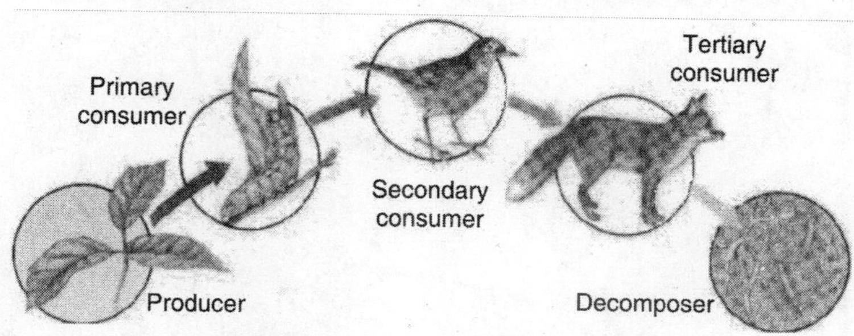

Fig. 3.3 Flow of energy in a food chain.

When a herbivore animal eats a plant (or when bacteria decompose it) and these organic compounds are oxidized, the energy liberated is just equal to the amount of energy used in synthesizing the substances (first law of thermodynamics), but some of the energy is released in the form of heat and not useful energy (second law of thermodynamics). If this animal is eaten by another one, along with transfer of energy from a herbivore to carnivore a further decrease in useful energy occurs as the second animal (carnivore) oxidizes the organic substances of the first (herbivore or omnivore) to liberate energy to synthesize its own cellular constituents. Such transfer of energy from organism to organism sustains the ecosystem and when energy is transferred from individual to individual in a particular community, as in a pond or a lake or a river, we come across the food chains. The number of steps in a food chain are always restricted to four or five, since the energy available decreases with each step. Many direct or indirect methods are employed to study food chain relationships in nature. They include gut content analysis, use of radioactive isotopes, precipitin test, etc.

In nature, basically two types of food chains are recognized—grazing food chain and detritus food chain.

1. **Grazing food chain:** This type of food chain (Fig. 9.4) starts from the living green plants, goes to grazing herbivores and on to the carnivores. Ecosystems with such type of food chain directly depend on the influx of solar radiation. Thus, this type of food chain depends on autotrophic energy capture and the movement of this energy to herbivores. Most of the ecosystems in nature follow this type of food chain. These chains are very significant from energy standpoint. The phytoplankton zooplanktons → fish sequence or the grasses → rabbit → fox sequence are the examples of grazing food chain. Further the producer → herbivore → carnivore chain is a predator chain. Parasitic chains also exist wherein smaller organisms consume larger ones without outright killing as in the case of the predators.
2. **Detritus food chain:** The organic wastes, exudates and dead matter derived from the grazing food chain are generally termed detritus. The energy contained in this detritus is not lost to the ecosystem as a whole; rather it serves as the source of energy for a group of organisms (detritivores that are separate from the grazing food chain, and generally called the detritus food chain. The detritus food chain represents an exceedingly important

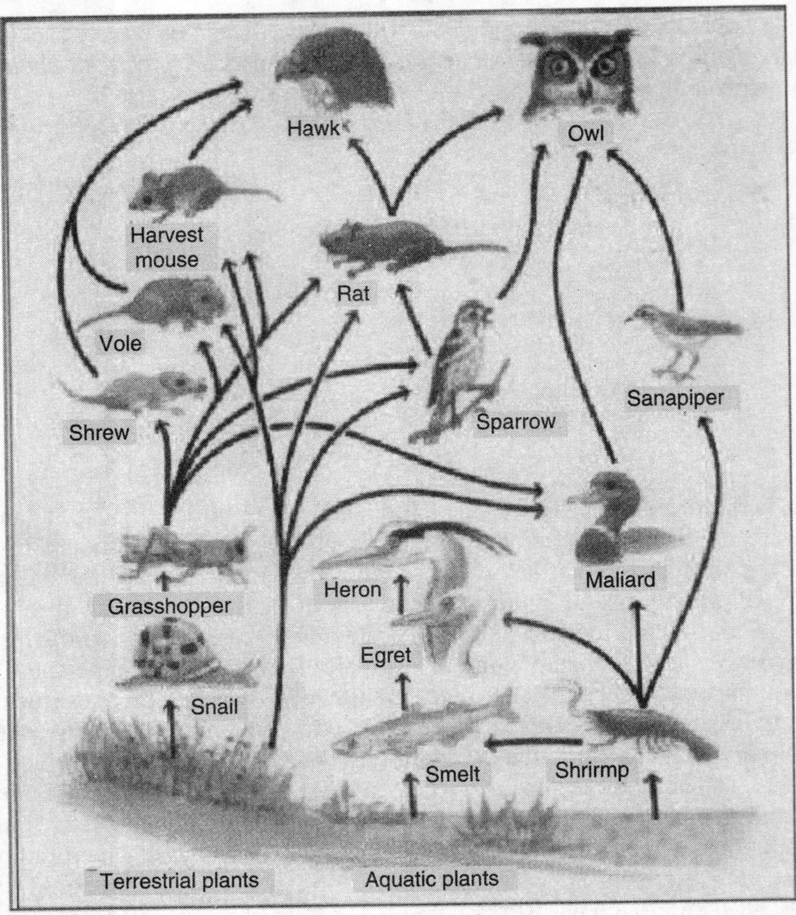

Fig. 3.4 Representation of food web.

component in the energy (low of an ecosystem). Indeed in some ecosystems, considerably more energy flows through the detritus food chain than through the grazing food chain. In the detritus food chain the energy flow remains as a continuous passage rather than as a stepwise flow between discrete entities. The organisms of the detritus food chain are many and include algae, bacteria, slime molds, actinomycetes, fungi, protozoa, insects, mites, crustacea, centipedes, molluscs, rotifers, annelid worms, nematodes and some vertebrates. Some species are highly specific in their food requirements and some can eat almost anything even detritus itself.

3. **Significance of food chain:** The food chain studies/help understand the feeding relationships and the interaction between organisms in any ecosystem. They also help us to appreciate the energy flow mechanism and matter circulation in ecosystem, and understand the movement of toxic substances in the ecosystem and the problem of biological magnification.

3.5 FOOD WEB

In nature simple food chains occur rarely. The same organism may operate in the ecosystem at more than one trophic level, i.e., it may derive its food from more than one source. Even the same organism may be eaten by several organisms of a higher trophic level or an organism may feed upon several different organisms of a lower trophic level. Usually, the kind of food changes with the age of the organism and the food availability. Thus, in a given ecosystem various food chains are linked together and interrelated with each other to form a complex network called food web. Generally, food webs are not too complex. As more and more species are involved in a web the connectance falls. Except in insect communities, omnivores are scare and when they occur they usually feed on species in adjacent trophic levels. Within habitats, food webs are rarely broken up into discrete compartments. The number of species of predators in a food web typically exceeds the number of species of prey by an average age of 1.3 predator species per prey species.

3.6 ECOLOGICAL PYRAMIDS

In the successive steps of grazing food chain-photosynthetic autotrophs, herbivorous heterotrophs, carnivores heterotrophs, decay bacteria, the number and mass of the organisms in each step is limited by the amount of energy available. Since some energy is lost as heat, in each transformation the steps become progressively smaller near the top. This relationship is sometimes called ecological pyramid. The ecological pyramids represent the trophic structure and also trophic function of the ecosystem. In many ecological pyramids, the producer form the base and the successive trophic levels make up the apex.

The ecological pyramids may be of the following three kinds:

1. **Pyramid of number:** It depicts the number of individual organisms at different trophic levels of food chain. This pyramid was advanced by Charles Elton (1927), who pointed out the great difference in the number of the organisms involved in each step of the food chain. The animals at the lower end (base of pyramid) of the chain are the most abundant. Successive links of carnivores decrease rapidly in number until there are very few carnivores at the top. The pyramid of number ignores the biomass of organisms and it also does not indicate the energy transferred or the use of energy by the groups involved. The lake ecosystem provides a typical example for pyramid of number.

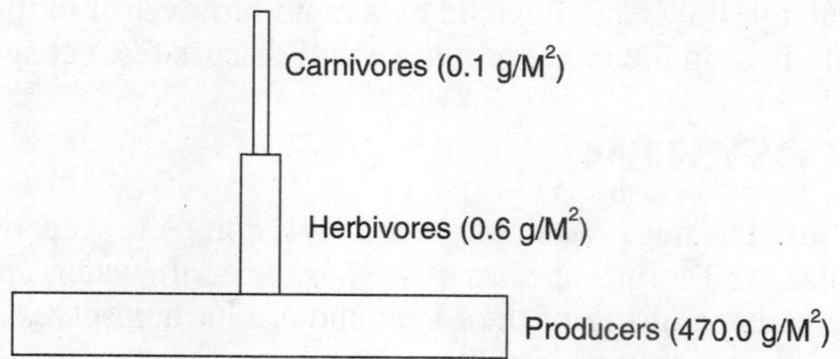

2. **Pyramid of biomass:** The biomass of the members of the food chain present at any time forms the pyramid of the biomass. Pyramid of biomass indicates decrease of biomass in each trophical level from base to apex. For example, the total biomass of the producers ingested by herbivores is more than the total biomass of the herbivores in an ecosystem. Likewise, the total biomass of the primary carnivores (or secondary consumer) will be less than the herbivores and so on.

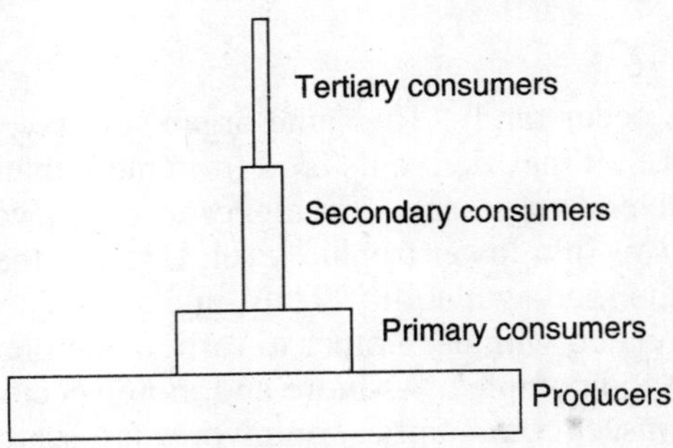

3. **Pyramid of energy:** When production is considered in terms of energy, the pyramid indicates not only the amount of energy flow at each level, but more important, the actual role the various organisms play in the transfer of energy. The base upon which the pyramid of energy is constructed is the quantity of organisms produced per unit lime, or in other words, the rate at which food material passes through the food chain. Some organisms may have a small biomass, but the total energy they assimilate and pass on, may be considerably greater than that of organisms with a much larger biomass.

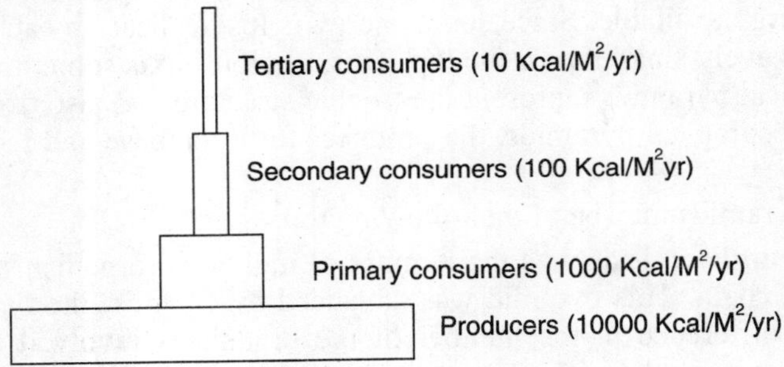

Energy pyramids are always slopping because less energy is transferred from each level than was stored into it. In cases such as in open water communities the producers have less bulk than consumers but the energy they store and pass on must be greater than that of the next level. Otherwise the biomass that producers support could not be greater than that of the producers themselves. This high energy flow is maintained by a rapid turnover of individual plankton, rather than an increase of total mass. In the next section we will discuss few ecosystem in detail.

3.7 FOREST ECOSYSTEM

A forest is more than just the trees. A single forest is a complete, functioning ecosystem that supports innumerable plant and animal species as well as the earth, water, and air subsystems. The subsystems provide the essence of life of the forest and are in themselves a by-product of forest systems, all of which are reciprocal and interdependent.

A forest ecosystem is a complex of living and non-living elements which interrelate. An ecosystem can be small (like your backyard) or large (like the planet earth). It depends on the range of individual species or group of species being discussed (e.g., a salmon stock or a forest type), geology (e.g., a mountain range or watershed), and other issues.

3.7.1 Types of Characteristic Features

3.7.1.1 *Temperature forest ecosystem*

The temperature forest ecosystem is very important on the earth. Temperate forests are in regions where the climate changes a lot from summer to winter. Tropical rainforests are in regions where the climate stays constant all year long. Temperate forests are almost always made of two types of trees, deciduous and evergreen.

Deciduous trees are trees that lose their leaves in the winter. Evergreens are trees that keep them all year long, like pine trees. Forests can either be one or the other, or a combination of both. A fourth kind of forest is a temperate rainforest. These are found in California, Oregon and Washington in the United States. These forests are made of redwoods and sequoias, the tallest trees in the world. The amount of rainfall in an area determines if a forest is present. If there is enough rain to support trees, then a forest will usually develop. Otherwise, the region will become grassland.

3.7.1.2 *The tropical rainforest ecosystem*

Tropical rainforests are one of the most important areas on the earth. These special ecosystems are home to thousands of animals and plants. Contrary to popular belief, rainforests are not only densely packed plants, but are also full of tall trees that form a ceiling from the sun above. This ceiling keeps small plants from growing. Areas where sunlight can reach the surface are full of interesting plants.

Structure of Forest Ecosystems: Different organisms exist within the forest layers. These organisms interact with each other and their surrounding. Each organism has a role or niche in sustaining the ecosystem. Some provide food for other organisms, others provide shelter or control populations through predation.

Producers: All living organisms intake energy in order to survive. In a forest ecosystem, trees and other plants get their energy from sunlight. Plants produce their own food, in the form of carbohydrates. Plants are, therefore, called the primary producers, since they produce the basic foodstuffs for other organisms within food chains and food webs. Photosynthesis is the chemical reaction that allows plants to produce their own food.

Consumers: Animals cannot produce their own food. They must consume food sources for the energy they need to survive. All animals, including mammals, insects, and birds are called consumers. Consumers rely on plants and other animals as a food source. Details of these animals in a forest ecosystem have been given earlier.

Primary consumes only eat plants and are referred to as herbivores. Second consumers are referred to as carnivores and feed on herbivores. Tertiary consumers are carnivores that feed on other carnivores. Omnivores eat both plant and animal matter.

Decomposers: Leaves, needles, and old branches fall to the forest floor as trees grow. Eventually all plants and animals die. These materials are decomposed by worms, microbes, fungi, ants, and other bugs.

Decomposers break these items down into their smallest primary elements to be used again. Decomposers are important in order to sustain the nutrient cycle of ecosystems.

Humans are parts of forest ecosystem

Humans are consumers. We get food and materials from forests. Because of this, we are a part of the forest ecosystem. Human consumption alters forest ecosystems. Human intervention may be necessary to sustain forest communities under the increased pressure of human use.

The components of an ecosystem are connected in a complex web. Many questions come forward while considering forest web, e.g.

How does the forest web sustain itself?

What starts the food cycle in the forests?

We will examine a variety of cycles that exist in forest ecosystems, and their functions.

3.7.2 Variety of Cycles

3.7.2.1 Energy cycle

The sun starts the food pyramid by fuelling the photosynthesis process. The sun's energy constantly changes form as plants convert light into plant biomass, which is then consumed by other organisms. Plants perform an important action of energy fixation, and establish the basis of all life.

3.7.2.2 Water cycle

Water is constantly cycling. The water cycle collects, purifies, and distributes water on the earth. Without the water cycle, life on the earth would be impossible. Trees and plants are part of this water cycle. Transpiration is the controlled evaporation progress by which plants lose H_2O through the pores in their leaf structures. A full-grown tree can transpire hundreds of gallons of water a day during growing season.

3.7.2.3 Nutrient cycle

We already know trees rely on nutrients like phosphorus and nitrogen for healthy growth and reproduction. Throughout a tree's life stages, they constantly use and return nutrients to the soil. Nutrient cycles regularly transform nutrients from the non-living environment.

Disturbances can be small to large scale events. The composition of the forest is altered by disturbance. Wind throw may remove several trees from the forest. Some species of trees are more susceptible to wind throw, like aspen.

Intense wildfires are a disturbance that can eliminate all trees in an area. Dry, hot weather, a build up of dead plant material and lightning start many natural wildfires. Wildfires are part of the natural succession process of forests.

Other climate disturbances also change forests. A winter with temperatures below zero for several consecutive days weakens some trees, making them more susceptible to diseases. At the same time, a hot dry summer in Wisconsin causes water stress in some trees. While this may not kill plants and trees right away it weakens them. This makes it harder to store food for the winter, or to fight off disease and pests.

Insects can alter the composition and structure of forests by removing entire species from an area. Herbivores such as deer browse tree bark, branches, and young seedlings. This browsing severely affects the tree's growth potential and limits the ability of the next generation to regenerate. Browsing by insects and herbivores weakens and exposes trees to natural and exotic diseases.

Although natural disturbances cause dramatic changes in forests, these events are important for forest health. These changes provide opportunities for the species to change over time.

3.7.3 Human Disturbance

Nature has changed by human disturbance. For over 10,000 years, humans have influenced the composition of forests by both the introduction and removal of species. When humans remove trees, this opens the over story to suppress trees, leading to quick changes in structure and function.

Human change often mimics natural forest changes but at a much faster pace. The uses of many management techniques are often meant to speed along the natural succession proceed of forests and alter the composition and structure. Examples of human disturbances are forest logging and thinning, fires, and the introduction of exotic species.

3.8 GRASSLAND ECOSYSTEM

3.8.1 Types and Characteristic Features

Over one quarter of the earth's surface is covered by grasslands. Grasslands are found on every continent except Antarctica, and they make up most of Africa and Asia. There are several types of grassland. They are distinguished by different names like plains, prairies, savannas and pampas.

Grasslands develop where there isn't enough rain for forests but too much rain for deserts. Grasslands are full of many types of grass. Fields of wheat are considered grasslands, even though they are often cultivated by people. Grass is special because it grows underneath the ground. During cold periods the grass can stay dormant until it warms up.

One can find at least one desert on every continent except Europe and Antarctica. Each desert is different in some way, but they all have one thing in common. In order for an area of land to be considered a desert, it must receive less than 10 inches of water a year.

How come deserts get such little water? Clouds are scarce in these regions and we all know that without clouds, there can't be rain, snow or any other precipitation. But clouds also serve another purpose – they block out some of the sun. The desert gets mighty hot during the day because the sun beats the sand. At night, the desert gets very cold, because there aren't clouds around to stop the heat from escaping to the atmosphere.

There are plenty of differences between different deserts of the world. Some deserts are made of very fine red sand, others consist of sand mixed with pebbles and rocks. The desert sand started out as rock, but years of weathering by wind and water has created dunes in the deserts. These sands are mostly minerals, and sometimes oil is found hidden deep within the rocks.

3.8.2 Structure and Function

The different components of a desert ecosystem are:

Abiotic Component: The abiotic component includes the nutrients present in the soil and the aerial environment. The characteristic feature of the abiotic component is lack of organic matter in the soil and scarcity of water.

Biotic Component: The various biotic components representing three functional groups are:

(*a*) ***Producer organisms:*** The producers are mainly shrubs or bushes, some grasses and a few trees. The most famous desert plant is the cactus. There are many species of cacti. The saguaro cactus is the tall, pole shaped cactus. The saguaro can grow up to 40 feet tall. It can hold several tonnes of water inside its soft tissue. Like all cacti, the saguaro has a thick, waxy layer that protects it from the sun.

Other succulents include the desert rose and the living rock. This strange plant looks like a spiny rock. Its disguise protects it from predators. The welwitschia is a weird looking plant. It has two long leaves and a big root. This plant is actually a type of tree and it can live for thousands of years.

There are many other kinds of desert plants. Some of them have thorns while others have beautiful flowers and deadly poisons. Even in the worst conditions, these plants continue to thrive.

(b) ***Consumers:*** These include animals such as insects and reptiles. Besides them, some rodents, birds and some mammalian vertebrates are also found.

Desert Insects and Arachnids: There are plenty of insects in the desert. One of the most common and destructive pests is the locust. A locust is a special type of grasshopper. They travel from place to place, eating all the vegetation they find. Locusts can destroy many crops in a single day.

Not all desert insects are bad, though. The yucca moth is very important to the yucca plant, because it carries pollen from the flower to the stigma. The darkling beetle has a hard, white, wing case that reflects the sun's energy. This allows the bug to look for food during the day.

There are also several species of ants in the desert. The harvester ants gather seeds and store them for use during the dry season. And the honey pot ants have a very weird habit. Some members of the colony eat large amounts of sugar, so that their abdomens get too large for them to move. The rest of the colony feeds off this sugar.

There are also arachnids in the desert. Spiders are the most notable arachnids, but scorpions also belong in this group. Some species of scorpions have poison in their sharp tails. They sting heir predators and their prey with the piercing tip.

The desert is also full of wild horses, foxes and jackals, which are part of the canine family. And we can't forget the cats. Lions are found all over the deserts of southern Africa. They get their water from the blood of their prey.

Camels – The Car of the Desert: Camels could be included in the mammal section. Camels are the cars of the desert. Without them, people would have great difficulty crossing the hot terrain. There are two types of camels: Bactrian and dromedary. The main difference between the two is the number of humps. Dromedaries have one hump and Bactrian have two. Both are used by people, but Bactrian are found in the wild.

Camels are great transportation because they use very little water. Camels can withstand very high temperatures without sweating. They also store fat in their humps for food. If a Bactrian camel travels a long distance without eating, its hump gets smaller.

(c) ***Decomposers:*** Due to poor vegetation the amount of dead organic matter is very less. As a result the decomposers are very few. The common decomposers are some bacteria and fungi, most of which are thermophillic.

3.9 AQUATIC ECOSYSTEM

An **aquatic ecosystem** is an ecosystem located in the body of water. Communities of organisms that are dependent on each other and on their environment live in aquatic ecosystems. The two main types of aquatic ecosystems are marine ecosystem and freshwater ecosystem. It also contains plants and animals that predominantly depend on a significant amount of water to be present for at least some part of the year.

3.9.1 Components of an Aquatic Ecosystem

Pond layers – Like in a forest, the top, middle, and bottom of a pond can be vastly different from each other, and even the layers in between. Under different temperature or light conditions the water in a pond can vary greatly in oxygen, clarity, and other factors that affect where plants and animals might live. The air above the pond and the land below the pond are important as well, as those provide space for animals to live, plants to root, and predators and prey to interact.

Diversity – while many people think of a pond as just a small lake with frogs and fish, there are thousands of different species of plants and animals living together in a natural pond. The more diverse a pond is (more species that it has) the stronger and healthier it is.

Microorganisms: Some of the most important plants and animals in a pond are so small that they are difficult to see without a microscope. They are called "microorganisms" (micro=small, organism=life form) and while a few may cause diseases, almost all are very beneficial and important to a pond ecosystem. While bigger animals may fly, walk, or swim away to other ponds, microorganisms are always present in large numbers.

Macroorganisms: Larger plants and animals that are easy to see on a pond are called "macroorganisms (macro=large). They are the plants and animals that we often notice first, and can more easily spread from pond to pond.

3.9.2 Aquatic Systems that can be Found at Riverbed

- Marsh – shallow water with non-woody plants growing above water level
- Swamp – like a marsh but with bushes and trees growing from the water as well
- Pond – a small and shallow body of water with plants growing above water level only on the edges; generally freezes solid during winter.
- River – a moving body of water that flows from one place to another.
- Stream – smaller than a river, may even dry up sometimes
- Puddles – body of water that lasts for a few days or more may attract aquatic life
- Spring – area where underground water is discharged onto the land surface forming a pond or stream

Other types of aquatic systems found on the earth are:

- Lake
- Ocean
- Creek
- Lagoon
- Bog
- Sea
- Glacier
- Tidal pool
- Geyser
- Fen
- Floodplain
- Bog
- Estuary
- Aquifer
- Salt lake

3.9.3 Types, Characteristic Features, Structure and Functions

3.9.3.1 Lakes

A lake is a body of water completely surrounded by land. Lakes can either be salty or fresh water. Most lakes are in places where glaciers used to exist. When a glacier moves forward, it carves away a deep valley and when the ice melts it forms a lake in the valley. Other lakes are formed in craters or when a river changes its course.

Lakes are short-lived surface features because the water can sink into the ground or evaporate into the sky. In order for a lake to remain same, it must be constantly fed by a river or rainfall.

Lakes are inland bodies of water that lack any direct exchange with an ocean. Lake ecosystems are made up of the physical, chemical and biological properties contained within these water bodies. Lakes may contain fresh or salt water (in arid regions). They may be shallow or deep, permanent or temporary. Lakes of all types share many ecological and biogeochemical processes and their study falls within the discipline of 'limnology'. Lakes are superb habitats for the study of ecosystem dynamics: interactions among biological, chemical and physical processes are frequently either quantitatively or qualitatively distinct from those on land or in air. Because the boundaries between water and land, and water and air are distinct, there is tight coupling among many ecosystem components.

Although lakes contain <0.01% of all the water on the earth's surface, they hold >98% of the liquid surface fresh water. Many organisms depend on fresh water for survival, and humans frequently depend on lakes for a great many 'goods and services' such as drinking water, waste removal, fisheries, agricultural irrigation, industrial activity, and recreation. For these reasons lakes are important ecosystems.

Lake ecosystems are influenced by their watersheds, the geological, chemical and biological processes that occur on the land and streams that lie uphill. The movement of chemicals, sediments, detritus, and of many organisms, is typically unidirectional from the watershed to the lake, but fish may migrate upstream, and aquatic insects may emerge and disperse on to land. A lake and its watershed are often considered to be a single ecosystem.

(i) Lake Thermal Structure: During summer, sunlight increases the temperature of lake surface water. Water at greater depths is less warm. Wind at the surface causes the top several metres of lake water to mix homogeneously to form a warm surface layer called the 'epilimnion'. Below the level of wind mixing, temperature drops rapidly through a zone called the thermocline, and below this is a region of homogeneously cool water called the 'hypolimnion'. This two-layer physical structure is called 'thermal stratification'. Summer stratification neither occurs in shallow water bodies, nor in high latitude or high altitude lakes where summers are short. In winter, surface water directly under the ice is about 0°C and deeper water is slightly warmer. Wind cannot mix water below the ice so winter stratification persists while the lake is frozen. Winter stratification does not occur in tropical or subtropical lakes, in many large temperate-zone lakes, or in many salt lakes. Lakes that stratify during summer and winter mix from top to bottom during spring and autumn in events called 'turnover'. Lakes that do not stratify in winter, and those that do not stratify in summer, mix continuously throughout their unstratified periods. Tropical lakes may stratify every day and mix every night.

Summer thermal stratification divides the lake environment into two distinct parts: an illuminated and warm epilimnion where phytoplankton (algae and photosynthetic cyanobacteria) carry out photosynthesis, and a dark and cool hypolimnion where this production is decomposed.

Decomposition also dominates under the ice of lakes that freeze over. When the amount of decomposing organic matter is large, anoxic conditions (no dissolved oxygen) can prevail.

(ii) Lake Habitats and Food Chains: In the pelagic zone (open-water region) of the lake, phytoplankton carries out photosynthesis at the base of the food web. These unicellular or simple colonial algae or cyanobacteria sink only very slowly and are easily re-suspended by wind-driven water movements. Very small phytoplankton and bacteria are consumed by unicellular zooplankton; larger phytoplanktons are consumed by larger zooplankton. Some taxa are generalists that filter most algae encountered (e.g. Daphnia) and can have a major impact on phytoplankton densities in lakes. Other taxa tend to select the more nutritious phytoplankton to consume.

Fine detritus (dead plankton) suspended in the pelagic zone is colonized by heterotrophic bacteria, which is then consumed by protists (ciliates and flagellates) and generalist grazers. The protists are in turn consumed by other protists or by copepods. This return of energy to the pelagic food chain is called the 'microbial loop'. Its ultimate importance in lake ecosystems remains a point of debate.

Other detritus produced in the epilimnion may be trapped at the thermocline or sink into the hypolimnion where it is decomposed by bacteria. Grazing zooplankton is consumed by predatory invertebrates (rotifers, cyclopoid copepods, some cladocerans, and insect larvae of the genus Chaoborus) or vertebrates (typically fish). Zooplanktivorous fish also consume predatory invertebrates. Piscivorous fish (consume other fish) sit atop the natural food webs of most lakes although in some cases there are piscivorous birds, otters, seals, crocodiles or alligators.

Humans act as top predators in a great many lakes worldwide. Submersed rooted plants (macrophytes) growing at the lake margin define the littoral zone, and provide habitat for attached algae (epiphytes), insects and other invertebrates, and fishes that use this area for breeding, cover and foraging. Some fish consume rooted plants, but most eat invertebrates or other fish. The littoral zone captures much of the chemicals, sediments and detritus washing in from the watershed and processes these materials before they reach the pelagic zone. Because macrophytes require light to grow up from the lake bottom each spring, the distance the littoral zone extends into the lake depends upon how steeply the lake bottom drops off near shore, and how turbid the lake water is with phytoplankton or suspended sediments.

The profundal zone is the bottom water and sediments of deep lakes where there is insufficient light for photosynthesis. In this region bacteria and fungi obtain energy by decomposing detritus, or by chemoautotrophy (breaking energy-rich chemical bonds in inorganic molecules). Insect larvae (typically dipterans) and annelid worms (oligochaetes) live in the soft bottom sediments and consume detritus. All organisms that live in, on, or in association with the lake bottom are called 'benthos'.

There are exchanges among all lake habitats. Nutrients and dissolved organic carbon ('DOC') molecules released by macrophytes in the littoral zone diffuse to the pelagic zone where they are used by algae and bacteria. Detritus from the epilimnion sinks to the hypolimnetic profundal zone where nutrients are released. Nutrients in the hypolimnion are returned to the epilimnion via diffusion, turbulent mixing across the thermocline, and at turnover. Planktonic animals migrate between the epilimnion and hypolimnion on a daily cycle, excreting nutrients as they travel. Fish move between the littoral and pelagic zones feeding and breeding in one place and excreting and defecating nutrients in another.

(iii) The Role of Nutrients: Identifying the determinants of algal growth in lakes is crucial both for understanding lake ecosystem functioning, and because extensive algal blooms are a nuisance that can be caused by human activity. Primary production in lake ecosystems depends on nutrients and light as essential resources. Phytoplankton take up nutrients dissolved in lake water; rooted macrophytes obtain nutrients from the sediments. Primary producers are potentially limited by carbon, nitrogen or phosphorus. Of these, carbon is the most common element in algal tissue and is also the most abundant in solution in lake water (CO_2, HCO_3^-, or CO_3^{2-}). Nitrogen (NO_3^-, NH_4^+, N_2) and phosphorus (PO_4^{3-}) are much less available, suggesting that phosphorus, followed by nitrogen, is most likely to limit algal production in lakes.

Limitation by other nutrients can occur, for example diatoms, algae characterized by hard cell walls containing silica, can be limited by silica availability. Light is also taken up and consumed like other algal resources. Light limitation can occur during algal blooms when cells close to the surface shade algae deeper in the water column, or when phytoplankton shade the lake bottom and prevent macrophyte growth.

1. **Phosphorus:** Phosphorus is essential in nucleic acids, phospholipids, adenosine triphosphate (ATP) etc., but is scarce in bio accessible forms in lakes. Whereas carbon and nitrogen enter lakes by exchange of CO_2 and N_2 from the atmosphere, the only natural source of phosphorus is weathering from the watershed of PO_4^{3-} ions, which dissolve only poorly in water. Thus, bio-available phosphorus occurs in extremely low concentrations in fresh water and is taken up quickly by phytoplankton. In water containing dissolved oxygen, PO_4^{3-} can combine with iron to form insoluble salts that sink to the lake bottom, making phosphorus still less available to algae. Under anoxic conditions in the hypolimnion or the sediments, PO_4^{3-} becomes soluble again (a process called 'internal loading') and can stimulate algal growth through mixing processes.

   The paramount importance of phosphorus as a resource for primary producers in lakes has been shown: (1) as a close relationship between phosphorus supply to lakes worldwide and the abundance of algae they contain; (2) through studies of lakes polluted with phosphorus-containing human waste and laundry detergents, where phosphorus removal has eliminated noxious algal blooms; and (3) through whole lake experiments in which enrichment with both phosphorus and nitrogen resulted in a bloom of cyanobacteria much higher than the growth of algae observed in a lake enriched only with nitrogen. The important role played by phosphorus in lake algal production is generally accepted today, and successful management of many lake ecosystems depends upon controlling phosphorus inputs.

2. **Nitrogen:** Algal cells require nitrogen to synthesize proteins. Typically nitrogen is available in lakes in higher concentrations than phosphorus. Most algae take up NH_4^+ ions (from decomposition) or NO_3^- ions (from bacterial nitrification of NH_4^+). However, many species of cyanobacteria can also use N_2 diffusing across the air-water boundary in a process called nitrogen fixation which forms NH_4.

   Lake ecosystems are not as frequently nitrogen limited as marine or terrestrial ecosystems. Lakes can only be nitrogen limited if the relative supply of phosphorus is greater than that of nitrogen. Adding wastewater from sewage to a lake can create this condition, which in turn favours the growth of nitrogen-fixing cyanobacteria because they have an alternative source of nitrogen. This is why lakes with excessive phosphorus inputs tend to have noxious blooms of cyanobacteria.

3.9.4 Trophic Control of the Food Chain

The enrichment of algal growth by nutrients, and the resulting enhanced production of grazing zooplankton, predatory invertebrates and fish, is called 'bottom-up control' because nutrients flow from the base of the food chain up through the trophic levels. However, an alternative regulation of algal abundance and primary production can come from 'top-down control'. When grazing zooplankton, especially Daphnia, is abundant, algal biomass can be suppressed by consumption even when nutrient concentrations are high. Large-bodied and abundant zooplankton occur in lakes that have a relatively low number of zooplanktivorous fish. And low zooplanktivorous fish density is found in lakes with high numbers of piscivorous fish. Thus, an alternative to reducing phosphorus inputs as a means of regulating the abundance of algae is to increase the density of stocked piscivorous fish. Managing fisheries has become a tool for controlling phytoplankton biomass in lakes.

In natural lakes both bottom-up and top-down processes operate. Their relative importance may vary among lakes or can change seasonally within a lake. For example, in many temperate-zone lakes in spring the disappearance of ice increases light availability and turnover which mixes nutrients throughout the water column. The result is a bottom-up driven algal bloom. Daphnia then grows in abundance to the point where it suppresses algal densities and causes the water to become quite transparent in a seasonal top-down event called the 'clear water phase'.

Within these broad generalizations, there is ecologically important variations among taxa. Only some phytoplankton can fix nitrogen. Not all phytoplankton species are equally edible for grazing zooplankton: cyanobacteria can be of poor nutritional quality (low in essential fatty acids), often contain toxins, and grazers have a reduced impact on these cells. Similarly, not all zooplankton species are equally vulnerable to predation. Fish are generally visually-orienting predators that feed during the day on the largest prey in the water. Small zooplankton and those that migrate into the dark hypolimnion during the day can escape detection but still graze on phytoplankton. Still other zooplankton grow spines that defend them against predation by invertebrates, and zooplanktivorous fish may grow sufficiently large that piscivores are unable to capture and consume them.

The rates and even the occurrence of many ecosystem processes, such as transfers of nutrients and energy from one trophic level to the next, depend not only on the physical and chemical environment, but also critically on the characteristics of the species present.

Rivers and Streams: Rivers are very important to the earth because they are major forces that shape the landscape. Also, they provide transportation and water for drinking washing and farming. Rivers can flow on land or underground in deserts and seas.

Rivers may come from mountain springs, melting glaciers or lakes. A river's contribution to the cycle is that it collects water from the ground and returns it to the ocean. The water we drink is about 3 billion years old because it has been recycled over and over since the first rainfall.

A delta is where a river meets the sea. Usually, the river flows more slowly at the delta than at its start because it deposits sediment. Sediment can be anything from mud, sand and any rock fragments. A special environment is created when the fresh water from the river mixes with the salty ocean water. This environment is called estuary.

The longest river is the Nile River in Africa, and the Amazon River in South America carries the most water. The muddiest river is the Yellow River in China.

Rivers provide many critical goods and services, including drinking water, power generation, nutrient recycling, organic matter retention, and habitat for many unique plants and animals and recreational activities. Coniferous forests and acidic soils cause many of the rivers to be tea-coloured, indicating high concentrations of dissolved organic carbon (DOC). Because of the surrounding thin soils and the low buffering capacity of the waters, these ecosystems are vulnerable to acid precipitation. As described in the section on Forests, fire suppression has reduced the frequency and extent of wildfires, but when fires do occur they have large impacts on the biological and chemical processes occurring at the land—water interface and in streams. Forest harvest and mining are now the major disturbances in the watersheds of these streams. Stream flow is generally stable in areas with coarse-textured sediments and large expanses of lakes and wetlands, while areas with exposed bedrock or thin soils have very flashy flow regimes.

Dominant disturbances result from riparian vegetation disruption or removal, nutrient and pesticide loading from agricultural fields, industrial discharge, and storm water discharge from urban areas.

3.10 THE SYNERGISTIC THREATS TO STREAM ECOSYSTEMS

Stream ecosystems are regulated by features and processes occurring at a range of spatial scales. At the largest scale, climate, geomorphology, and land use control channel morphology and stream hydrology, thermal regime and water chemistry, and biotic community structure. At finer scales, the availability of suitable habitat and food resources, as well as species interactions regulate organism populations. When superimposed on climate changes, anthropogenic disturbances including land use activities, nitrogen and acidic deposition, and species exchanges will undoubtedly alter the way in which stream ecosystems respond to change. Changes in climate variables relevant to streams include increased air temperature and a general drying of watersheds especially during summer and autumn, mainly due to increased air temperatures and higher rates of evaporation during a longer ice-free period. Past trends of longer ice-free periods, earlier spring flows, and increased frequency of mid-winter breakup events and ice jams are expected to persist. Despite a general drying, model predictions for the region suggest that over the next 100 years precipitation will increase during winter and spring, which would increase the magnitude of spring floods, especially when intense rain storms coincide with snowmelt when soils are still frozen. The magnitude of stream responses will vary greatly across the region mainly because of the relative contribution of groundwater versus surface water to flow regimes.

(i) ***Estuaries:*** An estuary is a unique environment where fresh water and salt water come together. Estuaries are found on the coast where a river or bay or other sources of fresh water has access to the open sea. A good example of an estuary is a salt marsh that can be found close to the coast. Another example is when a river feeds directly into the ocean. The largest estuary in the United States is the Chesapeake bay estuary on the east coast of the U.S.

Estuaries are affected by the tides. So, there can be changes of salinity, temperature and other physical properties in an estuarial system which means the organisms there must be very tolerant to change. Even with that consideration, estuaries are among the most fertile places in the world. All kinds of plants and animals live there. When looking at estuaries, scientists quickly realized that these areas were extremely nutrient-rich because of sediment deposit of rivers, creeks or streams feeding into the salt water environment. Unfortunately, estuaries haven't always been seen as valuable. In the past, they were seen as worthless and were used as dumps, or places for new land development (by filling in the marshy area). However, now the systems are being set up to establish, manage and maintain estuary reserves, and to provide for their long-term stewardship.

(ii) Pond Ecosystem: The definition of a pond is any body of water over 1m square that is inundated, "wet" for more than 8 months of the year.

That definition tells us quite a lot about a pond's ecosystem. The first is obvious, it doesn't have to be large to be useful, in fact even a bird bath will be visited by the water boatman if he is in need of a place to rest and eat.

If a pond does dry out it isn't a disaster, both plants and animals have adapted to this eventuality.

The pond is the most valuable as far as wildlife is concerned, if the pond is reasonably shallow (no more than 90 cm) and very well planted it will support a vast array of animals and plants.

Some potential inhabitants and visitors are frogs, newts, toads, foxes, hedgehogs, badgers birds, bats, grass snakes, slow worms, water voles, birds of all sizes, great diving beetles, water boatman, dragon fly larvae and adults, damsel flies, daphnia, cyclops, spiders, water scorpion (not dangerous), zoo plankton, wirlygig beetle, Pond skaters and Lace wings. The potential list of insects and other small things is huge.

An ecosystem is a dynamic complex unit of plant, animal, and microorganism communities and the nonliving environment, interacting as a functional unit. Remember that the organisms living in an ecosystem are broken down into categories: producers, consumers, and decomposers. A pond is a quiet body of water that is too small for wave action and too shallow for major temperature differences from top to bottom. It usually has a muddy or silty bottom with aquatic plants around the edges and throughout. However, it is often difficult to classify the differences between a pond and a lake, since the two terms are artificial and the ecosystems really exist on a continuum. Generally, in a pond, the temperature changes with the air temperature and is relatively uniform. Lakes are similar to ponds, but because they are larger, temperature layering or stratification takes place in summer and winter, and these layers turnover in spring and fall. Ponds get their energy from the sun. As with other ecosystems, plants are the primary producers. The chlorophyll in aquatic plants captures energy from the sun to convert carbon dioxide and water to organic compounds and oxygen through the process of photosynthesis. Nitrogen and phosphorus are important nutrients for plants. The addition of these substances may increase primary productivity. However, too many nutrients can cause algal blooms, leading to eutrophication.

3.10.1 Producers

Phytoplankton, literally "wandering plants," are microscopic algae that float in the open water and give it a green appearance. They carry out photosynthesis using carbon dioxide that is dissolved in water and releases oxygen that is used by the bacteria and animals in the pond.

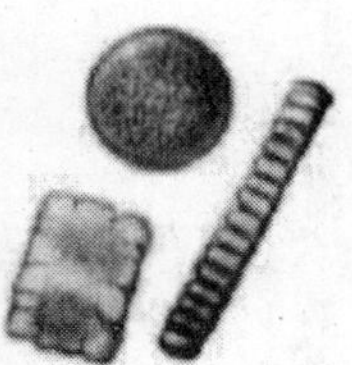

Periphytic algae are microscopic algae that attach themselves to substrates and give the rocks and sticks a greenish brown slimy appearance. They also carry out photosynthesis and produce oxygen, often near the bottom of the pond where it can be used by decomposers.

Submerged plants grow completely under water.

Floating plants include plants that float on the surface and plants that are rooted on the bottom of the pond but have leaves and/or stem that float.

Emergent plants are rooted in shallow water but their stems and leaves are above water most of the time.

Shore plants grow in wet soil at the edge of the pond.

3.10.2 Consumers

Zooplankton are microscopic animals that eat phytoplankton or smaller zooplankton. Some are single-celled animals, tiny crustaceans, or tiny immature stages of larger animals. Zooplankton float about in the open water portions of the pond and are important food for some animals.

Invertebrates include all animals without backbones. Macro-invertebrates are big enough to be seen with the naked eye. Some of them are only found in clean water.

Vertebrates are animals with backbones. In a pond these might include fish, frogs, salamanders, and turtles.

3.10.3 Decomposers

Animal waste and dead and decaying plants and animals form **detritus** on the bottom of the pond. **Decomposers,** also known as **detritovores,** are bacteria and other organisms that break down detritus into material that can be used by primary producers, thus returning the detritus to the ecosystem. As this material decomposes it can serve as a food resource for microbes and invertebrates. During decay microbes living on detritus can pull nutrients from the overlying water thus acting to improve water quality. In the process of breaking down detritus, decomposers produce water and carbon dioxide.

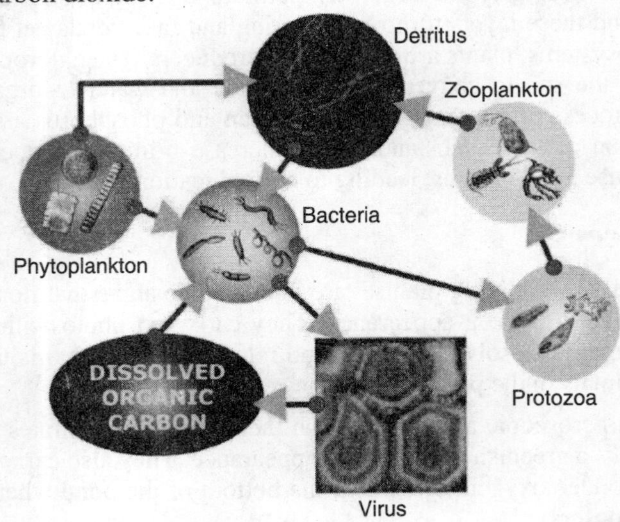

Learning Outcomes

The main learning outcomes from this chapter should be fundamental understanding of:

- Concept, Structure and functions of an ecosystem;
- Producers, consumers and decomposers; Energy flow in the ecosystem; Ecological succession; and Food chains, food webs and ecological pyramids.

- Also the readers will get comprehensive understanding of introduction, types, characteristic features, structure and functions of the following ecosystems: forest ecosystem; grassland ecosystem; desert ecosystem; aquatic ecosystems.

Readers are also advised to carry out following exercise:

Exercise

1. Write a 1000 words essay on "learning outcomes from the present chapter".
2. Give an example of food chain and web and create schematic diagram.
3. Write a critical review on threats faced by our ecosystem in general and then selected one (from exercise 2) in particular.

Chapter 4

Ecosystem Biodiversity and Conservation

4.1 INTRODUCTION

Biodiversity (an abbreviation of biological diversity) describes the diversity of life from genes to ecosystems and spans every level of biological organization. Biodiversity means different things to different people and there are many ways to index, measure, characterize, and represent its complex organization. Biodiversity includes the complex processes operating at and among these respective levels. Biodiversity plays an important role in as much as it does for human health. Populations and species migration, for example, are more sensitive indicators of that sustain and contribute towards the well-being of humanity. An understanding of biodiversity has practical applications for ecosystem-based conservation planners as they make ecologically responsible decisions in management recommendations to consultant firms, governments and industry.

The term BIODIVERSITY was first coined by the entomologist E.O. Wilson in 1986. A neologism from biology and diversity, it refers to the variety of life on the planet. There is no single standard definition for biodiversity.

Biodiversity may be defined as "the totality of different organisms, the genes they contain, and the ecosystems they form".

The Convention on Biological Diversity defines biodiversity as the "variability among living organisms from all sources including, among other things, terrestrial, marine, and other aquatic ecosystems and the ecological complexes of which they are a part; this includes diversity within species, between species and of ecosystems".

4.2 LEVEL OF BIODIVERSITY

Biodiversity may be considered at three levels: genetic diversity, species diversity, and ecosystem diversity.

Conservation and sustainable use of biodiversity is fundamental to ecologically sustainable development. Biodiversity is a part of our daily lives and livelihood, and constitutes resources upon which families, communities, nations and future generations depend. Every country has the responsibility to conserve, restore and sustainably use the biological diversity within its jurisdiction. Biological diversity is fundamental to the fulfilment of human needs. An environment rich in biological diversity offers the broadest array of options for sustainable economic activity, for sustaining human welfare and for adapting to change. Biodiversity is defined as the variability among living organisms from all sources, including terrestrial, marine and other aquatic ecosystems and the ecological complexes of which they are a part. This includes diversity within species, between species and of ecosystems.

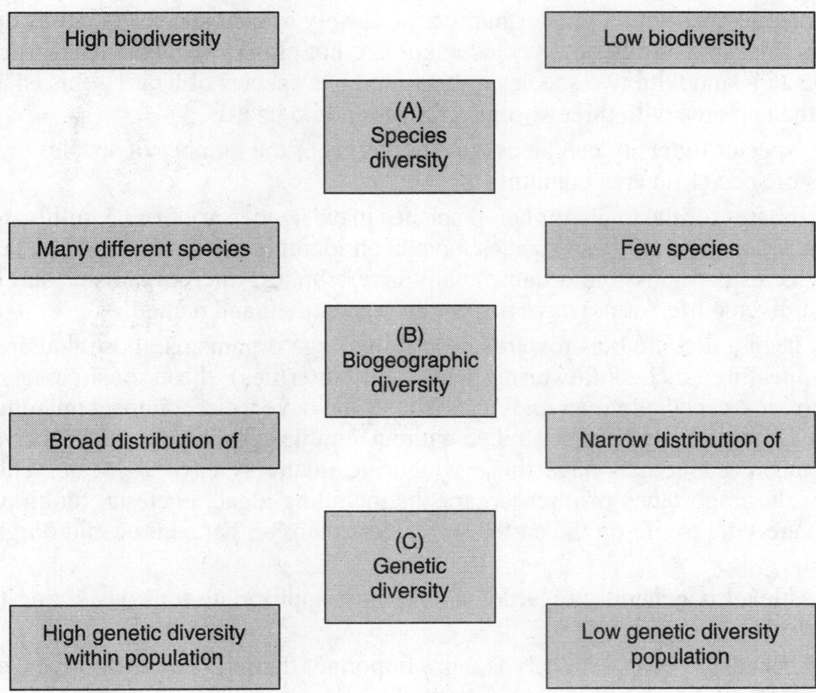

Fig. 4.1 Schematics for biodiversity classification.

4.2.1 Species Diversity

Species diversity refers to the variety of living species within a geographic area. A species may be defined as a group of organisms which are able to interbreed freely under natural conditions to produce viable offspring.

Species diversity may be measured using the following characteristics:

I. Species richness – the number of species within a particular sample area.

(In combination with)

Species evenness – this refers to the evenness in number of individuals of each species in the area, e.g., for two sample areas X and Y, there are two species, a, and b. In X, there are 92 individuals of species a, and only 8 of species b, while in Y, there are 50 individuals of each species.

If species richness only was used to account for species diversity in X, the diversity might seem lower than Y (although both have the same number of species), because almost all the individuals encountered would be from only one species. Species evenness in conjunction with species richness is thus a more useful indicator of species diversity, because it takes into account rarer species.

II. Relative abundance of species of various categories (the categories might include size classes, tropic levels, taxonomic groups, or morphological types).

For example, an area with a greater number of closely related species is not as diverse as the same area with the same number of species which are not closely related. An illustration of this point would be an island with two species of birds and one species of lizard – this island would be more diverse than an area with three species of birds and no lizards.

Therefore, species diversity can be assessed in terms of the number of species or the range of different types of species an area contains.

Current estimates for the total number of species in existence vary from 5 million to nearly 100 million. About 1.7 million of these species have been identified to date. A complete and current inventory of species is impossible because many invertebrates, microorganisms and lower plants (the three most diverse life-forms) have not yet been identified and named.

There has been a definite bias towards describing large organisms, those that are considered attractive or appealing (such as flowering plants and butterflies), those most closely resembling humans (vertebrates, especially mammals), and those that have a direct impact on human activities (such as pests). Organisms that can be studied without complex procedures or expensive equipment have also taken precedence, as have those which are relatively easy to locate. This, however, underestimates the importance of microorganisms including algae, bacteria, fungi, protozoa and viruses, which are vital to life on the earth. So far, less than 3-5 per cent of microorganisms have been described.

The species level is generally regarded as the most appropriate for considering the diversity between organisms.

This is not because species diversity is more important than the other two types, but because:

I. Species diversity is easier to work with.

   Species are relatively easy to identify by eye in the field, whereas genetic diversity requires laboratories, time and resources to identify, and ecosystem diversity needs many complex measurements to be taken over a long period of time.

II. Species are also easier to conceptualize and have been the basis of much of the evolutionary and ecological research that biodiversity draws on.

Species are well known and are distinct units of diversity. Each species can be considered to have a particular "role" in the ecosystem, so the addition or loss of single species may have consequences for the system as a whole. Conservation efforts often begin with the recognition that a species is endangered in some way, and a change in the number of species in an ecosystem is a readily obtainable and easily understandable measure of how healthy the ecosystem is.

4.2.2 Genetic Diversity

Genetic diversity refers to the differences in genetic make-up between distinct species, as well as the genetic variations within a single species. This is the least visible and, arguably, least studied level of biological diversity.

Genetic diversity is the variety present at the level of genes. Genes, made of DNA, are the building blocks that determine how an organism will develop and what its traits and abilities will be. This level of diversity can differ by alleles (different variants of the same gene, such as blue or brown eyes), by entire genes (which determine traits, such as the ability to metabolize a particular substance), or by units larger than genes such as chromosomal structure.

The amount of diversity at the genetic level is important because it represents the raw material for evolution and adaptation. More genetic diversity in a species or population means a greater ability for some of the individuals in it to adapt to changes in the environment. Less diversity leads to uniformity, which is a problem in the long term, as it is unlikely that any individual in the population would be able to adapt to changing conditions. As an example, modern agricultural practices use monocultures, which are large cultures of genetically identical plants. This is an advantage when it comes to growing and harvesting crops (for example, all the plants can be harvested at once), but can be a problem when a disease or parasite attacks the field, as every plant in the field will be susceptible. Monocultures are also unable to deal well with changing conditions.

Within species, genetic diversity often increases with environmental variability, which can be expected. If the environment often changes, different genes will have an advantage at different times or places. In this situation genetic diversity remains high because many genes are in the population at any given time. If the environment didn't change, then the small number of genes that had an advantage in that unchanging environment would spread at the cost of the others, causing a drop in genetic diversity.

Since the gene is the fundamental unit of natural selection, and thus of evolution, some scientists argue that the real unit of biodiversity is genetic diversity. However, species diversity is the easiest one to study.

4.2.3 Ecosystem Diversity

Ecosystem diversity encompasses the broad differences between ecosystem types, and the diversity of habitats and ecosystem processes within each ecosystem type.

Ecosystem diversity deals with species distributions and community patterns, the role and function of key species, and combines species functions and interactions. The term "ecosystem" here represents all levels greater than species: associations, communities, ecosystems, and the like. This is the least understood level of the three described here due to the complexity of the interactions. Trying to understand all the species in an ecosystem and how they affect each other and their surroundings while at the same time being affected themselves, is extremely complex.

4.2.3.1 Difficulties in examining ecosystem diversity

I. The enormous range of terrestrial and aquatic environments on the earth has been classified into a number of ecosystems. Major habitat types include tropical rainforests, grasslands, wetlands, coral reefs and mangroves. Measuring changes in the extent of ecosystems is difficult, because there is no globally agreed classification of ecosystems. Thus, ecosystems can be considered on different scales.

II. Transitions between them are usually not very sharp. A lake may have a very sharp boundary between it and the deciduous forest it is in, but the deciduous forest will shift much more gradually to grasslands or to a coniferous forest. This lack of sharp boundaries is known as "open communities" (as opposed to "closed communities," which would have sudden transitions) and makes studying ecosystems difficult, since even defining and demarcating them is problematic.

III. Species contained within a given ecosystem vary over time.

IV. The classification of the earth's immense variety of ecosystems into a manageable system is a major scientific challenge.

Studies of ecosystem diversity are carried out on different scales: from one ecosystem to an entire region containing many different ecosystems. Regions containing a great variety of ecosystems are rich in biodiversity, but individual ecosystems containing endemic species also make a significant contribution to global biodiversity.

Some of the world's richest habitats are tropical moist forests. Although they cover only 7 per cent of the world's surface, these areas contain at least 50 per cent, and possibly up to 90 per cent of all plant and animal species. Isolated islands such as Jamaica are often rich in endemic species.

Another aspect to this level of diversity is biogeographic diversity which refers to the distribution of species within habitats or ecosystems.

4.3 BIOGEOGRAPHICAL CLASSIFICATION OF INDIA

There are 10 biogeographical zones in India. They can be classified as under:

Trans-Himalayas: An extension of the Tibetan plateau, harbouring high-altitude cold desert in Laddakh (J&K) and Lahaul Spiti (H.P) comprising 5.7 % of the country's land mass.

Himalayas: The entire mountain chain running from north-western to north-eastern India, comprising a diverse range of biotic provinces and biomes, 7.2 % of the country's land mass.

Desert: The extremely arid area west of the Aravalli Hill range, comprising both the salty desert of Gujarat and the sand desert of Rajasthan. 6.9% of the country's land mass.

Semi-arid: The zone between the desert and the Deccan plateau, including the Aravalli Hill range. 15.6 % of the country's land mass.

Western Ghats: The hill ranges and plains running along the western coastline, south of the Tapti river, covering an extremely diverse range of biotic provinces and biomes. 5.8% of the country's land mass.

Deccan peninsula: The largest of the zone covering much of the southern and south central plateau with predominantly deciduous vegetation. 4.3 % of the country's land mass.

Gangetic plain: Defined by the Ganges river system, these plains are relatively homogenous. 11% of the country's land mass.

North-east India: The plains and non-Himalayan hill ranges of northeastern India, with a wide variation of vegetation 5.2% of the country's land mass.

Islands: The Andaman and Nicobar Islands in the Bay of Bengal, with a highly diverse set of biomes. 0.03% of the country's land mass.

Coasts: A large coastline distributed both to the west and east, with distinct differences between the two; Lakshadweep Islands are included in this with the per cent area being negligible.

4.4 VALUE OF BIODIVERSITY

The multiple uses of biodiversity value have been classified as follows:

(1) Consumptive use value

(2) Productive use value

S. No.	Biogeographic zone	Biotic province	Total area (Sq.km)
1	Trans-Himalayan	Upper regions	186200
2	Himalayan	North-west Himalayan West Himalayas Central Himalayas East Himalayas	6900 720000 123000 83000
3	Desert	Kutch Thar Ladakh	45000 180000 NA
4	Semi arid	Central India Gujarat – Rajwara	107600 400400
5	Western Ghats	Malabar coasts Western Ghat mountains	59700 99300
6	Deccan Peninsula	Deccan plateau south Central plateau Eastern plateau Chhota Nagpur Central highlands	378000 341000 198000 217000 287000
7	Gangetic plain	Upper Gangetic plains Lower Gangetic plains	206400 153000
8	North-east India	Brahmaputra valley North-eastern hills	65200 106200
9	Islands	Andaman Islands Nicobar Islands Lakshadweep Islands	6397 1930 180
10	Coasts	West coast East coast	6500 6500

Fig. 4.2 India's major biogeographic habitats.

(3) Social value
(4) Ethical value
(5) Aesthetic value
(6) Optional values
(7) Ecosystem service value
(8) Consumptive use value

These are direct use values where the biodiversity product can be harvested and consumed, e.g., fuel, food, drugs, fibre, etc.

Food: A large number of plants are consumed by human beings as food. About 90% of present-day crops have been domesticated from wild tropical plants.

Drugs and medicines: 75% of the world's population depends upon plants or plant extracts for medicines, e.g., Penicillin-Penicillium Quinine-Bark of cinchona Vinblastin and Vincristine-Catharanthus

Fuel: Forests have been used since ages for fuelwood. Fossil fuels are also products of fossilized biodiversity.

Productive use values

- These are commercially usable values where the product is marketed and sold. It may be lumber, wild gene resources.
- Animal products like tusks of elephants, musk from musk deer, and wool from sheep.
- Despite international ban on trade in products from endangered species, smuggled hide, fur, tusks worth millions of dollars are being sold.

Social value

- Values associated with the social life, customs, religion and psycho-spiritual aspects of people.
- Many of the plants are considered holy and sacred in our country (Tulsi, Peepal, Mango).
- Many animals like cow, snake, bull, peacock also have significant place in our psycho-spiritual arena and thus hold special social importance.

Ethical value

- Ethical values is also sometimes known as existence value.
- It involves ethical issues like "all life must be preserved" and the concept of "Live and let live".
- The ethical value means we may or may not use a species, but we feel the existence of the species is necessary.

Aesthetic value

- Great aesthetic value is attached to biodiversity.
- Stretches of barren lands with no signs of visible life is not a pleasant sight.
- Ecotourism is based on the aesthetic value of biodiversity. It is estimated to generate about 12 billion dollars of revenue annually.

Optional values

- These include the potential of biodiversity that are presently unknown and need to be explored.

Ecosystem service value

It refers to the services provided by ecosystem like:

(1) Prevention of soil erosion
(2) Prevention of floods
(3) Maintenance of soil fertility
(4) Cycling of nutrients and water
(5) Fixation of nitrogen
(6) Role as carbon sinks

Different categories of biodiversity value clearly indicate that ecosystem, species and genetic diversity have enormous potential and a decline in biodiversity will lead to huge economic, ecological and socio-cultural losses

Biodiversity value of some selected organisms in monetary terms

- A male lion living up to the age of 7 years can generate up to $ 515000 due to its aesthetic value as paid by tourists.
- In a lifetime a Kenyan Elephant can earn worth $ 1 million as tourist revenue.
- The mountain Gorillas in Rwanda are fetching up to $ 4 million annually through eco-tourism.
- Whale watching on Hervey Bay on Queensland's coast earns $ 12 million annually.
- Tourism to Great Barrier Reef in Australia earns $ 2 billion each year.
- A tree provides $ 19,62,150 worth of ecological services as oxygen, clean air, water recycling, fertile soil, erosion control, wildlife habitat, toxic gas moderation, etc., whereas its worth is only $ 590 if sold in market as timber.

4.5 BIODIVERSITY AROUND THE WORLD AND INDIA

- Roughly 1.8 million species are known till date.
- Most of the world's bio-rich nations are in the southern developing nations.
- The majority of the countries capable of exploiting biodiversity are northern regions – developed nations.
  - International agreements – World Heritage Convention attempts to protect and support such areas.
  - India is a signatory to the convention and has included areas covering Manas on the border between Bhutan and India, Kaziranga in Assam, Bharatpur in Rajasthan, Nandadevi in the Himalayas and the Sunderbans in the Ganges delta in West Bengal.

4.5.1 Indian Biodiversity

- Every country is characterized by its own biodiversity depending upon its climate.
- India has rich biological diversity of flora and fauna.
- 6% of the global species are found in India.
- The total number of species found in India is 150,000.
- Out of a total of 25 biodiversity hotspots in the world, India possesses 2, one in the northern region and one in the Western Ghats.

4.5.1.1 Regional or local biodiversity

Four types – based upon their spatial distribution.

- **Point Richness** refers to the number of species that can be found at a single point in a given space.
- **Alpha Richness** refers to the number of species found in a small homogeneous area.
- **Beta Richness** refers to the rate of change in species composition across different habitats.
- **Gamma Richness** refers to the rate of change across large landscape gradients.

4.5.1.2 India as mega diversity nation

- India is one of the 12 mega diversity countries in the world.
  - Australia
  - Brazil

- China
- Columbia
- Ecuador
- The United States
- India
- Indonesia
- Madagascar
- Mexico
- Peru
- The Democratic Republic of Congo

- The ministry of Environment and Forests, records show 47,000 plant species and 81,000 animal species, i.e., 7% and 6.5% of the global flora and fauna.
- A large proportion of the India's biodiversity is still unexplored.
- Due to diverse climatic conditions there is a complete rainbow spectrum of biodiversity in our country.

Table 4.1 Distribution of species in India.

Group-wise species distribution			
Plants	*Number*	*Animals*	*Number*
Bacteria	850	Lower groups	9979
Fungi	23,000	Mollusca	5042
Algae	2500	Arthropoda	57,525
Bryophytes	2564	Fishes	2546
Pteridophytes	1022	Amphibia	428
Gymnosperms	64	Reptilles	1228
Angiosperms	15,000	Birds	204
		Mammals	327

4.5.2 Reasons for India as a Mega-Diversity Nation

(1) Endemism

- Species which are restricted only to a particular area are known as endemic.
- India shows a good number of endemic species. About 62% amphibians and 50% lizards are endemic to India.
- Western Ghats are the site of maximum endemism.

(2) Centre of origin

- A large number of species are known to have originated in India.
- Nearly 500 species of flowering plants have their origin in India.
- India has been the centre or origin for 166 species of crop plants and 320 species of wild relatives of cultivated crops.

(3) Marine diversity

- The coastline of our country exhibits a rich biodiversity.
- Along 7500 km long coastline, in the mangroves, coral reefs, backwaters, etc., different species are found.
- The marine diversity is rich in mollusks, crustaceans and several species of mangrove plants and seagrasses are found.

4.5.2.1 Hotspots of biodiversity

- Areas which exhibit high species richness as well as high species endemism are called hotspots of biodiversity.
- The term was introduced by Myers (1988).

To qualify as a biodiversity, a region must meet two strict criteria:

(1) It must contain at least 0.5% or 1500 species of vascular plants as endemics,

(2) It has to have lost at least 70% of its primary vegetation.

According to Myers an area is designated as a hotspot when it contains atleast 0.5% of the plant species as endemics.

- The hotspots cover less than 2% of the world's land but are found to have 50% of the terrestrial biodiversity.
- The Indian hotspots are not only rich in floral wealth and endemic species of plants but also reptiles, amphibians, swallow tailed butterflies and some mammals.

4.5.2.2 Global hotspots of biodiversity

There are 25 hotspots at global level

1. Tropical Andes
2. Mesoamerican Forests
3. Caribbean
4. Brazil's Atlantic Forests
5. Darien of Panama Western Ecuador
6. Central Chile
7. California Floristic Province
8. Madagascar
9. Eastern Arc and Coastal Forest of
10. Tanzania
11. Western Africa Forests
12. Cape Floristic Province
13. Succulent Karoo
14. Mediterranean Basin
15. Caucasus
16. Sundal Land
17. Wallacea
18. Philippines
19. Indo-Burma Eastern Himalayas

20. South Central China
21. Western Ghats
22. South Western Australia
23. New Caledonia
24. New Zealand
25. Polynesia/Micronesia

4.5.2.3 *Salient features of Indian hotspots*

(a) Eastern Himalayas

- The Eastern Himalayan region encompasses Bhutan, North east India, and south, central and eastern Nepal
- Three largest herbivores –the Asian elephant, the greater one horned rhinoceros, and the wild water buffalo Carnivore-The tiger Birds-Vultures, Storks and hornbills
- Out of the world's recorded flora, 30%are endemic to India of which 35,000 are in the Himalayas

(b) Western Ghats

- It extends along a 17000 sq.km strip of forests in Maharashtra, Karnataka, TN and Kerala and has 40% of total endemic plants
- The major centers of diversity are Agastyamalai Hills and Silent valley-the New Amambalam reserve basin
- It is reported that only 6.8% of original forests are existing today while the rest has been deforested or degraded

4.6 CLASSIFICATION OF SPECIES

Extinct species – A species is said to be extinct when it is not seen in the wild for 50 years at a stretch. Eg Dodo, passanger pigeon

Endangered species – When the number has been reduced to a critical level or whose habitats have been drastically reduced and when not protected and conserved they are endangered species

Rare species – Species which are not endangered at present, but are at risk are categorised as rare species

Endemic species – Species which are restricted to a particular area is called endemic species endangered species:

- These are species which if not protected are likely to become extinct in near future.
- The International Union for Conservation of Nature and Natural Resources(IUCN) publishes the Red Data Book which includes the list of endangered species of plants and animals.
- In India nearly 450 plant species, 150 mammals and 150 species of birds have been listed as endangered species.

4.6.1 Few Endangered Species of India

1. Reptiles: Green sea turtle, tortoise, python
2. Birds: Peacock, great Indian hornbill, pelican, Siberian white crane

3. Carnivorous mammals: Indian wolf, red fox, tiger, lion, red panda, leopard, striped hyena, desert cat, dugong.
4. Primates: Capped monkey, golden monkey, Nilgiri langur, hoolock gibbon
5. Plants: species of orchids, rhododendrons Santalum, Cycasbeddonei, etc.

Zoological Survey of India reported that cheetah pink headed duck and mountain quail have already become extinct from India.

4.6.2 Endemic Species of India

- India has two biodiversity hotspots and thus possesses a large number of endemic species.
- Indian subcontinent has about 62% endemic flora, restricted mainly to Himalayas, Khasi Hills and Western Ghats.
- Indian endemic flora includes orchids and species like Sapria Himalayana, Uvaria Lirida, Nepenthes Khasiana, Pedicularis perroter, etc.

The Western Ghats are particularly rich in amphibians and reptiles:

- About 62% amphibians and 50% lizards are endemic to Western Ghats.
- Different species of Monitor Lizards (Varanus), reticulated python and indian salamandar and viviparous toad (Nectophryne) are some important species of our country.

4.7 THREATS TO BIODIVERSITY

- Extinction or elimination of a species is a natural process of evolution.
- The process of extinction has become particularly fast in the recent years of human civilization.
- One of the estimates puts the figure of extinction as 27/day.
- If the present trend continues we would lose 1/3rd to 2/3rd of our current biodiversity by the middle of 21st century.

Major causes for loss of biodiversity are:

1. Loss of habitat
2. Poaching
3. Man-wildlife conflicts

Loss of Habitat

- Destruction and loss of natural habitat is the single largest cause of biodiversity loss.
- Billions of hectares of forests and grasslands have been cleared over the past 10,000 years for conversion into agricultural lands, pastures, settlement areas or development projects.
- Thousand of species have perished due to loss of their natural habitat.
- A phenomenon known as habitat fragmentation have resulted in loss on many singing birds.
- As a result of human intervention marine biodiversity is under serious threat due to large scale destruction of the fragile breeding and feeding grounds.

Poaching

- Illegal trade of wildlife products by killing prohibited animals, is a threat to wildlife.

- Despite international ban on trade in products from endangered species, smuggling of wildlife items like furs, hides, tusks, etc. continues.
- The worse issue is that for every live animal that gets into market, about 50 additional animals are caught and killed.

Man-wildlife conflicts

- When wildlife causes immense damage and danger to man, the conflict occurs.
- In retaliation, the villagers electrocute or kill the animals which sometimes exceeds poaching.

4.7.1 Causes

(1) Dwindling of habitats and human encroachments into the forest areas.
(2) The wildlife corridors through which animals used to migrate seasonally have been used for human settlements and hence animals attack the settlements.
(3) The stopping of cultivation of paddy, sugarcane, etc., within the sanctuaries have led the animals to stray out.
(4) The compensation paid by the government in lieu of the damage caused to the crops is not adequate and the agonized farmers get revengeful and kill the wild animals.
(5) Usually the ill, weak and injured animals have a tendency to attack man.

4.7.2 Remedial Measures to Curb the Conflicts

- Tiger conservation project has made provisions for making available vehicles, tranquillizer guns, binoculars and radio sets, etc., to tactfully deal with any imminent danger.
- Adequate crop compensation and cattle compensation scheme must be started, along with substantial cash compensation for loss of human life.
- Solar powered fencing should be provided along with electric current-proof trenches to prevent animals from straying into fields.
- Cropping pattern should be changed near the forest borders and adequate fodder, fruit and water should be made available to the elephants within the forest zone.
- Wildlife corridors should be provided for mass migration of animals during unfavourable periods.
- WWF-TCP's initiative to curb the ritual of "akhand shikar" in Orrisa.

4.8 CONSERVATION OF BIODIVERSITY

- The enormous value of biodiversity due to their genetic, commercial medical, aesthetic importance emphasize the need to conserve biodiversity.
- There are two approaches of biodiversity conservation
  - In situ conservation
  - Ex situ conservation

In Situ conservation

One benefit of in situ conservation is that it maintains recovering populations in the surrounding where they have developed their distinctive properties. Another is that this strategy helps ensure the ongoing processes of evolution and adaptation within their environments. As a last resort, may be used on some or all of the population, when in situ conservation is too difficult, or impossible.

Wildlife and livestock conservation is mostly based on in situ conservation. This involves the protection of wildlife habitats. Also, sufficiently large reserves are maintained to enable the target species to exist in large numbers. The population size must be sufficient to enable the necessary to survive within the population, so that it has a good chance of continuing to and over time. This reserve size can be calculated for target species by examining the population density in naturally-occurring situations. The reserves must then be protected from intrusion or destruction by man, and against other catastrophes.

In, in situ conservation techniques are an effective way to improve, maintain and use traditional or native varieties of agricultural crops. Such methodologies link the positive output of scientific research with farmer experience and field work. First, the accessions of a variety stored at a germplasm bank and those of the same variety multiplied by farmers are jointly tested in the producer's field and in the laboratory, under different situations and stresses. Thus, the scientific knowledge about the production characteristics of the native varieties is enhanced. Later, the best tested accessions are crossed / mixed and multiplied under replicable situations. At last, these improved accessions are supplied to the producers. Thus, farmers are enabled to crop improved selections of their own varieties, instead of being lured to substitute their own varieties with commercial ones or to abandon their crop. This technique of conservation of agricultural biodiversity is more successful in marginal areas, where commercial varieties are not expedient, due to climate and soil fertility constraints. Or where the taste and cooking characteristics of traditional varieties compensate for their lower yields.

Basically the in situ conservation is achieved by the following methods:

- This is achieved by protection of wild flora and fauna in nature itself, e.g., biosphere reserves, national parks, sancturies, reserve forests, etc.
- The biosphere reserves conserve some representative ecosystems as a whole for long term in situ conservation. Gulf of mannar, Nilgiri, Sunderbans, Nanda Devi are biosphere reserves.
- A national park is an area dedicated for the conservation of wildlife along with its environment.
- Each park aims at conservation of some particular species of wildlife along with others.
- Wildlife sanctuaries are also protected areas where killing, hunting, shooting or capturing of wildlife is prohibited except under the control of highest authority.
- Private ownership rights are permissible and forestry operations are also permitted to an extent that do not affect the wildlife adversely.

At present there are 7 major biosphere reserves, 80 national parks, 420 wildlife sanctuaries and 120 botanical gardens in India covering 4% of geographic area.

Table 4.2 Important national parks in India.

Name	*Important wildlife*
Kaziranga	One horned Rhino
Gir national park	Indian Lion
Dachigam	Hangul
Bandipur	Eleplant

Contd...

Contd...

Periyar	Eleplant, Tiger
Kanha	Tiger
Corbett	Tiger
Dudwa	Tiger
Ranthambore	Tiger
Sariska	Tiger

Table 4.3 Wildlife sanctuaries.

Name	*Major wildlife*
Ghana bird sanctuary	300 species of birds
Hazaribagh sanctuary	Tiger, leopard
Sultanpur bird sanctuary	Migratory birds
Nal sarvovar bird sanctuary	Water birds
Abohar wildlife sanctuary	Black buck
Mudumalai wildlife sanctuary	Tiger, eleplant, leopard
Vedanthangal bird sanctuary	Water birds
Jaldapara wildlife sanctuary	Rhinoceros, eleplant, tiger
Wild ass sanctuary	Wild ass, wolf, nilgai, chinkara

Ex situ conservation

Ex situ conservation means literally offsite. It is the process of protecting a plant or animal outside of its natural habitat, for example, by removing part or whole of the threatened plant or animal and placing it in a new location, which may be a wild area or in the care of humans. While ex situ conservation comprises some of the oldest and best known conservation methods, it also involves newer, sometimes controversial laboratory methods.

The best method of maximizing a species chance of survival (when ex situ methods are required) is by relocating part of the population to a less threatened location. It is extremely difficult to mimic the environment of the original colony location given the large number of variables defining the original colony, species, etc. It is also technically challenging to uproot (in the case of plants) or trap (in the case of animals) the required without undue harm.

Ex situ conservation can be achieved in the following methods:

This type of conservation is mainly done for conservation of crop varieties, the wild relatives of crops and all the local varieties. In this the conservation in captivity under human care.

The objective is to conserve the total genetic variability of the crop species for future crop improvement or afforestation programmes.

4.8.1 Important Gene Bank/Seed Bank

(1) National Bureau of Plant Genetic Resources (NBPGR), New Delhi
Here agricultural and horticultural crops are cryo-preserved

(2) National Bureau of Animal Genetic Resources (NBAGR), Haryana.
Preserves semens of domesticated bovine animals.

(3) National Facility for Plant Tissue Culture Repository (NFPTCR) within NBPGR. Conservation of crop plants/trees by tissue culture.

4.9 CASE STUDY

4.9.1 Project Tiger

Project Tiger Scheme has been under implementation since 1973 as a centrally sponsored scheme of government of India.

- The main objective of Project Tiger is to ensure a viable population of tigers in India for scientific, economic, aesthetic, cultural and ecological values and to preserve for all time, areas of biological importance as a natural heritage for the benefit, education and enjoyment of the people.
- Main objectives under the scheme include wildlife management, protection measures and site specific eco development to reduce the dependency of local communities on tiger reserve resources.
- Initially, the Project started with 9 tiger reserves, covering an area of 16,339 sq.km., with a population of 268 tigers. At present there are 27 tiger reserves covering an area of 37761 sq.km., with a population of 1498 tigers.
- Project Tiger has put the tiger on an assured course of recovery from the brink of extinction, and has resurrected the floral and faunal genetic diversity in some of our unique and endangered wilderness ecosystems.

4.9.2 Project Elephant

Project Elephant (PE), is centrally sponsored scheme, was launched in February 1992 to provide financial and technical support to major elephant bearing states in the country for protection of elephants, their habitats and corridors

- The Project Elephant in India also aims to decrease the human-elephant battles and help in the welfare of domesticated elephants in India.

(1) Ecologically restoring the natural habitats and migratory routes of the elephants.

(2) Resolution of the increasing conflicts between man and elephants in important habitats and moderating the pressures of human and domestic stock activities in important elephant habitats.

(3) Developing scientific and planned management measures for conservation of elephants and controlling the population of wild Asiatic elephants, which are almost on the verge of extinction.

(4) Protecting the elephants from poachers and other unnatural causes of death and illegal ivory trade is also one of the major concerns of the Elephant Project in India.

(5) Researching on issues related to elephants and creating public awareness and education programs for it.

(6) Eco-development and veterinary care for the elephants.

(7) Project Elephant also aims at maintaining health care and breeding of tame elephants.

4.9.3 Current Status of Project Elephant

- The Project Elephant has still not led to as much increase in the number of elephants as was expected.
- Wildlife conservationists state that the progress has been slow and people in charge of the project are themselves not very clear of the causes of decline in the number of elephants.
- Project Elephant tries to ensure a free movement for the elephants and thus conserve large areas for them that are called 'elephant reserve range'.

Learning Outcomes

The main learning outcomes from this chapter should be fundamental understanding of: genetic, species and ecosystem diversity; Bio-geographical classification of India; Value of biodiversity- consumptive use, productive use, social, ethical, aesthetic and option values; Biodiversity at global, National and local levels; India as a mega-diversity nation; Hotspots of biodiversity; Threats to biodiversity- Habitat loss, poaching of wildlife, man-wildlife conflicts; Endangered and Endemic species of India; Conservation of biodiversity : In situ and Ex situ conservation of biodiversity.

Readers are also advised to carry out following exercise:

Exercise

1. Give an example of diversity at your regional level, e.g., district or state.
2. Write a 1000 words essay on "Project Tiger".
3. Write a critical review on the threats faced by our ecosystem and what can be done to conserve them.

Chapter 5

Environmental Pollution

5.1 INTRODUCTION

Pollution is the introduction of contaminants into the natural environment that causes adverse changes. It can take the form of chemical substances, energy (such as noise), heat or light. The components of the pollutions often called pollutants, can be either foreign substances or naturally occurring contaminations. Generally, pollution is classed as point source or non-point source pollution. Some form of pollutions among the existing many others are given below:

Air Pollution: The release of chemicals and particulates into the atmosphere. Common gaseous pollutants include carbon mono-oxide, sulphur dioxide, chlorofluorocarbons, nitrogen oxides, etc.

Light Pollution: This includes light trespass, over-illumination and problems occurring for astronomical interference.

Littering: The illegal throwing of manmade waste over public/private places.

Noise Pollution: Noise from speakers, traffic, aircraft, industries, etc.

Water Pollution can happen be fresh drinking water mixed with waste water, domestic sewage, chemical components and many others.

Radioactive Contamination: This type of pollution happens by contamination the affected areas by enriched radioactive material.

Soil Contamination occurs when chemicals are released by spill or underground leakage. Among most important contaminants are hydrocarbons, heavy metals, pesticides, etc.

In the present chapter some of the major environmental pollution sources will be discussed in detail.

5.2 WATER POLLUTION

Water is a unique substance, because it can naturally renew and cleanse itself, by allowing pollutants to settle down (through the process of sedimentation) or break down, or by diluting the pollutants to a point where they are not in harmful concentrations. However, this natural process takes time, and is difficult when excessive quantities of harmful contaminants are added to the water. Presently, humans are using more and more materials that are polluting the water sources. In nine of the last ten years, large blue-green algae blooms have appeared on the northern part of

Lake Winnipeg. These are caused by excess phosphorus in the water. Fertilizer use is 15 times higher today than it was in 1945. Beach closures are becoming increasingly common. The list of pollutants is long and the signs of water pollution surround us. Even though we are dumping contaminants into the small portion of water on the planet that is fit for drinking.

Water pollution occurs when any physical, biological or chemical change in water quality that adversely affects living organisms or makes the water unsuitable for desired uses. Water pollution takes place when energy and other materials are released, degrading the quality of the water for the use of living being. It includes all of the waste materials that cannot be naturally broken down by water. In other words, anything that is added to the water, above and beyond its capacity to break it down, is pollution. Pollution, in certain circumstances, can be caused by nature itself, such as when water flows through soils with high acidities. But more often than not, human actions are responsible for the pollutants that enter the water.

5.3 SOURCES OF WATER POLLUTION

As we have classified above there are two main sources of water pollution; point sources and non-point sources.

5.3.1 Point Sources

Include factories, wastewater treatment facilities, septic systems, and other sources that discharge pollutants into water sources.

5.3.2 Non-point Sources

These are more difficult to identify, because they cannot be traced back to a particular location. Non-point sources include run-off including sediment, fertilizers, chemicals and animal wastes from farms, fields, construction sites and mines. Landfills can also be a non-point source of pollution, if substances leach from the landfill into water supplies.

The United States Environmental Protection Agency (EPA) divides water pollution into the following six categories:

1. Biodegradable waste consists mainly of human and animal waste. When biodegradable waste enters a water supply, the waste provides an energy source (organic carbon) for bacteria. Organic carbon is converted to carbon dioxide and water, which can cause atmospheric pollution and acid rain; this form of pollution is far more widespread and problematic than other forms of pollutants, such as radioactive waste. If there is a large supply of organic matter in the water, oxygen-consuming (aerobic) bacteria multiply quickly, consume all available oxygen, and kill all aquatic life.
2. Plant nutrients, such as phosphates and nitrates, enter the water through sewage, and livestock and fertilizer run-off. Phosphates and nitrates are also found in industrial wastes. Though these chemicals are natural, 80 per cent of nitrates and 75 per cent of phosphates in water are human-added. When there is too much nitrogen or phosphorus in water (0.3 parts per million for nitrogen and 0.01 parts per million for phosphorus), algae begin to develop. When algae blooms, the water can turn green and cloudy, feel slimy, and smell bad. Weeds start to grow and bacteria spread. Decomposing plants use up the oxygen in the water, disrupting the aquatic life, reducing biodiversity, and even

killing aquatic life. This process, called eutrophication, is a natural process, but generally occurs over thousands of years. Eutrophication allows a lake to age and become more nutrient-rich; without nutrient pollution, this may take 10,000 years, but pollution can make the process occur 100 to 1000 times faster.

3. Heat can be a source of pollution in water. As the water temperature increases, the amount of dissolved oxygen decreases. Thermal pollution can be natural, in the case of hot springs and shallow ponds in the summer time, or human-made, through the discharge of water that has been used to cool power plants or other industrial equipment. Fish and plants require certain temperatures and oxygen levels to survive, so thermal pollution often reduces the aquatic life diversity in the water.
4. Sediment is one of the most common sources of water pollution. Sediment consists of mineral or organic solid matter that is washed or blown from land into water sources. Sediment pollution is difficult to identify, because it comes from non-point sources, such as construction, agricultural and livestock operations, logging, flooding, and city run-off. Each year, water sources in the United States are polluted by over one billion tonnes of sediment. Sediment can cause large problems, as it can clog municipal water systems, smother aquatic life, and cause water to become increasingly turbid. Turbid water can cause thermal pollution, because cloudy water absorbs more solar radiation.
5. Hazardous and toxic chemicals are usually human-made materials that are not used or disposed of properly. Point sources of chemical pollution include industrial discharges and oil spills. The oil pollution fact sheet includes more detailed information about oil spills, as well as other sources of oil pollution. Non-point sources of chemical pollution include run-off from paved roads and pesticide run-off. Many people think industries produce the greatest amount of chemical pollution. But domestic and personal use of chemicals can significantly contribute to chemical pollution. Household cleaners, dyes, paints and solvents are also toxic, and can accumulate when poured to drains or flushed down the toilet. In fact, one drop of used motor oil can pollute 25 litres of water. And, people who use pesticides in their gardens and lawns tend to use ten times more pesticide per acre than a farmer would.
6. Radioactive pollutants include wastewater discharges from factories, hospitals and uranium mines. These pollutants can also come from natural isotopes, such as radon. Radioactive pollutants can be dangerous, and it takes many years until radioactive substances are no longer considered dangerous.

It would be wise to add a seventh category of water pollution; pharmaceuticals and personal care products (often abbreviated PPCPs), including medications, lotions and soap, are being found in increasing concentrations in lakes and rivers. Scientists have discovered that many PPCPs act as hormone disrupters, which means that the synthetic hormones in the products interfere with the natural hormones in animals, especially fish that live in the water. There has not been enough research to determine the effects that PPCPs can have on humans, but there is evidence to suggest that these chemicals may be partially responsible for an increase in cancer and birth defects.

5.4 IMPACT OF DOMESTIC WASTE ON WATER POLLUTION

The categories of water pollution that domestic waste fits into are biodegradable waste, hazardous and toxic chemical pollutants and PPCPs. Generally, wastewater treatment facilities are equipped to effectively remove harmful substances generated from biodegradable waste. The hazardous and toxic chemicals that individuals release into the environment are more dangerous (and more preventable). Chemicals, such as cleaners, dyes, paints, pesticides and solvents, which are poured down drains, are a substantial and dangerous form of pollution. Wastewater treatment facilities are generally unequipped to remove PPCPs from wastewater; water pollution from PPCPs is a growing concern.

5.5 IMPACT OF INDUSTRIAL ACTIVITIES ON WATER POLLUTION

Industrial pollution comes in a variety of forms. There are many federal regulations regarding types and amounts of pollutants that can be emitted from industries, though in some countries, companies who are over their limit can buy "pollution credit" from companies who are under the targeted amount.

Heat pollution is commonly caused by industries, but many regions have passed legislation requiring that power plants and industries cool water before they release it. Construction, mining and logging operations can cause great amounts of sediment to pollute lakes and streams. While agricultural practices are responsible for the most sediment pollution, construction and mining can lose up to 70 tonnes of sediment per acre per year, which is 15 times higher than the normal cropland rate.

5.6 IMPACT OF AGRICULTURE ON WATER POLLUTION

The greatest agricultural contributions to water pollution are through nutrient and sediment pollution. Livestock waste and fertilizers contain nitrogen and phosphorus, which, if carried to lakes and streams through run-off, can cause significant problems resulting in excess algae growth.

In the last ten years, the number of livestock in Manitoba has increased by about 65 per cent, mostly in the form of pigs and cows. The livestock produce a large amount of waste, which many farmers use as fertilize on their fields. In the Winnipeg area, thousands of hectares of farmland have been designed for efficient run-off, which minimizes flooding. However, when the water runs off, it carries organic matter from the fertilizers straight into the creeks that feed Lake Winnipeg. Agricultural practices in Manitoba are estimated to be responsible for about 17 per cent of the phosphorus problem in Lake Winnipeg. There are several good management practices that can reduce the amount of agricultural water pollution, such as collecting animal wastes in a lagoon, or spraying pesticides in small amounts and at minimal run-off times.

Agricultural practices are the leading cause of sediment pollution, because bare lands are susceptible to large amounts of erosion. Erosion causes problems both for the water source and the farmland, which loses significant amounts of topsoil each year.

5.7 TYPES OF WATER POLLUTION (FRESH WATER MAINLY)

- Infectious agents
- Oxygen demanding wastes
- Plant nutrients and eutrophication
- Toxic inorganic materials
- Organic chemicals
- Sediment and suspended solids

- Thermal pollution/ thermal shock
- **Infectious Agents**
  - Most serious water pollutants in terms of human health
  - Ex. typhoid, cholera, dysentery, polio, hepatitis
  - 25,000,000 deaths each year (2/3 of child deaths, 80% of sickness in developing countries)
  - Comes from untreated human wastes and animal wastes
  - 2.5 billion people lack sanitation (more lack clean water)
- **Oxygen-demanding Wastes**
  - Healthy water has a high level of dissolved oxygen (> 8ppm)
  - Oxygen-poor water (<2ppm) only supports detritivores
  - Oxygen is added to the water by diffusion from air (effect of temperature) and photosynthesis
  - Oxygen is removed by respiration of plants and animals
  - The addition of sewage and wastes stimulates oxygen consumption by detritivores

5.8 MEASURING OXYGEN CONTENT

The health of water can be measured using oxygen content and majorly done in two ways given below;

- *BOD: Biological Oxygen Demand*

  –this measures the demand for oxygen that the detritivores place on the system

  –how much O_2 is used by organisms over a 5-day period
- *DO: Dissolved oxygen content*

  –how much oxygen is dissolved in the water

  –is affected by temperature and aeration

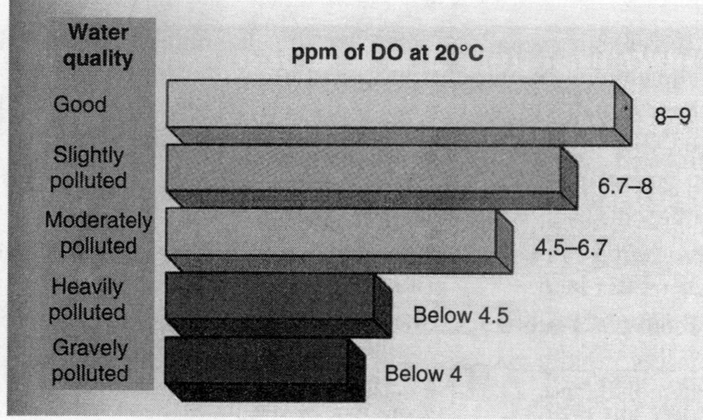

Fig. 5.1 Dissolved oxygen (DO) as indicator for water health.

Oxygen sag is another parameter measured to check the water quality in stream by measuring dissolved oxygen in a stream (Fig. 5.2)

- The oxygen sag is the pattern of dissolved oxygen in a stream that is being dumped into.
- The pattern of organisms is determined by the DO content.
- Know the different types of organisms and where they occur.
- The length of the oxygen sag will depend upon how fast the stream is flowing, and how turbid it is.

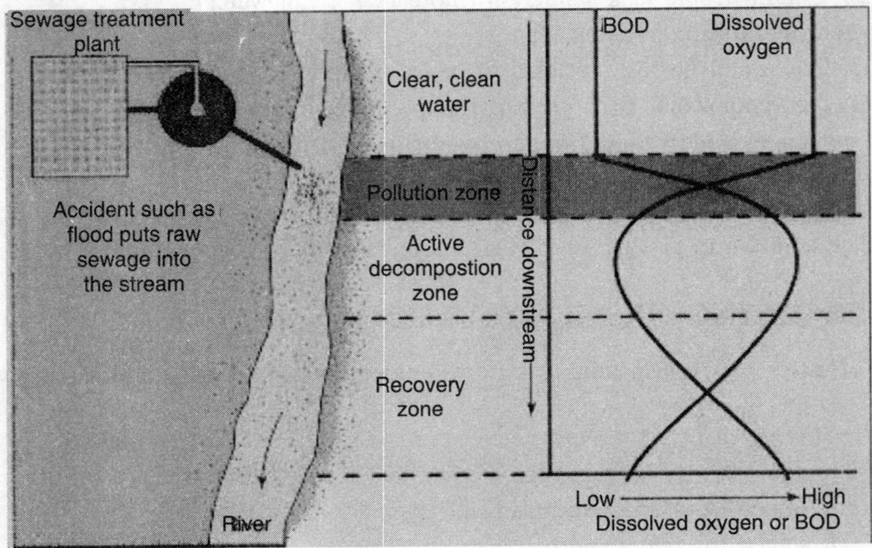

Fig. 5.2 Oxygen sag in a stream.

5.9 PLANT NUTRIENTS AND EUTROPHICATION

Water ecosystems (lakes) are usually limited by the amount of nutrients in them. Humans artificially increase the amount of nutrients in lakes through fertilizers, run-off. Over succession, lakes gradually increase in nutrients and productivity. The increase in nutrients leads to a series of stems culminating in eutrophication. Steps of eutrophication are given below:

- Nutrients are added to water.
- Increase in nutrients causes an algae bloom.
- As the algae bloom progresses, the algae begin to die, and organic material accumulates at the bottom of the lake.
- This material supports a boom in the decomposer populations.
- The decomposers rapidly rob the lake of its oxygen, suffocating most other organisms in the process.

5.10 OTHER SOURCES OF WATER POLLUTION

Along with the above-mentioned sources of water pollution, there are several other means which deliver significant amount of water pollution and their short summary in terms of source and origin is given below.

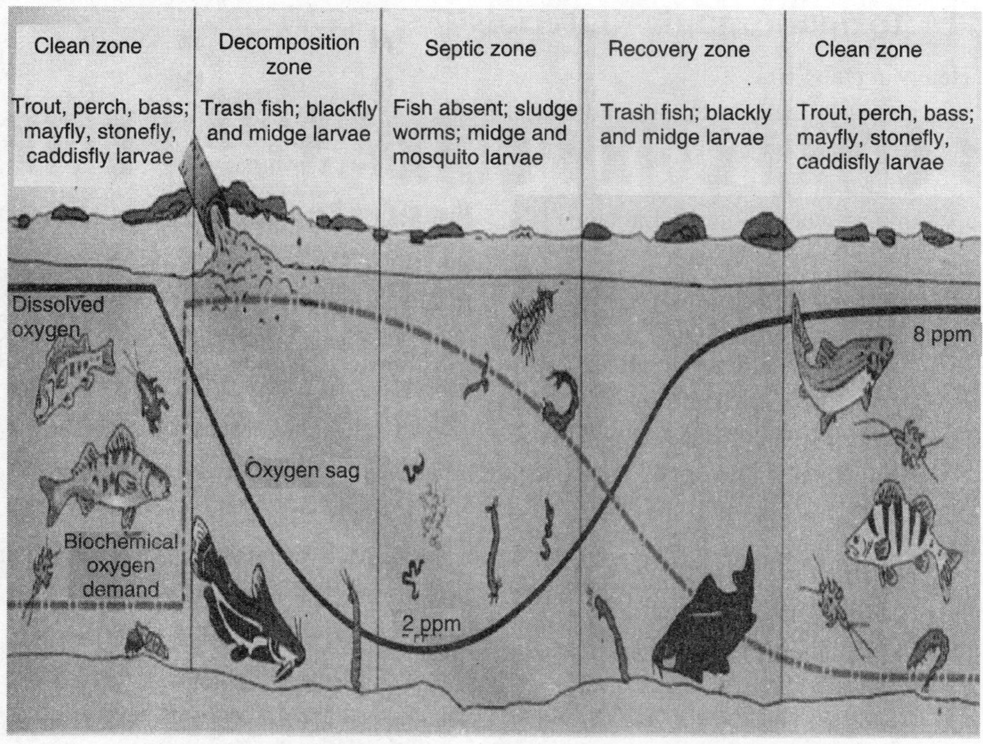

Fig. 5.3 Water pollution zones while putting waste.

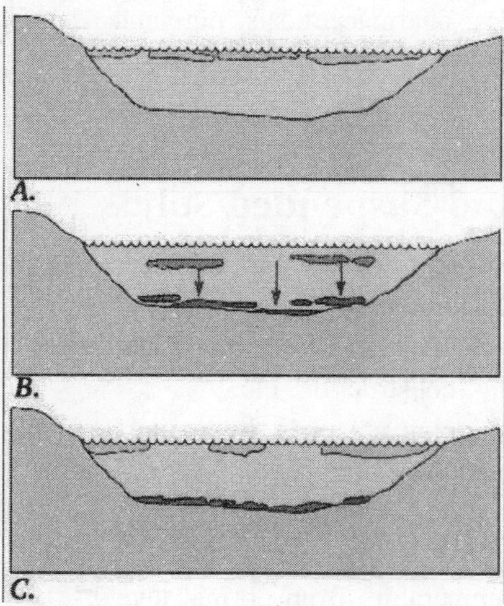

Fig. 5.4 Schematic Diagram showing effects of global bloom on water quality. (A) Abundant growth of Algae in sunlight shallow water when nutrients are abundant. (B) In colder weather, algae die and sink to the bottom. (C) The next growing season, more algae thrive at the surface while older material decays at the bottom, increasing BOD ad releasing more nutrients to fuel new growth.

5.10.1 Toxic Inorganic Materials

- Heavy metals:
  - –mercury, lead, cadmium and nickel
  - –lead pipes
  - –gold mining
  - –mining wastes, mine drainage
  - –tin on boat bottoms
- Nonmetallic salts:
  - –Occurs in desert soils
  - –As water evaporates, high levels of toxins are left behind
  - –Also from road run-off
    - e.g., selenium and arsenic
  - –Also table salt in very high concentrations
    - e.g., Salton Sea
- Acids and Bases
  - –Produced during leather tanning, metal smelting, plating, petroleum distillation and organic chemical formation
  - –Coal mining produced sulfuric acid
    - Acid Mine Drainage
  - –Acid rain (HNO_3 and H_2SO_4)

5.10.2 Organic Chemicals

- Pesticides, oils, plastics, pharmaceuticals, pigments, detergents, cleaning solutions, and paints
- DDT, etc.
- String fellow Site

5.10.3 Sediment and Suspended Solids

- Largest pollutant by volume in most parts of the world
- Erosion has increased sediment levels
- 25 billion metric tonnes of topsoil from runoff and erosion
- 50 billion from grazing, construction, etc.
- Fills reservoirs, fills shipping channels, less suitable for life, recreation
- Small levels of sediment are good

5.10.4 Thermal Pollution

- Raising or lowering temperature from normal levels.
- Water temperatures are usually stable so organisms are poorly adapted to rapid change.
- Oxygen solubility decreases as temperature increases.
- Most happens in industrial cooling.

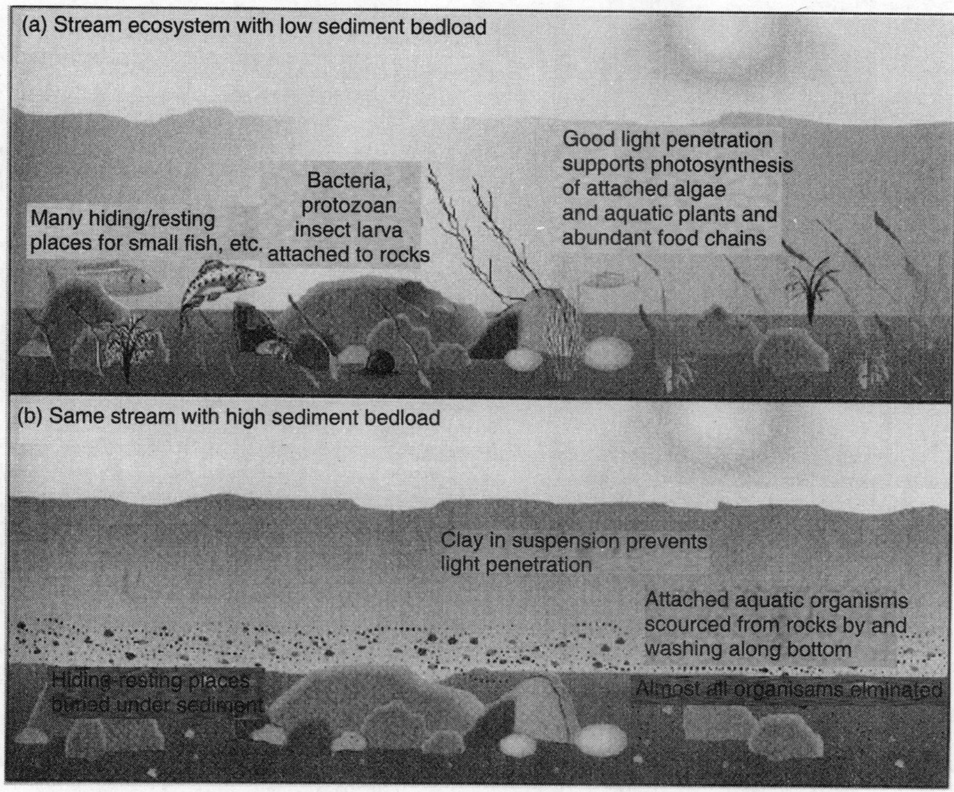

Fig. 5.5 Effect of sediments.

- Can be good for raising species that wouldn't be there otherwise.
- But can be harmful → manatees

5.11 LAND POLLUTION

Soils are the major support systems of human life. They provide anchorage for roots, hold water long enough for plants to make use of it, and hold nutrients that sustain life. Soils are home to myriad microorganisms that accomplish a suite of biochemical transformations - from fixing atmospheric nitrogen to the decomposition of organic matter and to armies of microscopic animals as well as the familiar earthworms, ants and termites. In fact, most of the land's biodiversity lives in the soil.

Soils, like landscapes, are often manmade. Farmers make agricultural soils which, if they are successful, support sustainable farming systems; sometimes, a farmer fails to make such a soil but all the operations of good soil management are directed towards this goal. Our footprint is stamped even more firmly on urban development that makes different and very special demands on soils; by land use changes that distort infiltration and drainage, and in climatic change, against which the soil provides the only buffer that we know how to manage.

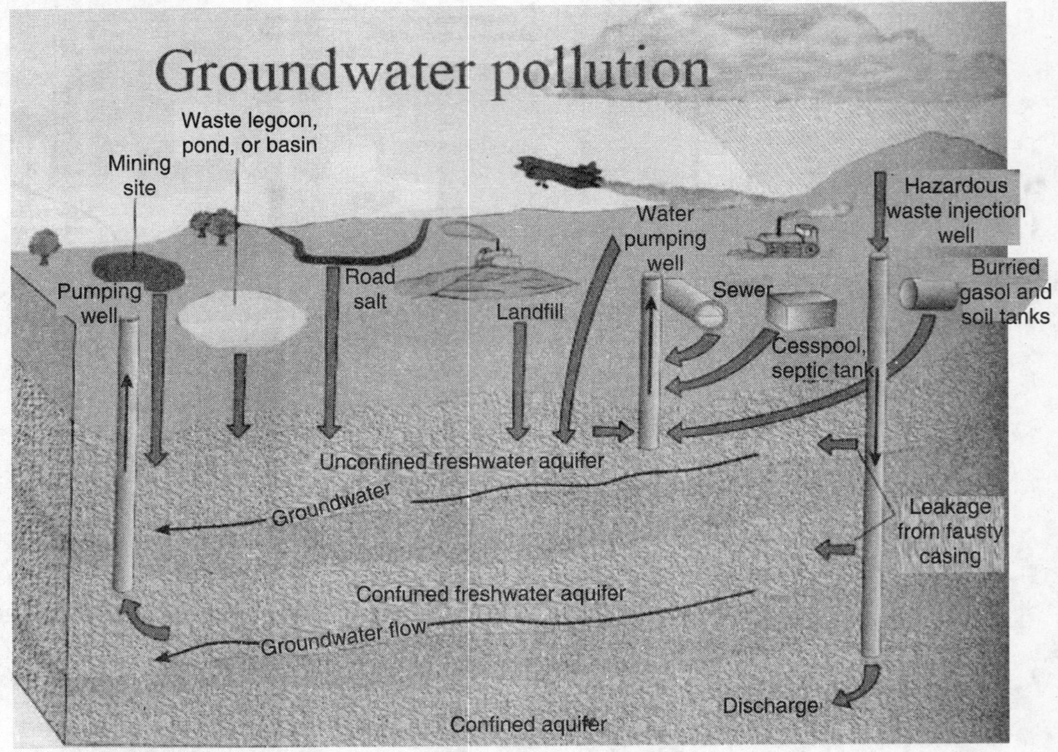

Fig. 5.6 Groundwater pollution.

Different soil landscapes respond to management (and mismanagement) in various ways. For this reason, particular kinds of production or construction are favoured in some places and not in others. Soil surveys identify and characterise soil landscapes; other branches of soil science are then applied to maximise the natural advantages or avoid the difficulties, e.g., by irrigating the dry, draining the wet, fertilizing the poor, and devising strong foundations in weak soil.

Not so long ago, soil, water, fuel and mineral resources were simply resources. Of course, economies and societies are built upon soils and most earth science activity is devoted to obviously useful activities that support the economy. In the case of soil science, this means supporting many different activities, including agricultural production, civil engineering, water supplies, water and air quality, sanitation and waste disposal to achieve the sustainable use of this finite and delicate system.

The one ultimate source of all fresh water is rain. Depending on land cover and soil condition, it is either intercepted or evaporated, infiltrated into the soil - or lost as damaging runoff. Rapid runoff brings floods, erosion of fertile soil, and riverbank erosion - damaging aquatic ecosystems, and clogging reservoirs and waterways. Depending on the thickness, permeability and water-retaining capacity of the soil, the infiltrated water may be held in the soil and used by plants, or drained to recharge groundwater and stream flow.

Whether the water becomes “hazard” or “resource” depends on the way it is partitioned at the soil surface and in the soil profile. That is to say, it depends on the kind of soil, its use and management. Water resources may be destroyed, or increased threefold, depending on how the

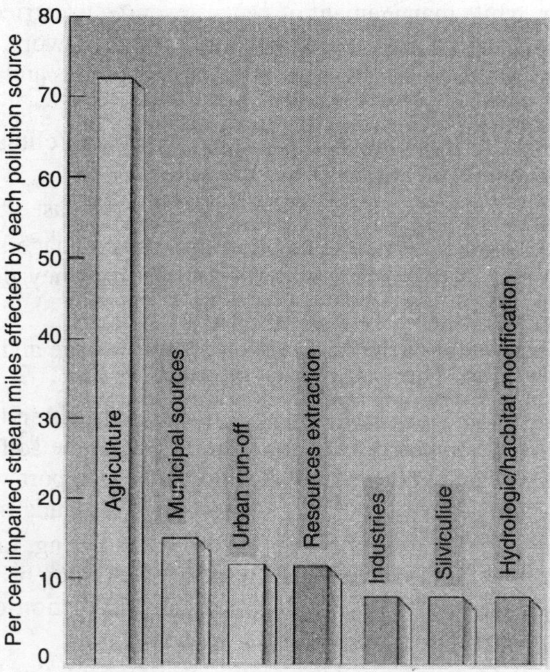

Fig. 5.7 Relative polluters of river.

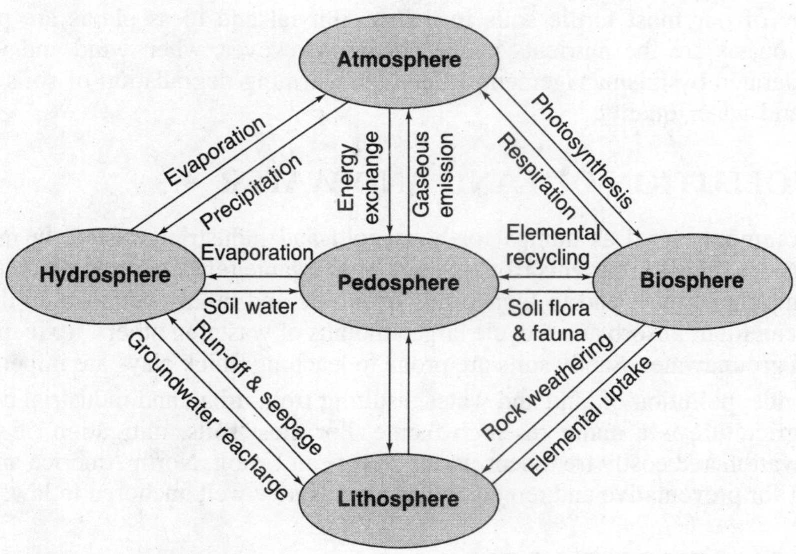

Fig. 5.8 Cycle in land pollution.

soil is managed. However, while management of water at source is carried out largely by farmers and grazers in rural areas of catchments, the beneficiaries of their work live overwhelmingly in urban areas downstream. Sustainable management of water resources and mitigation of associated hazards requires:

- Better understanding of each unique water delivery system (climate, soil, terrain, surface and groundwaters, and land use)
- Management applied at the scale of a whole watershed, not just at selected points or farms
- Payment by downstream beneficiaries to the upstream managers which allows them to manage both land and water more comprehensively than they do now.

Among these global systems, soils are integral to

- **Climate:** through mediation of the water cycle, carbon storage and emission of greenhouse gases (water vapour, CO_2, NOx and methane).
- **Water cycle:** Soils make up a key link and buffer system within the world's hydrological cycle. About 60% of fresh water is "green" water, held in the soil and available to plants. Soils also regulate streams and groundwater flows that support wetlands, irrigation, and domestic and industrial water supplies – sometimes thousands of miles downstream.
- **Waste and nutrient cycles:** Nutrients released by weathering, or fixed from the air, are recycled; toxins are neutralised. Disturbance of the cycle may bring about eutrophication or pollution of soils and water or, on the other hand, depletion of nutrients, threatening livelihoods across the world.
- **Erosion:** Loss of soils cover can lead to the stripping of the living skin, with all its irreplaceable functions, muddying waters and dumping sediment where it is not wanted – on fertile soils elsewhere, in streams, reservoirs and harbours. Erosion is not all bad; many of our most fertile soils in deltas, alluvial and loess plains are products of past erosion, as are the nutrients in the oceans. However, when wind and water erosion is accelerated by mismanagement, it leads to alarming degradation of soils and diminished air and water quality.

5.12 POLLUTION OF LAND AND WATER

Soils are commonly used as dumps for household and industrial wastes. In many intensively farmed areas, leaching of nutrients from manure or inorganic fertilizers and effluent from livestock or processing plants may lead to high levels of nitrate and other chemicals in the groundwater. Some soils can filter, absorb and recycle large amounts of waste; in others, toxic materials leach to streams and groundwater. Sandy soils are prone to leaching; thick clays are impermeable.

Worldwide, pollution of land and water resulting from urban and industrial development and intensive agriculture is a major research theme. For most soils, mitigation of severe pollution means excavation and costly treatment. In the European Union, North America and Australia, the requirement for preventative and remedial measures is now well anchored in law.

5.13 POLLUTION IN SOIL

Naturally, some soils are dry when hard, very sticky when wet, poorly drained, gravelly and stony or have very low nutrient contents or toxic amounts of aluminium or salt. Ever since settled agriculture began, people have altered such soils – although some soils remain more problematical than others.

Acid sulphate soils are the nastiest soils in the world. Undisturbed, they are not a problem. If drained, they generate sulphuric acid – 10 cubic metres of sulphidic soil may generate 1.5 tonnes of sulphuric acid and release a cocktail of aluminium, heavy metals and arsenic into drainage and floodwaters.

The acid corrodes steel and concrete; pollutes streams and estuaries, killing fish and causing disease. The effects of aluminium, heavy metal and arsenic in the food chain are not well understood, but are certainly unwelcome.

These soils occur mainly in coastal swamps, and there has always been an incentive to reclaim this land – to exploit their expected, though rarely realised fertility – or, more recently, to make way for urban and leisure development. Generations of people depending on these soils have been impoverished and poisoned by their drinking water. The engineering and environmental consequences have usually been disastrous.

Gradually, and only in some favourable cases, local people found empirical solutions. Science arrived late in the day. The process of fixing sulphates in waterlogged soils and subsequent oxidation upon drainage was elucidated by J M van Bemmelen in the 1880s, following the initial failure of the largest land reclamation project of its day, the Haarlemmermeerpolder in 1852. It took more than a century to build up a coherent body of scientific knowledge to solve practical problems and predict the occurrence and severity of these soils worldwide.

Soil pollution is defined as the build-up in soils of persistent toxic compounds, chemicals, salts, radioactive materials, or disease causing agents, which have adverse effects on plant growth and animal health.

Soil is the thin layer of organic and inorganic materials that covers the earth's rocky surface.

The organic portion, which is derived from the decayed remains of plants and animals, is concentrated in the dark uppermost topsoil. The inorganic portion made up of rock fragments, was formed over thousands of years by physical and chemical weathering of bedrock. Productive soils are necessary for agriculture to supply the world with sufficient food.

There are many different ways that soil can become polluted, such as:

- Seepage from a landfill
- Discharge of industrial waste into the soil
- Percolation of contaminated water into the soil
- Rupture of underground storage tanks
- Excess application of pesticides, herbicides or fertilizer
- Solid waste seepage

The most common chemicals involved in causing soil pollution are:

- Petroleum hydrocarbons
- Heavy metals
- Pesticides
- Solvents

5.13.1 Types of Soil Pollution

- Agricultural soil pollution
  - (i) Pollution of surface soil
  - (ii) Pollution of underground soil

- Soil pollution by industrial effluents and solid wastes
  (i) Pollution of surface soil
  (ii) Disturbances in soil profile
- Pollution due to urban activities
  (i) Pollution of surface soil
  (ii) Pollution of underground soil

5.13.2 Causes of Soil Pollution

Soil pollution is caused by the presence of man-made chemicals or other alterations in the natural soil environment. This type of contamination typically arises from the rupture of underground storage links, application of pesticides, and percolation of contaminated surface water to subsurface strata, oil and fuel dumping, leaching of wastes from landfills or direct discharge of industrial wastes to the soil. The most common chemicals involved are petroleum hydrocarbons, solvents, pesticides, lead and other heavy metals. The occurrence of this phenomenon is correlated with the degree of industrialization and intensity of chemical usage.

A soil pollutant is any factor which deteriorates the quality, texture and mineral content of the soil or which disturbs the biological balance of the organisms in the soil. Pollution in soil has adverse effect on plant growth.

Pollution in soil is associated with

- Indiscriminate use of fertilizers
- Indiscriminate use of pesticides, insecticides and herbicides
- Dumping of large quantities of solid waste
- Deforestation and soil erosion

5.13.3 Indiscriminate Use of Fertilizers

Soil nutrients are important for plant growth and development. Plants obtain carbon, hydrogen and oxygen from air and water. But other necessary nutrients like nitrogen, phosphorus, potassium, calcium, magnesium, sulphur and more must be obtained from the soil. Farmers generally use fertilizers to correct soil deficiencies. Fertilizers contaminate soil with impurities, which come from the raw materials used for their manufacture. Mixed fertilizers often contain ammonium nitrate (NH_4NO_3), phosphorus as P_2O_5, and potassium as K_2O. For instance, As, Pb and Cd present in traces in rock phosphate mineral get transferred to super phosphate fertilizer. Since the metals are not degradable, their accumulation in the soil above their toxic levels due to excessive use of phosphate fertilizers, becomes an indestructible poison for crops.

The over use of NPK fertilizers reduces quantity of vegetables and crops grown on soil over the years. It also reduces the protein content of wheat, maize, grams, etc., grown on that soil. The carbohydrate quality of such crops also gets degraded. Excess potassium content in soil decreases Vitamin C and carotene content in vegetables and fruits. The vegetables and fruits grown on over-fertilized soil are more prone to attacks by insects and disease.

5.13.4 Indiscriminate Use of Pesticides, Insecticides and Herbicides

Plants on which we depend for food are under attack from insects, fungi, bacteria, viruses, rodents and other animals, and must compete with weeds for nutrients. To kill unwanted populations living in or on their crops, farmers use pesticides. The first widespread insecticide use began at the end of World War II and included DDT (dichlorodiphenyltrichloroethane) and gammaxene. Insects soon became resistant to DDT and as the chemical did not decompose readily, it persisted in the environment. Since it was soluble in fat rather than water, it biomagnified the food chain and disrupted calcium metabolism in birds, causing eggshells to be thin and fragile. As a result, large birds of prey such as the brown pelican, ospreys, falcons and eagles became endangered. DDT has been now been banned in most western countries. Ironically, many of them including USA, still produce DDT for export to other developing nations whose needs outweigh the problems caused by it.

The most important pesticides are DDT, BHC, chlorinated hydrocarbons, organophosphates, aldrin, malathion, dieldrin, furodan, etc. The remnants of such pesticides used on pests may get adsorbed in the soil particles, which then contaminate root crops grown in that soil. The consumption of such crops causes the pesticide remnants to enter human biological systems, affecting them adversely.

An infamous herbicide used as a defoliant in the Vietnam War called Agent Orange (dioxin), was eventually banned. Soldiers' cancer cases, skin conditions and infertility have been linked to exposure to Agent Orange.

Pesticides not only bring toxic effect on humans and animals but also decrease the fertility of the soil. Some of the pesticides are quite stable and their biodegradation may take weeks and even months.

Pesticide problems such as resistance, resurgence, and health effects have caused scientists to seek alternatives. Pheromones and hormones to attract or repel insects and using natural enemies or sterilization by radiation have been suggested.

5.13.5 Dumping of Solid Wastes

In general, solid waste includes garbage, domestic refuse and discarded solid materials such as those from commercial, industrial and agricultural operations. They contain increasing amounts of paper, cardboards, plastics, glass, old construction material, packaging material and toxic or otherwise hazardous substances. Since a significant amount of urban solid waste tends to be paper and food waste, the majority is recyclable or biodegradable in landfills. Similarly, most agricultural waste is recycled and mining waste is left on site.

The portion of solid waste that is hazardous such as oils, battery metals, heavy metals from smelting industries and organic solvents are the ones we have to pay particular attention to. These can in the long run, get deposited to the soils of the surrounding area and pollute them by altering their chemical and biological properties. They also contaminate drinking water aquifer sources. More than 90% of hazardous waste is produced by chemical, petroleum and metal-related industries and small businesses such as dry-cleaners and gas stations contribute as well.

Solid waste disposal was brought to the forefront of public attention by the notorious Love Canal case in USA in 1978. Toxic chemicals leached from oozing storage drums into the soil underneath homes, causing an unusually large number of birth defects, cancers and respiratory, nervous and kidney diseases.

5.13.6 Deforestation

Soil Erosion occurs when the weathered soil particles are dislodged and carried away by wind or water. Deforestation, agricultural development, temperature extremes, precipitation including acid rain, and human activities contribute to this erosion. Humans speed up this process by construction, mining, cutting of timber, overcropping and overgrazing. It results in floods and causes soil erosion.

Forests and grasslands are excellent binding materials that keep the soil intact and healthy.

They support many habitats and ecosystems, which provide innumerable feeding pathways or food chains to all species. Their loss would threaten food chains and the survival of many species. During the past few years quite a lot of vast green land has been converted into deserts. The precious rainforest habitats of South America, tropical Asia and Africa are coming under pressure of population growth and development (especially timber, construction and agriculture). Many scientists believe that a wealth of medicinal substances including a cure for cancer and AIDS, lie in these forests. Deforestation is slowly destroying the most productive flora and fauna areas in the world, which also form vast tracts of a very valuable sink for CO_2.

5.13.7 Pollution Due to Urbanisation: Pollution of Surface Soils

Urban activities generate large quantities of city wastes including several biodegradable materials (like vegetables, animal wastes, papers, wooden pieces, carcasses, plant twigs, leaves, cloth wastes as well as sweepings) and many non-biodegradable materials (such as plastic bags, plastic bottles, plastic wastes, glass bottles, glass pieces, stone/cement pieces). On a rough estimate Indian cities are producing solid city wastes to the tune of 50,000 – 80,000 metric tonnes every day. If left uncollected and decomposed, they are a cause of several problems such as:

- Clogging of drains: Causing serious drainage problems including the burst / leakage of drainage lines leading to health problems.
- Barrier to movement of water: Solid wastes seriously damage the normal movement of water thus creating problem of inundation, damage to foundation of buildings as well as public health hazards.
- Foul smell: Generated by dumping the wastes at a place.
- Increased microbial activities: Microbial decomposition of organic wastes generate large quantities of methane besides many chemicals to pollute the soil and water flowing on its surface.
- Hospital wastes create many health problems: These have dangerous pathogen, dangerous medicines and injections syringes, etc.

5.13.8 Pollution of Underground Soil

Underground soil in cities is likely to be polluted by:

- Chemicals released by industrial wastes
- Decomposed and partially decomposed materials of sanitary wastes

Many dangerous chemicals like cadmium, chromium, lead, arsenic, selenium are likely to be deposited in underground soil. Similarly, underground soil polluted by sanitary wastes generates many harmful chemicals. These can damage the normal activities and ecological balance in the underground soil. Some of the major causes of underground soil pollution are:

- Polluted water discharged from factories
- Run-off from pollutants (paint, chemicals, rotting organic material) leaching out of landfills
- Oil and petroleum leaks from vehicles washed off the road by the rain into the surrounding habitat

- Chemical fertilizer run-off from farms and crops
- Acid rain (fumes from factories mixing with rain)
- Sewage discharged into rivers instead of being treated properly
- Over-application of pesticides and fertilizers
- Purposeful injection into groundwater as a disposal method
- Interconnections between aquifers during drilling (poor technique)
- Septic tank seepage
- Lagoon seepage
- Sanitary/hazardous landfill seepage
- Cemeteries
- Scrap yards (waste oil and chemical drainage)
- Leaks from sanitary sewers

5.13.9 Effects of Soil Pollution

Soil pollution has diverse effects on many associated stakeholders. Many of major affected areas due to soil pollution are summarised below:

Agricultural

- Reduced soil fertility
- Reduced nitrogen fixation
- Increased erodibility
- Larger loss of soil and nutrients
- Deposition of silt in tanks and reservoirs
- Reduced crop yield
- Imbalance in soil fauna and flora

Industrial

- Dangerous chemicals entering underground water
- Ecological imbalance
- Release of pollutant gases
- Release of radioactive rays causing health problems
- Increased salinity
- Reduced vegetation

Urban

- Clogging of drains
- Inundation of areas
- Public health problems
- Pollution of drinking water sources
- Foul smell and release of gases
- Waste management problems

5.13.10 Long-Term Environmental Effects of Soil Pollution

When it comes to the environment itself, the toll of contaminated soil is even drier. Soil that has been contaminated should no longer be used to grow food, because the chemicals can leech into the food and harm people who eat it.

If contaminated soil is used to grow food crops, the land will usually produce lower yields than it would if it were not contaminated. This, in turn, can cause even more harm because a lack of plants on the soil will cause more erosion, spreading the contaminants onto land that might not have been tainted before.

In addition, the pollutants change the makeup of the soil and the types of microorganisms that live in it. If certain organisms die off in the area, the larger predator animals will also have to move away or die because they've lost their food supply. Thus it's possible for soil pollution to change whole ecosystems

Effects of soil pollution in brief:

- Pollution runs off into rivers and kills the fish, plants and other aquatic life
- Crops and fodder grown on polluted soil may pass the pollutants on to the consumers
- Polluted soil may no longer grow crops and fodder
- Soil structure is damaged (clay ionic structure impaired)
- Corrosion of foundations and pipelines
- Impairs soil stability
- May release vapours and hydrocarbons into buildings and cellars
- May create toxic dusts
- May poison children playing in the area

5.14 CONTROL OF SOIL POLLUTION

The following steps have been suggested to control soil pollution. To help prevent soil erosion, we can limit construction in sensitive areas. In general, we would need less fertilizer and fewer pesticides if we could all adopt the three R's: reduce, reuse, and recycle. This would give us less solid waste.

5.14.1 Reducing Chemical Fertilizer and Pesticide Use

Applying biofertilizers and manures can reduce chemical fertilizers and pesticides use.

Biological methods of pest control can also reduce the use of pesticides and thereby minimize soil pollution.

5.14.2 Reusing of Materials

Materials such as glass containers, plastic bags, paper, cloth, etc., can be reused at domestic levels rather than being disposed of, reducing solid waste pollution.

5.14.3 Recycling and Recovery of Materials

This is a reasonable solution for reducing soil pollution. Materials such as paper, some kinds of plastics and glass can and are recycled. This decreases the volume of refuse and helps in the conservation of natural resources. For example, recovery of one tonne of paper can save 17 trees.

5.14.4 Reforesting

Control of land loss and soil erosion may be attempted through restoring forest and grass cover to check wastelands, soil erosion and floods. Crop rotation or mixed cropping can improve the fertility of the land.

5.14.5 Solid Waste Treatment

Proper methods should be adopted for management of solid waste disposal. Industrial wastes can be treated physically, chemically and biologically until they are less hazardous. Acidic and alkaline wastes should be first neutralized; the insoluble material if biodegradable should be allowed to degrade under controlled conditions before being disposed.

As a last resort, new areas for storage of hazardous waste should be investigated such as deep well injection and more secure landfills. Burying the waste in locations situated away from residential areas is the simplest and most widely used technique of solid waste management. Environmental and aesthetic considerations must be taken into consideration before selecting the dumping sites.

Incineration of other wastes is expensive and leaves a huge residue and adds to air pollution.

Pyrolysis is a process of combustion in absence of oxygen or the material burnt under controlled atmosphere of oxygen. It is an alternative to incineration. The gas and liquid thus obtained can be used as fuels. Pyrolysis of carbonaceous wastes like firewood, coconut, palm waste, corn combs, cashew shell, rice husk paddy straw and saw dust, yields charcoal along with products like tar, methyl alcohol, acetic acid, acetone and a fuel gas.

Anaerobic/aerobic decomposition of biodegradable municipal and domestic waste is also being done and gives organic manure. Cow dung which releases methane into the atmosphere, should be processed further in 'gobar gas plants' to produce 'gobar gas' and good manure.

5.15 NATURAL LAND POLLUTION

Land pollution occurs massively during earthquakes, landslides, hurricanes and floods. All cause hard to clean mess, which is expensive to clean and may sometimes take years to restore the affected area. These kinds of natural disasters are not only a problem in that they cause pollution but also because they leave many victims homeless.

5.16 MARINE POLLUTION

Pollution occurs when concentrations of various chemical or biological constituents exceed a level at which a negative impact on ecosystem or human health can occur. Pollution results primarily from human activities. More specifically, in the marine context "pollution of marine environment means the introduction by man, directly or indirectly, of substances or energy into the marine environment, including estuaries, which results or is likely to result in such a deleterious effect as to harm living resources and marine life, hazardous to human health, hindrance to marine activities, including fishing and other legitimate uses of the sea, impairment of quality for use of seawater and reduction of amenities". The sources of pollution include sewage, urban run-off, industrial wastes, coastal development, and shipping activities. The chemical or biological constituents creating pollution are known as contaminants. More specifically, they can be divided into inorganic contaminants, such as metals; organic contaminants such as pesticides, PCB or petroleum hydrocarbons; biological contaminants such as coliform bacteria and other pathogens.

The list of potential contaminants is quite exhaustive; our understanding of the potential impacts of trace metals, bacteria, or hydrocarbons is well established. However, the long-term effects of man-made organic compounds such as pesticides, PCB and other plastics, etc., are not yet fully understood.

Pollution is not a recent phenomenon. It occurred since the man concentrated around a port area.

5.16.1 Effects of Marine Pollution

The major effects of marine pollution are:

- Disturbance in the balance of the ecosystem
- Loss of flora and fauna
- Migration of fisheries
- Biomagnification of undesirable material
- Aesthetic degradation
- Accelerated corrosion

5.16.2 Sources of Marine Pollution

The oceans have long been considered to have limitless capacity to receive and absorb all types of wastes. Since the 1950's many scientists began to warn that this limitless capacity is running out and the very survival of the marine environment was in danger. Therefore, it is imperative to control the sources of pollution.

The origins of various sources of pollution are listed below:

Table 5.1 Sources of marine pollution.

Source	*Percentage*
Land	45
Air	33
Transportation	12
Dumping	9
Offshore	1

As observed from the above table marine transportation routinely contributes to only 12% of the total marine pollution out of which major oil spills contribute to only 2%. Therefore, routine ship movements cause more than 10% of the pollution. However, land based activities are the major sources and have to be significantly controlled to save the marine ecosystem.

5.16.3 Dredging

Among all land based activities that cause marine pollution, dredging of ports and harbours is probably the most significant in terms of propagation of pollutants. The process of dredging causes the disruption of bottom sediments. More often than not, bottom sediments of ports and harbours, are contaminated by chemicals, minerals, hydrocarbons and domestic wastes. Toxic materials released from the disturbed soils may go into solution and cause severe mortality to marine estuarine life. Redeposition of suspended materials could possibly eliminate bottom flora and fauna. Organic material in the suspended material may deplete available oxygen in the surrounding waters and can create temporary stress conditions for aquatic animals. If suspended sediments are sufficiently concentrated, penetration of light in the water column is significantly reduced, thereby causing damage to light depending photosynthetic algae, corals and other aquatic organisms. Explosives used to break rock formations destroy bottom habitants. Pressure radiation due to blasting, often injures or disables marine life at considerable distances.

The problems associated with disposal of dredged materials are a matter of serious concern in both in industrial and developing countries. Traditionally, dredged materials have been disposed of in open water, without any kind of restrictions. Increased concern with the environmental impact on the marine environment, restrictions have been placed on such disposals, especially if they are contaminated. Areas of sediment depositions typically contain fine grained sand which acts as "sinks" for many compounds. Many organic contaminants have a low solubility in water and therefore are associated with suspended solids and are transported along with the sediments. Since these sediment zones are areas often require maintenance dredging, in most cases dredged material contain hazardous contaminants. Usually, the contaminants are due to land based activities such as industrial effluents, fertilisers, and pesticides, urban sewage, etc.

As the contaminants originate from land based activities, the disposal of contaminated dredged materials do not address the source of the problem, it is only a symptom. The problem of dredged material disposal can be considerably reduced, if effluents from land based sources are controlled. Further to control limit the propagation of contaminants can be achieved by proper selection of dredging process, equipment and disposal sites. Another mitigative measure that could be taken into consideration is to avoid or minimise dredging during fish migration, breeding and spawning seasons.

In areas where, chemical contaminants of the bottom sediments exceed threshold values, selection of appropriate dredging equipment is to be considered. These include:

- Oozer pump suction dredger to avoid the release of material at the point of excavation
- Closed clamshell bucket dredger or trailing cutter suction dredger with no overflow
- Proper offloading equipment on a disposal barge to pump to disposal sites without leakage

Unconfined open water disposal has been traditional method of dredged spoils disposal because it is the most economical option. Even, if a "safe" disposal site has been identified; strong upper water column or low bottom currents can carry significant quantities of low-density materials away to distant locations. This off-site transportation can lead to re-filling of the dredged areas, silting fish spawning areas and the redistribution of contaminated areas.

For the contaminated dredged material the following disposal options must be considered:

- **Thin Layer Disposal:** In this method, dredged spoils are spread very thinly over a wide area of a pre-identified disposal site. The thickness of the layer is calculated so that leachate of contaminants does not exceed threshold values that could be dangerous to marine flora and fauna.
- **Confined Land Disposal:** In case, the contaminants are high, suitable land based clay/ polymer lined disposal sites may be considered. This is an expensive method and a rough estimate is that it costs five times more than unconfined open water disposal. This process is suitable for small quantities of highly contaminated dredge spoils.
- **Open Water Disposal with Capping:** In this process, dredged spoils are disposed of in the open sea followed by "capping" with inert material. Typically, contaminated sediments are buried under one meter of "clean sediment". In this process, the contaminants are immobilised by the capping material.

- **Specialised Treatment:** When the materials are small and the contaminants are highly toxic, for example radioactive wastes, the spoils must be brought to shore based facilities for specialised treatment before disposal.

5.16.4 Ship Discharges

Discharges and spills from ships and shore based facilities result in the environmental degradation of marine life. In ports, the situation is even more critical due to the fact that most ports have relatively poor exchanges with the open sea. This results in the pollutants having long residence times in the port area. Untreated sewage discharges stagnating in port areas can cause serious epidemics to humans. It may also result in serious degradation of port waters. Discharge of garbage in port waters results in unaesthetic conditions due to accumulation of plastics, glass and metal containers. Frequently, plastic bags and sheets block intakes of cooling water pipelines of industrial units and floating crafts.

Ship discharges include:

- Ballast water from cargo tanks, oil tankers
- Tank washings from oil tankers
- Bilge waste from machinery spaces
- Residues from fuel and lubricating oil purification units, spent lubricating oils, and sludge from tankers
- Sanitary wastes
- Garbage

Although many large ships have sewage treatment, and oil/water separation facilities, smaller ships and other port crafts lack these facilities. To mitigate this problem all ports must have connections from ship to shore-based facilities to receive and treat sanitary wastes, oily waste and garbage. For vessels anchored outside port, a floating reception facility such as towed or self-propelled barge may be considered.

5.16.5 Ballast Water Discharges

Discharge of ballast water in ports may result in coliform and pathogens from the port of origin to propagate to the port of destination. Increases in ship speeds, especially those of container vessels, have resulted in increase of survival rates of these pathogens. A proposed annexe to the MARPOL regulations makes it mandatory for ships to exchange their ballast water several times in course of their voyage.

5.16.6 Spills: Detection and Cleanup

Although major oil spills constitute an estimated two per cent of the total marine pollution, they cause in addition to environmental damage, a serious aesthetic degradation of the shoreline. The major causes of the spill are:

- Marine causalties such as collisions, grounding, fires, etc.
- Hull failure
- Failure of equipment such as pipelines, flanges, hoses, etc.
- Incorrect operating procedures

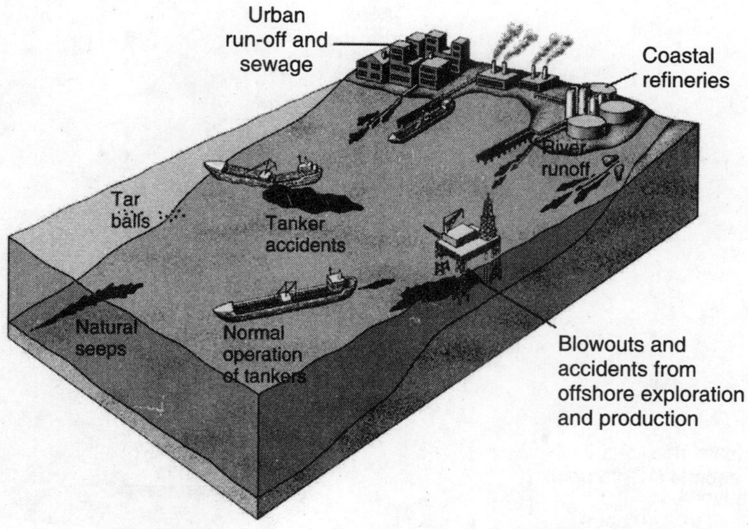

Fig. 5.9 Marine pollution source.

The avoidance of spill is a key requirement. Many ports in developing countries are particularly lax in enforcing regulations both from calling ships and their own port craft. To alleviate this problem, strict discipline must be observed. All ports must have an emergency contingency plan clearly outlining authority and responsibility for dealing with such situations. A mechanism for alerting and reporting must be established. Oil spill cleanup equipment must be available with all ports to deal with minor to medium spills. These equipment must include containment booms, recovery equipment and suitable dispersants. All operators should be trained to handle the equipment. These equipment should be routinely checked and regular exercises must take place to test the efficiency reporting, alerting and operating mechanisms.

5.16.7 Anti-fouling Paints

By their very nature anti-fouling paints inhibit the growth and/or destroy a wide variety of marine organisms. Severe damage to the aquatic life is caused by the application of paints containing tri-butyl-tin. Careless application of this paint causes more damage than necessary. Removal of old paint from vessels and floating craft in the waterfront also causes similar damage. New ship building/repair facilities should be located where marine life is minimal.

5.16.8 Run-off from Raw Material Storage

Raw materials such as coal, mineral ores, sulphur, salt, etc., are usually stored in the open in port areas prior to loading or after unloading from ships. Due to their large volumes, it is uneconomical to store them otherwise. Due to prevalent weather conditions such as rain and wind, frequently, significant amounts of these materials find themselves in harbour waters. Over a period of time, enormous quantities are built up to such levels that cause acute toxicity and fatalities to aquatic life.

To alleviate this problem, bunds are to be built along the periphery of the storage areas along with suitable drains. These drains may flow to a centralised location, where water can be treated before disposal to harbour waters.

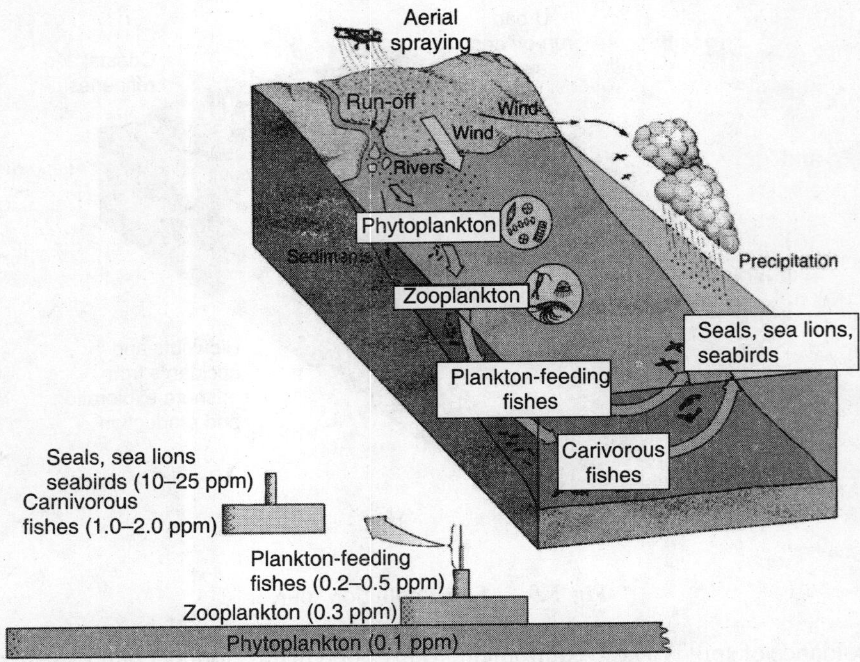

Fig. 5.10 Biomagnification of inert materials.

5.16.9 Waterfront Drainage

Most port jetties are designed so that they have slight slope towards the harbor waters. Ostensibly, this is to avoid flooding and/or water stagnation due to heavy rainfall. Inadvertently, this type of jetty design allows unimpeded flow of contaminants into harbour waters. The principal constituents of these contaminants are oils, diesel, hydraulic oils, and spillage of cargo on the wharf.

As far as possible, direct drainage from the wharves and piers must be avoided. These areas are to be paved and sloped so that they direct flows to catch basins, where the water collected is treated before disposal.

5.16.10 Hazardous Cargoes

A hazardous material can generally be defined as a substance which has an inherent instability, which under certain conditions can lead to fire, explosions or highly corrosive, toxic, or are radioactive. Some examples of hazardous chemicals are liquefied natural gas (LNG), acids, explosive chemicals, radioactive materials, ammonia, etc. During transportation and storage of these materials, it is quite possible that serious accidents may happen which not only endanger marine life but also human lives, directly.

Some of the techniques that are to be adopted to reduce the risk of accidents during transportation of such materials are listed below:

- Precise information must be obtained regarding the type, quantity, nature and condition of material.
- Proper segregation of these materials must be maintained.

- These materials are to be appropriately packed and the integrity of the packing must be maintained during transportation.
- Ships carrying hazardous cargoes must be constructed and operated at higher levels of safety.
- Entry and departure to and from ports are to be carefully monitored to avoid risks due collisions.

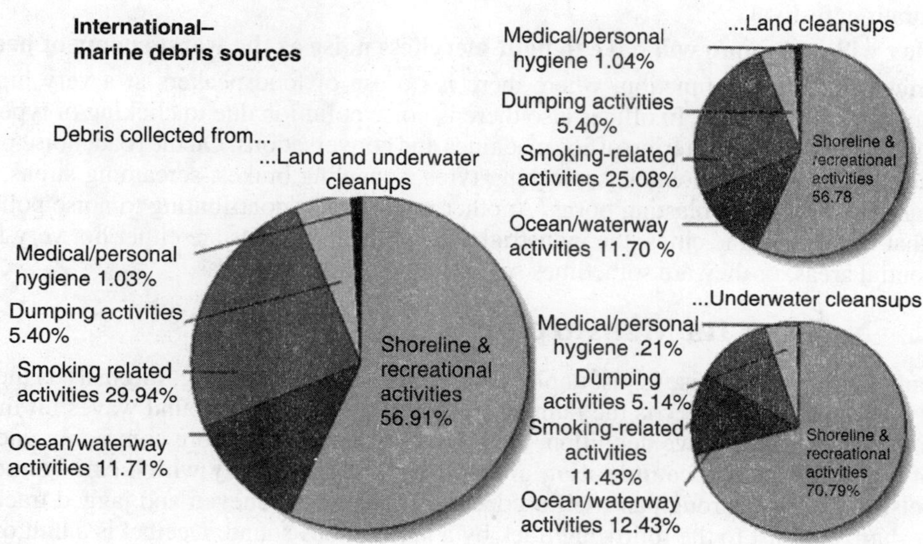

Fig. 5.11 Source of international marine pollution.

5.17 NOISE POLLUTION

Noise pollution did not create much public concern due to ignorance about the treacherous effect of noise on both workers in industry in particular and the public in the community in general. It is, therefore, imperative to assess the environment in which the noise is being heard by using suitable bases of judgement and awareness to determine whether or not a definite nuisance exists.

Noise is an important environmental pollutant like noxious gases that befoul our air, water and soil. It destroys bridges and produces cracks in buildings. The noise can cause skin and mental diseases.

It has been revealed that noise is a technology generated problem and that the overall noise doubles every ten years keeping pace with our social and industrial progress. This geometric progression-wise growth of noise could be mind-boggling in view of the ever increasing pace of technological growth. According to **Robert Koch** a Nobel Prize Winner German bacteriologist.

According to him **"Noise Like Smog is a Slow Agent of Death."**

5.17.1 Sources of Noise Pollution

Noise is a major factor of environmental pollution; on the one hand, industrialization, scientific and technological developments have contributed a great deal to the progress of society, on the other, these are main causes of environmental pollution, including noise pollution. As the day rises, the noise level in the different parts of the city increases in and around workplaces and homes. The peak noise levels are reached in the twilight hours as traffic reaches a peak. In India, the problem caused by noise pollution is more aggravated in view of the fact that there is hardly any celebration, festival.

"A day will come man will have to fight merciless noise as the worst enemy of health."

Marriage or religious function, where there is no use of loudspeakers at a very high pitch continuously for a long time. In offices also there is noise pollution due to clicking of typewriters, bells, ringing telephones, clattering office machines and conversations. On the road, noise pollution exists due to growing automobiles, screeching tyres, squealing brakes, screaming sirens, blaring televisions and radios, and blasting horns. Another major factor contributing to noise pollution in India is that in many of the cities, the industrial and commercial units are either not very far from the residential areas, or they are sometimes set up in the residential areas.

5.17.2 Noise & Its Measurement

In layman's terminology, noise is just unpleasant sound. Broadly speaking, **intensity** is the energy of sound waves and **frequency** is the number of times per second the sound waves hit the ear of man. On the basis of the above definition, the loudness which is the more commonly understood source of noise, is in fact a **combination of intensity and frequency which can be measured in decibels.** If we view through an oscilloscope, noise makes an uneven and jagged track on the screen in sharp contrast to the following track by a harmonious sound. Decibel is a unit of sound, name after **Alexander Graham Bell.**

Intensity – The level of sound is usually expressed in terms of the sound pressure level (SPL) in decibels, which is defined as

$$SPL = 20 \log 10\ P/PO\ dB$$

where P is the pressure variation measured in N/m^2 and PO is the standard reference pressure taken as $2 \times 10 - 5\ N/m^2$.

Human beings generally have the ability to hear sounds in the frequency range 20 to 20,000 Hz (1 Hz = 1 cycle per second). Human hearing is most sensitive to frequencies in the range 500 to 6000 Hz and less sensitive both at lower and higher frequencies.

A decibel can be defined as an abstract unit. It is to be remembered that the threshold at the normal hearing is 20 – 25 decibels and of normal conversation is 60 decibels. It has also been noticed that speech interference occurs at 75 decibels and definite annoyance begins at 80 decibels. The motor activities disturbs at 90 decibels and physiological disturbance occurs beyond 120 decibels definite pain occurs at 140 decibels

5.17.3 Effects of Noise on Human Beings

(a) **Auditory Effect:** The human ear is a very sensitive instrument. If the hearing mechanisms are damaged in any way either by excessive noise levels or by diseases which affect the brain, the auditory nerve or the auditory ossicles, then hearing will be impaired. Intense noise levels, for example, are encountered in many industries and they can cause temporary or progressively permanent loss of hearing.

(b) Physiological Effect: Noise is likely to harm the physiological and psychological well-being of people. Physiological changes include:

Effect Main Observations

- Cardiovascular response: Changes in heart beat rate and blood-pressure.
- Metabolism: Increased metabolic rate, increased pulse rate.
- Respiration: Increased respiratory rate, pulse frequency
- Eye pupil dilation: There is significant increase in pupil size above 55 dB.

(c) Non-Auditory Effects: In addition to its auditory effects (temporary and permanent threshold shifts, noise inducted deafness, etc.); noise can also produce many non-auditory effects. The exposure to noise:

- May interfere with verbal communication
- Cause annoyance and distraction
- Reduce working efficiency and work output
- Cause fatigue

5.17.4 Sleep Interference

The World Health Organization Task Group on Environmental Health Criteria for Noise has recommended a noise level of 35 (A) to preserve the restorative process of sleep.

5.17.5 Speech Interference

Speech reception is the most important and also the most complex use of the auditory system. Noise can either mask the speech to make it inaudible or by making only some frequencies leaving it audible but of reduced intelligibility.

A decibel can be defined as an abstract unit. It is to be remembered that the threshold at the **normal hearing is 20-25 decibels** and of **normal conversation is 60 decibels.** It has also been noticed that **speech interference occurs at 75 decibels** and **definite annoyance begins at 80 decibels**. The **motor activities disturbs at 90 decibels** and **physiological disturbance occurs beyond 120 decibels. Definite pain occurs at 140 decibels** and though no human being has reportedly died of noise, experimental results have shown that mice died at 175 decibels of noise. According to environmentalist **Thomas G. Aylesworth,** "Constant noise may cause our blood-vessels to contract, our skin to become pale, our muscles to constrict and adrenaline to be short into our blood-stream". This is the reason why factory workers develop abnormal heartbeat rates and suffering from insomania, nervousness and impaired motor coordination.

5.17.6 Noise Control

It is very important to control noise pollution by using many means. Some of the adopted methods used are discussed herein:

A. Industrial Noise: Control of noise (i.e. its prevention and reduction) is a system related problem.

This system is composed of the following components.

- The source of noise
- The path of sound propagation
- The receiver of noise.

- The source is part of the system that produces the acoustic energy. The source should be considered a group of noise generators which present different physical characteristics randomly distributed in space and time.
- The acoustic energy produced by the source is transmitted to the environment in which it is propagated, and which may be a solid structure or air.
- The third component of the system (i.e. the receiver of sound) may be a worker operating his machines.

Noise prevention and reduction measures should be aimed at the following goals:

(a) Controlling source of noise

(b) Precluding the propagation, amplification and reverberation of noise

(c) Isolating the workers.

Evidently, reduction of noise at the source is the most rational method of noise control. If all machines and all prime movers were sufficiently silent, there would not be many problems of excessive noise left for solution.

B. Community Noise: The traffic volume should be reduced by diversion of traffic, and use of horn should be banned.

- Residences should not be allowed to grow nearby industrial areas.
- There should be plenty of trees and bushes in open spaces, houses and lanes.
- At the time of festive occasions, as practised in some western countries, community fireworks could be displayed in lieu of individual celebrations, which will be a good way of enjoying the spirit of the occasion while keeping it safe and pollution free, thus lowering the noise, which mars the celebrations. Also, as a measure of awareness drive, slums, schools, and residential colonies covering all zones should be visited to address both youngsters and elders on the need of adhering to the relevant decibels and smoke emission levels in a bid to make festivals noise and pollution free, besides monitoring noise levels and ambient air quality.

C. Noise Pollution Standards: The Noise Pollution (Regulations & Control) Rules 2000 prescribe the ambient air quality standards in respect to noise in industrial, commercial, residential and silence zone areas as below:

Table 5.2 Noise pollution standards in India.

S.No	*Area*	*Noise level (Leq in dBA)*	
		Day	*Night*
1.	Industrial	75	70
2.	Commercial	65	55
3.	Residential	55	45
4.	Silent Zone	50	40
Day time - 0600 hrs to 2200 hrs		Night time – 2200 to 0600 hrs	

5.18 FRAMEWORK STANDARDS FOR NOISE MEASUREMENT AT MACHINES

The determination of the sound power is conducted internationally according to the ISO 3740 series, especially ISO 3744 and ISO 3746; the ISO standards were revised in 1989. In these standards the enveloping measurement surface procedure is used to determine the sound power level. The subjects of the standards are in particular the selection of the measurement surface, the arrangement of measuring points, the determination of the environmental and background noise correction, the principles for the selection of the mounting and operating conditions and for determining measurement uncertainties.

The emission sound pressure level is determined internationally according to ISO 11201, ISO 11202, ISO 11203, and ISO 11204. These standards address, as a complement to ISO 3744, selection of the measuring point or points for the workplace or workplace area and the environmental correction for one or more measuring points. Since the emission sound pressure level is taken as a criterion for the type and scope of the noise declaration in Europe, the precise determination and reproducibility should be stressed. A method which ensures the most correct possible environmental correction for a measuring point has now been developed and it is described in ISO 11204. ISO 11203 lays down procedures for calculating the emission sound pressure level from the sound power level. This procedure could be quite effective for smaller machines for which no precise workplace has been fixed (e.g. household appliances). ISO 11201 and ISO 11202 contain simplified procedures which are only applicable to a limited extent.

The type of noise declaration and a simplified verification procedure are given in ISO 4871.

The verification procedures for noise emission already exist as International Standards (ISO 7574) Most of the framework standards for determination, declaration and verification of the noise emission have been completed. With the noise emission declaration (Table 5.3) for a machine, which is in Europe mandatory for the manufacturer or importer, there is the possibility for the first time of evaluating low noise level or noise quality. This creates conditions for achieving noise reduction at source.

Noise emission declarations are also absolutely essential for forecasts by calculation (ISO/TR 11690-3) of the noise impact (rating level) at workplaces and in the vicinity for the noise control planning of new workshops.

Table 5.3 Example of a noise declaration.

Machine model number, operating conditions, and other indentifying information:		
Type-990, Model 11-TC, 50 Hz, 230 V rated load.		
Declared dual-number noise emission values in accordance with ISO-4871		
	Operating mode 1	*Operating mode 2*
Measured A-weight sound power level, L_{WA} (ref 1pW) in decibels)	88	95
Uncertanity, K_{WA}, in decibels	2	2
Measured A-weighted emission sound pressure level, L_{pA} (ref 20 mPa) at the operator's position in decibles	78	86
Ucertainty, K_{pA}, in decribels	2	2
Values determined according to noise test code given in ISO XXXX, using the basic standards ISO YYYY and ISO ZZZZ **Note:** The sum of a measured noise emission value and its associated uncertainty represented an upper boundary of the range of values which is likely to occur in measurements.		

5.18.1 Standards for Noise Reduction Devices and Materials

Since noise control devices and materials can of course also be used subsequently and in addition, standards were drawn up separately with noise-related requirements and measuring procedures for such devices/materials: for noise control enclosures (ISO 11546, ISO 11957), for noise absorbers (ISO 11654, ISO 10534), for silencers (ISO 7235, ISO 11691, ISO 11820, ISO 14163), noise baffles (ISO 11821, ISO 10053).

Table 5.4 Typical noise levels of some point sources.

Source	*Noise levels dB(A)*	*Source*	*Noise level, dB(A)*
Air compressors	95–104	Quiet garden	30
Lathe machine	87	Ticking clock	30
Milling machine	112	Type institute	60
Oxy-acetylene cutting	96	Printing press	80
Pulveriser	92	Sports car	80–95
Riveting	95	Trains	96
Power operated portable saw	108	Trucks	90–100
Steam turbine (12,50 kW)	91	Car horns	90–105
Pneumatic chiseling	118	Jet takeoff	120

5.18.2 Impacts of Noise

Why bother about noise? Often neglected, noise induces a severe impact on humans and on living organisms. Some of the adverse effects are summarised below.

- **Annoyance:** It creates annoyance to the receptors due to sound level fluctuations. The aperiodic sound due to its irregular occurrences causes displeasure to hearing and causes annoyance.
- **Physiological effects:** The physiological features like breathing amplitude, blood pressure, heart-beat rate, pulse rate, blood cholesterol are affected.

Loss of hearing: Long exposure to high sound levels cause loss of hearing. This is mostly unnoticed, but has an adverse impact on hearing function.

Human performance: The working performance of workers/humans is affected as they lose their concentration.

Nervous system: It causes pain, ringing in the ears, feeling of tiredness, thereby affecting the functioning of human system.

Sleeplessness: It affects sleep thereby inducing the people to become restless and lose concentration and presence of mind.

Damage to material: The buildings and materials may get damaged by exposure to infrasonic / ultrasonic waves and even get collapsed.

Learning Outcomes

The main learning outcomes from this chapter should be fundamental understanding of:

Type of pollutions; *Water pollution*: Hydrosphere, classification, Sources & effects of pollutants, D.O, B.O.D, C.O.D, Eutrophication; *Land pollution:* Lithosphere, source & effect of pollutants; *Marine pollution:* Classification, sources and Effects of Pollutants; *Noise pollution*: Introduction, Rating Systems

Readers also advised to carry out the following exercise:

Exercise

1. Write a 200 words essay on "learning outcomes from this chapter".
2. Create a poster reflecting, how pollution can affect the life of livings beings.
3. Give a 10 minute presentation on marine pollution in your class or among your colleagues.

Chapter 6

Environment Impact Assessment

6.1 INTRODUCTION

Environmental impact assessment (EIA) is a systematic process to identify, predict and evaluate the environmental impacts of a proposed project. It also includes beneficial and adverse effects of a developmental project on socio-economical systems, cultural systems and human health services. It is a term used to describe the total process of assessing the environmental effects of a development project. According to UNEP, "Environmental impact assessment is a tool used to identify the environmental, social and economic impacts of a project prior to decision-making." It is used for taking decision regarding projects and developments. It is done prior to any development of a project taking place. It produces an environmental statement which usually includes: A non-technical summary, a description of the project: location, design, scale, size, etc., a description of the receiving environment, description of significant effects and the mitigating measures to be taken to minimize the effects of the project. The International Association for Impact Assessment (IAIA) describes it as "the process of identifying, predicting, evaluating and mitigating the biophysical, social, and other relevant effects of development proposals prior to major decisions being taken and commitments made." EIA predicts how the project could harm to people, their homeland, their livelihoods, and the other nearby developmental activities. After predicting potential impacts, the EIA identifies measures to minimize the impacts and suggests ways to improve the project viability.

6.1.1 Aims and Objectives of an EIA

The aims and objectives of an EIA can be classified as: immediate aim or short-term objectives and ultimate aim or long-term objectives. The immediate aim of an EIA is to identify the potentially significant environmental impacts and risks of a proposed project. It helps the decision makers in improving design of a project and taking necessary actions. The ultimate aim of an EIA is to promote sustainable development of a project by ensuring that it will not disturb the natural resources, ecological functions, livelihood of the nearby people.

Immediate objectives of EIA:

- To improve the design of a project so that its benefits can be achieved and sustained without causing adverse impacts.
- To ensure that resources are used appropriately and efficiently. It helps in reducing the overall costs of the project.
- To identify the environmental impacts of a proposal at an early stage and take necessary actions accordingly.

- To facilitate informed decision making, including setting the environmental terms and conditions for implementing the proposal.

Long-term objectives of EIA:

- To protect human health and safety.
- To avoid irreversible changes and serious damage to the environment.
- To safeguard valued resources, natural areas (water, soil, air, biological system) and ecosystem components.
- To enhance the social aspects of the proposal.

6.1.2 History of EIA

Environmental impact assessment is "a powerful environmental safeguard in the project planning process" (Therivel and Morris 2001: 10). The process was developed in the USA and became a part of the National Environmental Policy Act (NEPA) in 1969. Initially, it was practised in countries like Canada, Australia, New Zealand (1973-74), Columbia (1974) and the Philippines (1978). Currently, it is practised in more than 100 countries. In 1989, the World Bank adopted EIA for major development projects, in which the borrower country had to undertake an EIA under the Bank's supervision.

Table 6.1 Evolution and history of EIA.

	Development of EIA
Pre-1970	Project review based on the technical/engineering and economic analysis. Limited consideration given to environmental consequences.
Early/mid-1970s	EIA introduced by NEPA in 1970 in US. Basic principle: Guidelines, procedures including public participation requirement instituted. Standard methodologies for impact analysis developed (e.g. matrix, checklist and network). Canada, Australia and New Zealand became the first countries to follow NEPA in 1973-1974. Unlike Australia, which legislated EIA, Canada and New Zealand established administrative procedures. Major public inquires help shape the process's development.
Late 1970 and early 1980s	More formalized guidance. Other industrial and developing countries introduced formal EIA requirements (France, 1976; the Philippines, 1977), began to use the process informally or experimentally (Netherlands, 1978) or adopted elements, such as impact statements or reports, as part of development applications for planning permission (German states [lander], Ireland). Use of EA by developing countries (Brazil, the Philippines, China, Indonesia) Strategic Environment Assessment (SEA), risk analysis included in EA processes.

Contd...

Contd...

	Greater emphasis on ecological modeling, prediction and evaluation methods. Provision for public involvement.
Mid-1980s to end of decade	Coordination of EA with land use planning processes.
	In Europe, EC Directive on EIA establishes basic principles and procedural requirements for all member states. Increasing efforts to address cumulative effects. World Bank and other leading international aid agencies establish EA requirements. Spread of EIA process in Asia.
1990s	Requirement to consider trans-boundary effects under Espoo convention. Increased use of GIS and other information technologies. Sustainability principal and global issues receive increased attention. India also adopted the EIA formally. Formulation of EA legislation by many developing countries. Rapid growth in EA training.

Source: International Study of the Effectiveness of Environmental Assessment, final report, environmental assessment in a changing world, Prepared by Barry Sadler, June 1996.

6.1.3 EIA Process in India

In India, it started in 1976-77. Till 1980, the subjects of environment and forests were the concern of the Department of Science and Technology and Ministry of Agriculture respectively. Later, the issues were formally attended by the Department of Environment which was established in 1980. In 1980, clearance of large projects from the environmental angle became an administrative requirement to the extent that the planning commission and the central investment board sought proof of such clearance before according financial sanction. It was then transferred to the Ministry of Environment & Forest in 1985. Till 1994, environmental clearance from the Central Government was an administrative decision and lacked legislative support. A major legislative measure for the purpose of environmental clearance was taken in 1994. The Union Ministry of Environment and Forests (MoEF), Government of India, under the Environmental (Protection) Act 1986, promulgated an EIA notification known as "Environment impact Assessment Notification 1994" making Environmental Clearance (EC) mandatory for expansion or modernization of any activity or for setting up new projects listed in Schedule 1 of the notification. Since then there have been 12 amendments made in the EIA notification of 1994. For developmental proposals, environmental clearance can be taken either at the state level or at the central level depending on certain criteria concerning the characteristics of the project. However, (regardless of where the final environmental clearance is obtained from), firstly, for most of the projects consent must be obtained from the state pollution control board or pollution control committees in the case of union territories.

Recently, the MoEF notified new EIA legislation in September 2006. The notification makes it mandatory for various projects such as mining, thermal power plants, river valley, infrastructure (road, highway, ports, harbours and airports) and industries including very small electroplating or foundry units to get environment clearance. However, unlike the EIA Notification of 1994, the new legislation has put the onus of clearing projects on the state government depending on the size/capacity of the project.

List of projects requiring environmental clearance from the central government

1. Nuclear power and related projects such as heavy water plants, nuclear fuel complex, rare earths.
2. River valley projects including hydel power, major irrigation and their combination including flood control.
3. Ports, harbours, airports (except minor ports and harbours).
4. Petroleum refineries including crude and product pipelines.
5. Chemical fertilizers (nitrogenous and phosphatic other than single superphosphate).
6. Pesticides (technical).
7. Petrochemical complexes (both olefinic and aromatic) and petro-chemical intermediates such as DMT, Caprolactam, LAB etc. and production of basic plastics such as LLDPE, HDPE, PP, PVC.
8. Bulk drugs and pharmaceuticals.
9. Exploration for oil and gas and their production, transportation and storage.
10. Synthetic rubber.
11. Asbestos and asbestos products.
12. Hydrocyanic acid and its derivatives.
13. (a) Primary metallurgical industries (such as production of iron and steel, aluminium, copper, zinc, lead and ferro alloys). (b) electric arc furnaces (mini steel plants).
14. Integrated paint complex including manufacture of resins and basic raw materials required in the manufacture of paints.
15. Viscose staple fibre and filament yarn.
16. Storage batteries integrated with manufacture of oxides of lead and lead antimony alloys.
17. All tourism projects between 200m—500 metres of high water line and at locations with an elevation of more than 1000 metres with investment of more than Rs 5 crore.
18. Thermal power plants.
19. Mining projects (major minerals) with leases more than 5 hectares.
20. Highway projects except projects relating to improvement work including widening and strengthening of roads with marginal land acquisition along the existing alignments provided it does not pass through ecologically sensitive areas such as National Parks, Sanctuaries, Tiger Reserves, Reserve Forests**
21. Tarred roads in the Himalayas and or forest areas.
22. Distilleries.
23. Raw skins and hides
24. Pulp, paper and newsprint.
25. Dyes.

26. Cement.
27. Foundries (individual)
28. Electroplating
29. Metaamino phenol

6.2 BASIC PRINCIPLES OF EIA

1. **Purposive:** The process should involve a proper planning and decisions that result in protection of environment and human beings.
2. **Rigorous:**The process should apply techniques and methodologies appropriately in mitigating the problems.
3. **Practical:** The information and outputs provided by the assessment process are readily usable in decision making and planning.
4. **Relevant:** The process should provide sufficient, reliable and usable information for development planning and decision making.
5. **Cost effective:** The assessment process and its outcomes will ensure environmental protection at the least cost to the society.
6. **Efficient:** It should impose the minimum burden in terms of time and money.
7. **Adaptive:** It should be able to adapt to deal efficiently with any proposal and decision making situation.
8. **Focused:** The process should focus on important environmental effects and key issues.
9. **Participative:** It should provide opportunities to involve the interested and affected people and include their suggestions during documentation and decision making.
10. **Interdisciplinary:** The process should ensure that the appropriate techniques and experts in the relevant bio-physical and socio-economic disciplines are employed, including use of traditional knowledge as relevant.
11. **Credible:** Assessment is undertaken with professionalism and objectivity. It should be subjected to independent checks and verification.
12. **Accountability:** The decision-makers are responsible to all parties for their action and decisions under the assessment process.
13. **Transparency:** All assessment decisions and their basis should be open and accessible.
14. **Certainty:** The process and timing of the assessment should be agreed in advanced and followed by all participants.
15. **Systematic:** The process should be done in a systematic way.

6.2.1 Types of EIA

There are two types of EIA: Rapid EIA and Comprehensive EIA.

Rapid EIA: In rapid EIA, data and information are collected for only one season (other than monsoon). It is carried out for the projects that are likely to cause minimum adverse effects on environment. The time frame for undertaking rapid EIA is very short about 3 months. It is for speedier appraisal process. It is acceptable if it does not compromise on the quality of decision making.

Comprehensive EIA: In comprehensive EIA, data and information are collected for three seasons (other than monsoon). It is carried out for the projects that are likely to cause serious adverse effects on environment. It is conducted over a year so, it is a very long process.

6.2.2 Process of EIA

The fundamental components of EIA involve following stages:

- **Screening:** To determine which projects or developments require a full or partial impact assessment study. To decide whether or not a proposal should be subject to the EIA process and, if so, at what level of detail. This is the first step of project evaluation.
- **Scoping:** To identify which potential impacts are relevant to assess (based on legislative requirements, international conventions, expert knowledge and public involvement), to identify alternative solutions that avoid, mitigate or compensate adverse impacts on biodiversity (including the option of not proceeding with the development, finding alternative designs or sites which avoid the impacts, incorporating safeguards in the design of the project, or providing compensation for adverse impacts), and finally to derive terms of reference for the impact assessment. It ensures that the study addresses all the issues of importance to the decision-makers. It selects primary impacts for the EIA depending on the basis of magnitude, geographical extent, importance to decision makers, etc.

These two steps, i.e., screening and scoping is done before starting an EIA.

- **Identification:** It includes the identification of those impacts which should be assessed in detail. This phase identifies the impacts of project on surroundings, human beings.
  1. Compile a list of key impacts (e.g., changes in air quality, noise levels, wildlife habitats, species diversity, landscape views, social and cultural systems, settlement patterns and employment levels from other EIA s for similar projects).
  2. Name all the project's sources of impacts (e.g. smoke emissions, water consumption, construction jobs) using checklists of questionnaires, then list possible receptors in the environment (e.g. crops, communities using same water for drinking, migrant of labour) by surveying the existing environment and consulting with interested parties.
  3. Identify impacts themselves through the use of checklist, matrices, networks, overlays, models and simulations.
- **Prediction:** Prediction scientifically characterizes the impacts, causes and effects and its secondary and synergetic consequences for the environment and the local community. Prediction draws on physical, biological, socio-economic and anthropological data techniques. In quantifying impacts, it may employ mathematical models, physical models, socio-cultural models, economic models, experiments or expert judgments.
- **Assessment and evaluation of impacts:** To predict and identify the likely environmental impacts of a proposed project or development, including the detailed elaboration of alternatives. It evaluates the predicted adverse impacts to determine whether they are significant enough to warrant mitigation. Thus judgment of significance can be based on one or more of the followings.
  1. Comparison with laws, regulations or accepted standards.
  2. Consultation with the relevant decision makers.

3. Reference to preset criteria such as protected sites features of species.
4. Acceptability to the local community or the general public.

- **Mitigation:** To develop measures to avoid, reduce or compensate for impacts, making good any environmental damage. Possible mitigation measures include:
  1. Changing project sites, routes, processes, raw materials, operating methods, disposal methods, disposal routes or locations, timing or engineering designs.
  2. Introducing pollution controls, waste treatment monitoring, phased implementation, landscaping, personal training, special social services or public education.
  3. Offering (as compensation) restoration of damaged resources, money to affected persons, concessions on other issues, or off-site programmes to enhance some other aspects of the environment or quality of life for the community. All mitigation measures cost something and this cost must be quantified too.

  Thus, various measures are then compared, trade-offs between alternative measures are weighed, and the EIA study team proposes one or more action plans, usually combining a number of measures. Several analytical techniques are available for this purpose as given below:
  1. Cost benefit analysis in which all quantifiable factors are converted to monetary values, and actions are assessed for their effect on project costs and benefits.
  2. Explaining what course of action would follow from various broad 'value judgments' (e.g., that social impact is more important than resources).
  3. A simple matrix of environmental parameters versus mitigation measures, contain brief description of the effects of each measure.
  4. Pair-wise comparisons, whereby the effects of an action are briefly compared with the effects of each of the alternative actions are briefly compared with the effects of each of the alternative actions, one pair at a time.

- **Reporting the Environmental Impact Statement (EIS) or EIA report:** Including an environmental management plan (EMP) and a non-technical summary for the general audience. An EIA report should contains:
  1. An executive summary of the EIA findings.
  2. A description of the proposed development projects.
  3. The major environmental and natural resource issues that needed clarification and elaboration.
  4. The project's impacts on the environment (in comparison with a base line were identified andpredicated).
  5. A discussion of options for mitigating adverse impacts and for shaping the project to suit its proposed environment, and an analysis of the trade-offs involved in choosing between alternative actions.
  6. An overview of gaps or uncertainties in the information.
  7. A summary of the EIA for the general public.

- **Public participation:** The public must be informed and consulted on a proposed development after the completion of EIA report. Public participation can be assured by: Consulting the public directly affected by the proposed project and the voluntary groups like NGOs or pressure groups having a concern with a specific aspect of the environment.

Conducting direct interviews with the sample from public or by sending questionnaire to the people from public. Publishing the summary of EIA report for objections and suggestions from people. In India, public hearing of development projects has been made mandatory for environmental clearance by the amendment to the EIA notification of April 10, 1997.

- **Review of the Environmental Impact Statement (EIS):** Based on the terms of reference (scoping) and public (including authority) participation.
- **Decision-making:** On whether to approve the project or not, and under what conditions, it should be approved. The decision-maker also has the option to defer approval until certain conditions are met or to require a proponent to redesign the project so that the environmental effects are minimized.
- **Monitoring, compliance, enforcement and environmental auditing:** Monitor whether the predicted impacts and proposed mitigation measures occur as defined in the EMP. Verify the compliance of proponent with the EMP, to ensure that unpredicted impacts or failed mitigation measures are identified and addressed in a timely fashion.

6.2.3 Public Hearings Process in India

In India public hearing of development projects has been made mandatory for environmental clearance by the Amendment to the EIA Notification of April 10, 1997. All the affected persons, including residents residing in and around the project site or the site of displacement or site of alleged adverse environmental impact can participate in public hearing. Some features for public hearings are as follows:

Notice for public hearing: The state pollution control board must issue notice for environmental public hearing by publishing it in at least two newspapers circulated in the region around the project. The date, time and place of public hearing should be mentioned in the newspaper article. The notice must be given at least 30 days prior to the public hearing.

Involvement of the public: People can submit their written suggestions, views, comments and objection to the state pollution control board within 30 days from the date of publication of the notice.

Access to the documents: The public are entitled to have access to the executive summary containing the salient features of the project, both in English as well as the local language. They can view the EIA report.

Public hearing panel and its members: The following persons may be included in the panel for the public hearing:

1. SPCB representative.
2. District collector/ nominee.
3. State government representative dealing with the project.
4. Representative of the concerned department of the state government.
5. Not more than 3 representative of the local bodies like panchayat/ municipality.
6. Not more than 3 senior citizens of the area nominated by the district collector.

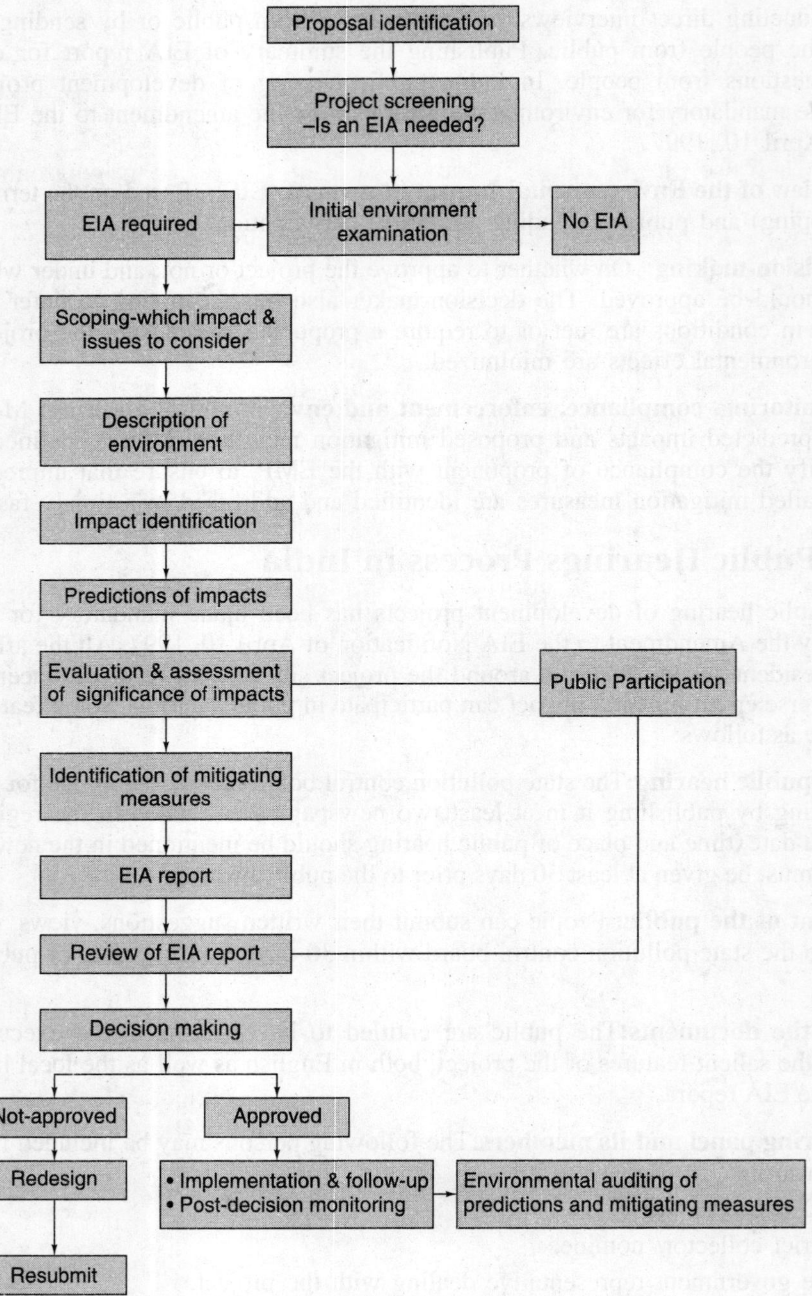

Fig. 6.1 Schematics of EIA approvals of projects.

6.3 METHODOLOGIES FOR PREDICTING IMPACT IN AN EIA

There are various methods for identification of impacts in a systematic way. Evaluation methods have been developed to bring together the various environmental effects (and impacts) associated with a proposal and its alternatives. The methods enable comparison of impacts and alternatives to be made. They also lead directly to an output which summarizes all the environmental impacts and provides the basis for the evaluation of the proposal (and alternatives) by decision-makers.

1. Adhoc method
2. Checklists
3. Matrices
4. Networks
5. Overlays
6. Battelle method
7. Simulation modeling

Ad hoc method: This method provides minimal guidance for total impact assessment while suggesting the broad areas of possible impacts. This gives the qualitative assessment, based on subjective or intuitive assessment. It does not address secondary impacts.

Checklist Method: It is a very common, easy and simple method. It helps in organizing work and identifying important environmental issues. Various types of checklists are:

I. **Simple checklists** list the components or aspects, usually of the environment, that might be considered by the assessor, but no other assistance are provided to guide the impact identification process.

II. **Descriptive checklists** provide additional assistance by indicating, for example, the specific variables to be measured to characterize each component.

III. **Scaling checklists** It includes simple devices for assessing importance or significance of suspected impacts. This might be through the use of letter or numeric scales, assigned after comparison with criteria supplied in the checklist, to indicate the importance of an impact. Another approach is to use threshold values, based on statutory criteria (e.g. for water quality standards) or on derived measures (e.g. visitor carrying-capacity for a given locality). The suspected impact can be estimated in broad terms and given a value to represent its significance. On that basis, a start can be made on comparing and ranking alternative project options.

IV. **Questionnaire checklist** is a form of scaling checklist but uses a series of carefully directed questions to elicit information about possible impacts and their likely importance.

6.3.1 Checklist of Impact Categories for Land Development Projects (Summarized From Schaenam 1976)

1. **Local economy**
   Public fiscal balance
   Employment
   Wealth
2. **Natural environment**
   Air quality

Water quality
Noise
Wildlife and vegetation
Natural disasters

3. **Aesthetics and cultural values**
Attractiveness
View opportunities
Landmarks

4. **Public and private services**
Drinking water
Hospital care
Crime control
Feeling of security
Fire protection
Recreation – public facilities
Recreation — informal settings
Education
Transportation – mass transit
Transportation – pedestrian
Transportation — private vehicles
Shopping
Energy services
Housing

5. **Other social impacts**
People displacement
Special hazards
Sociability/friendliness
Privacy
Overall contentment with neighbourhood

6.4 ENVIRONMENT IMPACT EVALUATION METHODS

6.4.1 Matrices

In matrices, project activities are cross-tabulated with environmental components. It can be made simpler or can be developed into a form in which a large amount of information is given. It is a simple and inexpensive method. Some of the limitations are inability to handle indirect impacts and temporal aspects, a potential rigidity of categories, and a difficulty to get an overview when many variables are included. Different types of matrices are:

I. **Descriptive matrices:** The various impacts associated with each activity are defined and short descriptions are provided.

II. **Symbolized matrices:** It uses symbols, such as words, abbreviations or plus/minus signs, to identify the impacts.

III. **Numeric matrices:** It derives facts to assist in showing the degree of impact or to help in making a comparison.

IV. **Scaled matrices:** It provides different scales, such as scales of 1 to 10 called Leopold matrices, or scales of –5 to +5 to indicate both positive and negative impacts. Leopold matrix is the best matrix methodology. It is a two-dimensional matrix cross-referencing the activities linked to the project that are supposed to have an impact on man and the environment and the existing environmental and social conditions that could possibly be affected by the project.

6.4.2 Overlays

Overlays may be used to identify geographic areas where the environmental load or the sensibility of the environment to a load is high. Here maps of different properties and environmental loads are superimposed.Overlay maps use a series of transparencies for identification, prediction, assignment of relative significance to and communication of impacts, usually on a larger than local scale. In the overlays both qualitative and quantitative data can be considered. Computer technology and programs, such as GIS, can be applied.

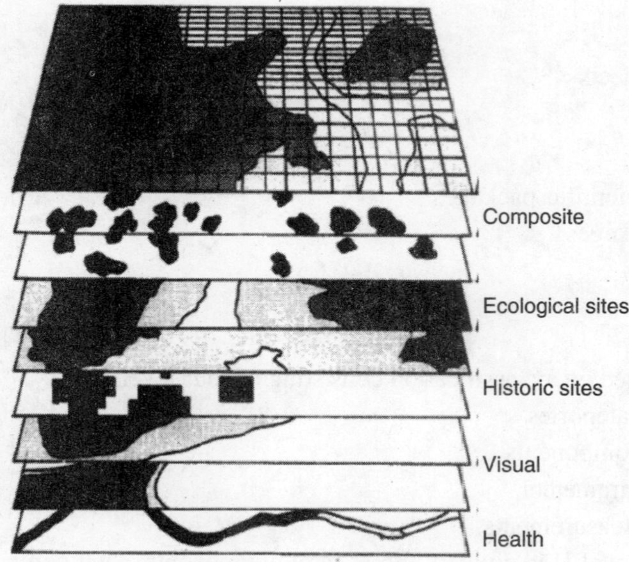

Fig. 6.2 Use of overlays.

6.4.3 Networks

Networks are directional diagrams developed to clearly consider the secondary or higher order impacts on the biophysical and socio-economic environment.This method provides a chain type of approach to identification of second and third order effects. Two types of impacts are usually considered in this method; primary impacts arise directly, and secondary impacts arise indirectly from the project activity.

6.4.4 The Battle Environmental Evaluation System

The principle lies in splitting the environmental impacts in four major categories: ecology, pollution, aesthetics and human interest. These categories are divided into thematic data as shown:

Ecology

(a) Species and populations
(b) Habitats and communities
(c) Ecosystems

Pollution

(a) Air pollution
(b) Water pollution
(c) Soil pollution
(d) Land pollution
(e) Noise pollution

Aesthetic

(a) Land
(b) Water
(c) Air
(d) Biota
(e) Man-made objects
(f) Composition

Human Interests

(a) Educational/scientific packages
(b) Historical packages
(c) Cultures
(d) Mood/atmosphere
(e) Life patterns

The system is based on a classification consisting of four levels:

- Level I : Categories
- Level II : Components
- Level III : Parameters
- Level IV : Measurements.

Each category (Level I) is divided into several components, each component (Level II) into several parameters, and each parameter (Level III) into one or more measurements. The EES identifies a total of four categories, eighteen components and seventy-eight parameters.The EES methodology is based on the assignment of an importance unit to each parameter. Collectively, these "importance units" are referred to as "parameter importance units" or PIUs. A total of 1000 PIUs distributed among the 78 parameters based on value judgments [of the system development team]. For each parameter, i, its $(PIU)_i$ represents a weight w_i. Each PIU_I or w_i requires a specific quantitative measurement. The methodology converts different measurements into common units by means of a scalar or "value function." Values of $V_i = V_{i,0}$ are obtained for conditions 'without' the project (current condition) and $V_i = V_{i,1}$ for conditions 'with' the project (future condition). The environmental impact E_I is evaluated as follows:

$$E_I = \Sigma\,[V_{i,1}\,w_i] - \Sigma\,[V_{i,0}\,w_i]$$

For $i = 1$ to n, where n = number of parameters (78).

For $E_I > 0$, the situation 'with' the project is better than 'without' the project, indicating that the project has positive environmental benefits. Conversely, for $E_I < 0$, the situation 'with' the project is worse than 'without' the project, indicating that the project has negative environmental benefits, i.e., certain negative impacts. A large negative value of E_I indicates the existence of substantial negative impacts.

6.4.5 Simulation Modeling

Systems analysis and modeling are among the few techniques that allow consideration of multi-dimensional problems that involve multiple (and usually conflicting) objectives, multiple criteria, multiple purposes and users, as well as interest groups. The use of model simulation emphasises the need for model validation, i.e. to show that the model really describes the processes that will occur in nature. Modeling has been used extensively in developed countries, but its use for impact assessment in developing countries has been rather limited because of constraints on resources, especially in expertise and data.

6.4.6 Environmental Auditing

Environmental auditing (or audit) is a management tool used by industry to evaluate its environmental performance. An audit assess the actual environmental impact, the accuracy of prediction, the effectiveness of environmental impact mitigation and enhancement measures, and the functioning of monitoring mechanisms. The audit should be undertaken upon a project run in operation, for some time, and is usually performed once or twice in the entire project cycle. Environmental auditing should compare monitoring results with information generated during the pre-project period. Comparison can be made with similar projects or against standard norms. Relating actual impacts with predicted impacts, help in evaluating the accuracy and adequacy of EIA predictions.There are two levels of environmental audits, i.e., environmental impact audit and environmental management audit. Environmental impact audit involves comparing the impacts predicted in an EIS with those that actually occur after implementation of the project and environmental management audit involves checks against adherence to plans, mitigation measures and general compliance of terms and conditions.

Different steps of an environment auditing are:

Pre-audit activities: It includes:

- Selection and scheduling of facility to audit
- Selection of audit team
- Contact with facility and planning of the audit.

Site activities: It is divided in 5 steps:

- First understanding of internal controls
- Assessment of internal controls

- Gathering of audit evidence
- Evaluation of audit findings
- Report of findings to facility.

Post-audit activities: This includes:

- Production of a draft report
- Production of a final report
- Preparation and implementation of an action plan
- Monitoring of action plan.

6.4.7 Environment Management Plan

An environmental management plan (EMP), is usually prepared as part of EIA reporting. It translates recommended mitigation and monitoring measures into specific actions that will be carried out by the proponent. Depending upon particular requirements, the plan may be included in, or appended to, the EIA report or may be a separate document. The EMP will need to be adjusted to the terms and conditions specified in any project approval. It will then form the basis for impact management during project construction and operation. Although there is no standard format, the EMP should contain the following:

- Summary of the potential impacts of the proposal
- Description of the recommended mitigation measures
- Statement of their compliance with relevant standards
- Allocation of resources and responsibilities for plan implementation
- Schedule of actions to be taken
- Programme for surveillance, monitoring and auditing
- Contingency plan when impacts are greater than expected.

6.5 BENEFITS OF EIA

By considering environmental effects and mitigation measures early in the project planning cycle, environmental assessment can support better decision making and result in many benefits, such as:

- Facilitates informed decision making by providing clear, well structured dispassionate analysis of the effect and consequences of proposed projects.
- Pre-emption or early withdrawal of unsound proposals.
- Assists in the selection of alternatives, including the selection of the best practicable and most environment-friendly option.
- Influences both project selection and design by screening out environmentally unsound projects, as well as modifying feasible projects — mitigation of negative environmental and social impacts.
- Guides formal approval, including the establishment of terms and conditions of project implementation and follow-up.
- Results in best practice prediction and mitigation of adverse effects of projects.
- Serves as an adaptive, organizational learning process, in which the lessons of experience are included into the policy, institutional and project design enhancement of positive aspects.

- Incorporates stakeholder analysis.
- Mitigation of negative environmental and social impacts.
- Reduced time and costs of approvals of development applications.
- Ensuring compliance with environmental standards.

Limitations

Many EIAs lack a central, conceptual or analytical theme or recognizable investigative design to guide the collection and interpretation of data. Since EIA involves many experts and other stakeholders, it actually costs more than the traditional planning process. Sometimes EIA also causes time delays for verification findings and is considered not contributing to economic development.Generally, three problems are concerned by experts in EIA worldwide practices. They are: institutional arrangements problem; low quality of EIA practices; and lack of implementation of EIA in decision making process.

Learning Outcomes

The main learning outcomes from this chapter should be fundamental understanding of:

Environmental impact assessment, procedure and methods; introduction of projects in India, benefits of EIA and limitations.

Readers are also advised to carry out following exercise:

Exercise

1. Perform full cycle environmental impact assessment of construction project of multi-storey apartment building.
2. Summarize EIA methods in a schematic diagram.

Chapter 7

Human Population and Environment

7.1 POPULATION GROWTH, VARIATION AMONG NATIONS

Our global human population will cross the 7 billion mark by 2015. The needs of this huge number of human beings cannot be supported by the earth's natural resources, without degrading the quality of human life.

In the near future, fossil fuel from oil fields will run dry. It will be impossible to meet the demands for food from existing agro systems. Pastures will be overgrazed by domestic animals and industrial growth will create ever-greater problems due to pollution of soil, water and air. Seas will not have enough fish. Larger ozone holes will develop due to the discharge of industrial chemicals into the atmosphere, which will affect human health. Global warming due to industrial gases will lead to a rise in sea levels and flood all low-lying areas, submerging coastal agriculture as well as towns and cities. Water 'famines' due to the depletion of fresh water will create unrest and eventually make countries go to war. The control over regional biological diversity, which is vital for producing new medicinal and industrial products, will lead to grave economic conflicts between biotechnologically advanced nations and the bio-rich countries. Degradation of ecosystems will lead to extinction of thousands of species, destabilizing natural ecosystems of great value. These are only some of the environmental problems related to an increasing human population and more intensive use of resources that we are likely to face in future. These effects can be averted by creating a mass environmental awareness movement that will bring about a change in people's way of life. Increase in production per capita of agricultural produce at a global level ceased during the 1980s. in some countries, food shortage has become a permanent feature. Two of every three children in South Africa are underweight.

In other regions famines due to drought have become more frequent. Present development strategies have not been able to successfully address these problems related to hunger and malnutrition. On the other hand, only 15% of the world's population in the developed world is earning 79% of income. Thus the disparity in the extent of per capita resources that are used by people who live in a 'developed' country as against those who live in a 'developing' country is extremely large. Similarly, the disparity between the rich and the poor in India is also growing.

The increasing pressures on resources place great demands on the in-built buffering action of nature that has a certain ability to maintain a balance in our environment. However, current development strategies that essentially lead to short-term gains have led to a breakdown of our earth's ability to replenish the resources on which we depend.

7.1.1 Global Population Growth

The world population is growing by more than 90 million per year, of which 93% is in developing countries. This will essentially prevent their further economic 'development'. In the past, population growth was a gradual phenomenon and the earth's ability to replenish resources was capable of adjusting to this increase. In the recent past, the escalation in growth of human numbers has become a major cause of our environmental problems.

Present projections show that if our population growth is controlled, it will still grow to 7.27 billion by 2015. However, if no action is taken it will become a staggering 7.92 billion.

Human population growth increased from:

1 to 2 billion, in 123 years.

2 to 3 billion, in 33 years.

3 to 4 billion, in 14 years.

4 to 5 billion, in 13 years.

5 to 6 billion, in 11 years.

It is not the census figures alone that need to be stressed, but an appreciation of the impact on natural resources of the rapid escalation in the rate of increase of human population in the recent past. The extent of this depletion is further increased by affluent societies that consume per capita more energy and resources, that less fortunate people. This is of great relevance for developing a new ethic for a more equitable distribution of resources.

In the first half of the 1900s human numbers were growing rapidly in most developing countries such as India and China. In some African countries the growth was also significant. In contrast, in the developed world population growth had slowed down. It was appreciated that the global growth rate was depleting the earth's resources and was a direct impediment to human development. Several environmental ill-effects were linked with the increasing population of the developing world. Poverty alleviation programmes failed, as whatever was done was never enough as more and more people had to be supported on Earth's limited resources. In rural areas population growth led to increased fragmentation of farmland and unemployment. In the urban sector it led to inadequate housing and an increasing level of air pollution from traffic, water pollution from sewage, and an inability to handle solid waste. By the 1970s most countries in the developing world had realized that if they had to develop their economics and improve the lives of their citizens they would have to curtail population growth.

Though population growth shows a general global decline, there are variations in the rate of decline in different countries. By the 1990s the growth rate was decreasing in most countries such as China and India. The decline in the 90s was greatest in India. However, fertility continues to remain high in sub-Saharan African countries.

There are cultural, economic, and political and demographic reasons that explain the differences in the rate of population control in different countries. It also varies in different parts of certain countries and is linked with community and/ or religious thinking. Lack of government initiatives for family welfare programme and a limited access to a full range of contraceptive measures are serious impediments to limiting population growth in several countries.

7.2 POPULATION EXPLOSION – FAMILY WELFARE PROGRAMME

In response to our phenomenal population growth, India seriously took up an effective family planning programme which was renamed the family welfare programme. Slogans such as '*Hum do hamare do*' indicated that each family should not have more than two children. It however has taken several decades to become effective.

At the global level by the year 2000, 600 million, or 57% of women in the reproductive age group, were using some method of contraception. However, the use of contraceptive measures is higher in developed countries – 68%, and lower in developing countries — 55%. Female sterilization is the most popular method of contraception used in developing countries at present. This is followed by the use of oral contraceptive pills and, intrauterine devices for women, and the use of condoms for men. India and China have been using permanent sterilization more effectively than many other countries in the developing world.

The best decision for the method used by a couple depends on a choice that they make for themselves. This must be based on good advice from doctors or trained social workers who can suggest the full range of methods available for them to choose from.

Informing the public about the various contraceptive measures that are available is of primary importance. This must be done actively by government agencies such as health and family welfare, as well as education and extension workers. It is of great importance for policy-makers and elected representatives of the people – ministers, MPs, MLAs at central and state levels – to understand the great and urgent need to support family welfare. The media must keep people informed about the need to limit family size and the ill effects of a growing population on the world's resources.

The decision to limit family size depends on a couple's background and education. This is related to government policy, the effectiveness of family welfare programmes, the educational level, and information levels in mass communication. Free access to family welfare information provided through the health care system, is in some cases unfortunately counteracted by cultural attitudes. Frequently misinformation and inadequate information are reasons why a family does not go in for limiting its size.

The greatest challenge the world now faces is how to provide its exploding human population with the resources it needs. It is evident that without controlling human numbers, the earth's resources will be rapidly exhausted. In addition, economically advanced countries and rich people in poorer countries use up more resources than they need.

As population expands further, water shortage will become acute. Soil will become unproductive. Rivers, lakes and coastal waters will be increasingly polluted. Water related diseases already kill 12 million people every year in the developing world. By 2025, there will be 48 countries that will starved for water. Air will become increasingly polluted. Air pollution al- ready kills 3 million people every year.

The first 'green revolution' in the 1960s produced a large amount of food but has led to several environmental problems. Now, a new green revolution is needed, to provide enough food for our growing population, that will not damage land, kill rivers by building large dams, or spread at the cost of critically important forests, grasslands and wetlands.

The world's most populous regions are in coastal areas. These are critical ecosystems and are being rapidly destroyed. Global climate change is now a threat that can affect the very survival of high population density coastal communities. In the sea, fish populations are suffering from excessive fishing. Once considered an inexhaustible resource, over-fishing has depleted stocks extremely rapidly. It will be impossible to support further growth in coastal populations on existing fish reserves.

Human populations will inevitably expand from farmlands into the remaining adjacent forests. Many such encroachments in India have been regularized over the last few decades. But forest loss has long-term negative effects on water and air quality and the loss of biodiversity is still not generally seen as a major deterrent to human well-being. The extinction of plant and animal species resulting from shrinking habitats threatens to destroy the earth's living web of life.

Energy use is growing, both due to an increasing population, and a more energy dependent lifestyle that increasingly uses consumer goods that require large amounts of energy.

Production, packagingand transport: Our growing population also adds to the enormous amount of waste.

With all these linkages between population growth and the environment, family welfare programs have become critical to human existence.

Planning for the future: How governments and people from every community meet challenges such as limiting population size, protecting the natural environment, change their consumer oriented attitudes, reduce habits that create excessive waste, elevates poverty and creates an effective balance between conservation and development will determine the worlds future.

The Urban Challenge: Population increases will continue in urban centres in the near future. The UN has shown that by 2025 there will be 21 "megacities" most of which will be situated in developing countries. Urban centres are already unable to provide adequate housing, services such as water and drainage systems, growing energy needs, or better opportunities for income generation.

7.2.1 Methods of Sterilization

India's family welfare programme has been fairly successful but much still needs to be achieved to stabilize our population. The most effective measure is the one most suited to the couple once they have been offered all the various options that are available. The family welfare programme advocates a variety of measures to control population. Permanent methods or sterilization are done by a minor surgery. Tubectomy in females is done by tying the tubes that carry the ovum to the uterus. Male sterilization or vasectomy is done by tying the tubes that carry the sperm. Both are very simple procedures, done under local anesthesia, are painless and patients have no post-perative problems. Vasectomy does not cause any loss in the male's sexual ability but only arrests the discharge of sperm.

There are several methods of temporary birth control. Condoms are used by males to prevent sperms from fertilizing the ovum during intercourse. Intrauterine devices (Copper Ts) are small objects which can be placed by a doctor in the uterus so that the ovum cannot be implanted, even if fertilized. They do not disturb any functions in the woman's life or work. Oral contraceptive tablets (pills) and injectable drugs are available that prevent sperms from fertilizing the ovum.

There are also traditional but less reliable methods of contraception such as abstinence of the sexual act during the fertile period of the women's cycle and withdrawal during the sexual act.

7.2.2 Urbanization

In 1975 only 27% of the people in the developing world lived in urban areas. By 2000 this had grown to 40% and by 2030 well informed estimates state that this will grow to 56%. The developed world is already highly urbanized with75% of its population living in the urban sector.

CASE STUDY

Urban Environments

Nearly half the world's population now lives in urban areas. The high population density in these areas leads to serious environmental issues.

Today, more than 290 million people live in towns and cities in India. There were 23 metros in India in 1991, which grew to 40 by 2001.

Urban population growth is both due to migration of people to towns and cities from the rural sector in search of better job options as well as population growth within the city.

As a town grows into a city it not only spreads outwards into the surrounding agricultural land or natural areas such as forests, grasslands and wetlands but also grows skywards with high-rise buildings. The town also loses its open spaces and green cover unless these are consciously preserved. This destroys the quality of life in the urban area.

Good urban planning is essential for rational land use planning, for upgrading slum areas, improving water supply and drainage systems, improving water supply and drainage systems, providing adequate sanitation, developing effective waste water treatment plants and an efficient public transport system.

Unplanned and haphazard growth of urban complexes has serious environmental impacts. Increasing solid waste, improper garbage disposal and air and water pollution are frequent side effects of urban expansions.

While all these issues appear to be under the preview of local municipal corporations, better living conditions can only become a reality if every citizen plays an active role in managing the environment. This includes a variety of "dos and don'ts" that should become an integral part of our personal lives.

Apart from undertaking actions that support the environment every urban individual has the ability to influence a city's management. He or she must see that the city's natural green spaces, parks and gardens are maintained, river and water fronts are managed appropriately, roadside tree cover is maintained, hill slopes are afforested and used as open spaces and architectural and heritage sites are protected. Failure to do this leads to increasing an urban problem which eventually destroys a city's ability to maintain a healthy and happy lifestyle for its dwellers. All these aspects are closely linked to the population growth in the urban sector. In many cities growth outstrips the planner's ability to respond to this in time for a variety of reasons.

Table 7.1

Megacities	*Population for 2001 (in millions)*	*Population for 2015 (in millions)*
Mumbai	16.5	22.6
Kolkata	13.3	16.7
Delhi	20.9	20.9

Small urban centres too will grow rapidly during the next decades and several rural areas will require reclassification as urban centres. India's urban areas will grow by a projected 297 million residents. In India people move to cities from rural areas in the hope of getting a better income. This is the 'pull' factor. Poor opportunities in the rural sector thus stimulate migration to cities. Loss of agricultural land to urbanization and industry, the inability of governments to sustainably develop the rural sector, and a lack

Megacities – over 10 million inhabitants, 1950 – there was only 1 – New York. During 1975 – there were 5 and in 2001 – there were 15 (with Mumbai, Kolkata, and Delhi, being added to the list from India) and in coming 2015 – there will be 21 megacities.

In 2000 there were 388 cities with more than 1 million inhabitants. By 2015 these will increase to 554, of which 75% are in developing countries.

The poor supporting infrastructure in rural areas, pushes people from the agricultural and natural wilderness ecosystems into the urban sector. As our development strategies have focused attention mostly on rapid industrial development and relatively few development options are offered for the agricultural rural sector, a shift of population is inevitable.

As population in urban centres grows, they draw on resources from more and more distant areas. The "ecological footprint" corresponds to the land area necessary to supply natural resources and disposal of waste of a community. At present the average ecological footprint of an individual at the global level is said to be 2.3 hectares of land per capita. But it is estimated that the world has only 1.7 hectares of land per individual to manage these needs sustainably. This is thus an unsustainable use of land.

The pull factor of the urban centres is not only due to better job opportunities, but also better education, health care and relatively higher living standards. During the last few decades in India, improvements in the supply of clean water, sanitation, waste management, education and health care has all been urban-centric, even though the stated policy has been to support rural development. Thus, in reality, development has lagged behind in the rural sector that is rapidly expanding in numbers, For people living in wilderness areas in our forests and mountain regions, development has been most neglected. It is not appropriate to use the development methods used for other rural communities for tribal people who are dependent on collecting natural resources from the forests. A different pattern of development that is based on the sustainable extraction of resources from their own surroundings would satisfy their development aspirations. In general, the growing human population in the rural sector will only opt to live where they are if they are given an equally satisfying lifestyle.

The wilderness – rural-urban linkage

The environmental stresses caused by urban individuals cover an 'ecological footprint' that goes far beyond what one expects. The urban sector affects the land at the fringes of the urban area and the areas from which the urban centre pulls in agricultural and natural resources.

Urban centres occupy 2% of the worlds' land but use 75% of the industrial wood. About 60% of the world's water is used by urban areas of which half irrigates food crops for urban dwellers, and one third goes to industry and the rest is used for household use and drinking water.

The impact that urban dwellers have on the environment is not obvious to them as it happens at distant places which supports the urban ecosystem with resources from agricultural and even more remote wilderness ecosystems.

Urban Poverty and the Environment

The number of poor people living in urban areas is rapidly increasing. A third of the poor people in the world live in urban centres. These people live in hutments in urban slums and suffer from water shortages and unsanitary conditions. In most cases while a slum invariably has unhygienic surroundings, the dwellings themselves are kept relatively clean. It is the 'common' areas used by the community that lacks the infrastructure to maintain a hygienic environment.

During the 1990s countries that have experienced an economic crisis have found that poor urban dwellers have lost their jobs due to decreasing demands for goods, while food prices have risen. Well paid and consistent jobs are not as easily available in the urban centers at present as in the past few decades.

One billion urban people in the world live inadequate housing, mostly in slum areas, the majority of which are temporary structures. However, low income groups that live in high rise buildings can also have high densities and live in poor unhygienic conditions in certain areas of cities. Illegal slums often develop on government land, along railway tracks, on hill slopes, riverbanks, marshes, etc. that are unsuitable for formal urban development. On the riverbanks floods can render these poor people homeless. Adequate legal housing for the urban poor remains a serious environmental concern.

Urban poverty is even more serious than rural poverty, as unlike the rural sector, the urban poor have no direct access to natural resources such as relatively clean river water, fuelwood and non-wood forest products. The urban poor can only depend on cash to buy the goods they need, while in the rural sector they can grow a substantial part of their own food. Living conditions for the urban poor are frequently worse than for rural poor. Both outdoor and indoor air pollution due to high levels of particulate matter and sculpture dioxide from industrial and vehicle emissions lead to high death rates from respiratory diseases. Most efforts are targeted at outdoor air pollution. Indoor air pollution due to the use of fuelwood, waste material, coal, etc. in 'chulas' is a major health issue. This can be reduced by using better designed 'smokeless' chulas, hoods and chimneys to remove indoor smoke.

With the growing urban population, a new crisis of unimaginable proportions will develop in the next few years. Crime rates, terrorism, unemployment, and serious environmental health related issues can be expected to escalate. This can only be altered by stabilizing population growth on a war footing.

7.3 ENVIRONMENT AND HUMAN HEALTH

Environment related issues that affect our health have been one of the most important triggers that have led to creating an increasing awareness of the need for better environmental management. Changes in our environment induced by human activities in nearly every sphere of life have had an influence on the pattern of our health. The assumption that human progress is through economic growth is not necessarily true. We expect urbanization and industrialization to bring in prosperity, but on the down side, it leads to diseases related to overcrowding and an inadequate quality of drinking water, resulting in an increase in waterborne diseases such as ineffective diarrhea and airborne bacterial diseases such as tuberculosis. High-density city traffic leads to an increase in respiratory diseases like asthma. Agricultural pesticides that enhanced food supplies during the green revolution have affected both the farm worker and all of us who consume the produce. Modern medicine promised to solve many health problems, especially associated with infectious diseases through antibiotics, but bacteria found ways to develop resistant strains, frequently even changing their behaviour in the process, making it necessary to keep on creating newer antibiotics. Many drugs have been found to have serious side-effects. At times the cure is as damaging as the disease process itself.

Thus development has created several long-term health problems. While better health care has led to longer life spans, coupled with a lowered infant mortality, it has also led to an unprecedented growth in our population which has negative implications on environmental quality. A better health status of society will bring about a better way of life only if it is coupled with stabilizing population.

7.3.1 Environmental Health

Enviromental health as defined by WHO, comprises those aspects of human health, including quality of life, that are determined by physical, chemical, biological, social, and psycho-social factors in the environment. It also refers to the theory and practice of assessing, correcting, controlling, and preventing those factors in the environment that adversely affect the health of present and future generations.

Our environment affects health in a variety of ways. Climate and weather affect human health. Public health depends on sufficient amounts of good quality food, safe drinking water, and adequate shelter. Natural disasters such as storms, hurricanes, and floods still kill many people every year. Unprecedented rainfall triggers epidemics of malaria and water borne diseases.

Global climate change has serious health implications. Many countries will have to adapt to uncertain climatic conditions due to global warming. As our climate is changing, we may no longer know what to expect. There are increasing storms in some countries, drought in others, and a temperature rise throughout the world. The El Niño winds affect weather worldwide. The El Niño event of 1997/98 had serious impacts on health and well-being of millions of people in many countries. It created serious drought, floods, and triggered epidemics. New strategies must be evolved to reduce vulnerability to climate variability and changes.

Economic inequality and environmental changes are closely connected to each other. Poor countries are unable to meet required emission standards to slow down climate change. The depletion of ozone in the stratosphere (middle atmosphere) also has an important impact on global climate and in turn human health, increasing the amount of harmful ultraviolet radiation that reaches the earth's surface. This results in diseases such as skin cancer.

CASE STUDY: Bhopal Gas Tragedy

The sitting of industry and relatively poor regulatory controls leads to ill health in the urban centres. Accidents such as the Bhopal gas tragedy in 1984 where Union Carbide's plant accidentally released 30 tonnes of methyl isocyanide, used in the manufacture of pesticides, led to 3330 deaths and 1.5 lakh injuries to people living in the area.

Development strategies that do not incorporate ecological safeguards often lead to ill health. Industrial development without pollution control and traffic congestion affects the level of air pollution in many cities. On the other hand, development strategies that can promote health invariably also protect the environment. Thus environmental health and human health are closely interlinked. An improvement in health is central to sound environmental management. However this is rarely given sufficient importance in planning development strategies. Some of major example of linkages are given below:

- Millions of children die every year due to diarrhoea from contaminated water or food. An estimated 2000 million people are affected by these diseases and more than 3 million children die each year from waterborne diseases across the world. In India, it is estimated that every fifth child under the age of 5 dies due to diarrhoea. This is a result of inadequate environmental management and is mainly due to inadequate purification of drinking water. Wastewater and/or sewage entering water sources without being treated leads to continuous gastrointestinal diseases in the community and even sporadic large epidemics. Large numbers of people in tropical countries die of malaria every year and millions are infected. An inadequate environmental management of stagnant water, which forms breeding sites of Anopheles mosquitoes, is the most important factor in the spread of malaria. The resurgence of malaria in India is leading to cerebral malaria that affects the brain and has a high mortality.
- Millions of people, mainly children, have poor health due to parasitic infections, such as amoebiasis and worms. This occurs from eating infected food, or using poor quality water for cooking food. It is estimated that 36% of children in low-income countries and 12% in middle income countries are malnourished. In India, about half the children under the age of four are malnourished and 30% of newborns are significantly underweight.
- Hundreds of millions of people suffer serious respiratory diseases, including lung cancer and tuberculosis, from crowded homes and public places. Motor vehicle exhaust fumes, industrial fumes, tobacco smoke and cooking food on improper 'chulas', contribute to respiratory diseases.
- Millions of people are exposed to hazardous chemicals in their workplace or homes that lead to ill health due to industrial products where controls are not adhered to.
- Tens of thousands of people in the world die due to traffic accidents due to inadequate management of traffic conditions. Poor management at the accident site and inability to reach a hospital within an hour causes a large number of deaths, especially from head injuries.
- Basic environmental needs such as clean water, clean air and adequate nutrition which are all related to environmental goods and services do not reach over 1000 million people living in abject poverty.

- Several million people live in inadequate shelters or have no roof over their heads especially in urban settings. This is related to high inequalities in the distribution of wealth and living space.
- Population growth and the way resources are being exploited and wasted, threaten environmental integrity and directly affects health of nearly every individual.
- Health is an outcome of the interactions between people and their environment. Better health can only come from a more sustainable management of the environment.

Important strategic concerns

- The world must address people's health care needs and sustainable use of natural resources, which are closely linked to each other.
- Strategies to provide clean potable water and nutrition to all people is an important part of a healthy living environment.
- Providing clean energy sources that do not affect health is a key to reducing respiratory diseases.
- Reducing environmental consequences of industrial and other pollutants such as transport emissions can improve the status of health.
- Changing patterns of agriculture away from harmful pesticides, herbicides and insecticides which are injurious to the health of farmers and consumers by using alternatives such as integrated pest management and non-toxic biopesticides can improve health of agricultural communities, as well as food consumers.
- Changing industrial systems into those that do not use or release toxic chemicals that affect the health of workers and people living in the vicinity of industries can improve health and environment.
- There is a need to change from using conventional energy from thermal power that pollutes air and nuclear power that can cause serious nuclear disasters to cleaner and safer sources such as solar, wind and ocean power, that do not affect human health. Providing clean energy is an important factor that can lead to better health.
- The key factors are to control human population and consume less environmental goods and services which could lead to 'health for all'. Unsustainable use of resources by an ever growing population leads to unhealthy lives. Activities that go on wasting environmental goods and destroying its services by producing large quantities of non degradable wastes, leads to health hazards.
- Poverty is closely related to health and is itself a consequence of improper environmental management. An inequitable sharing of natural resources and environmental goods and services, is linked to poor health.

The world's consumption of non-renewable resources is concentrated in the developed countries. Rich countries consume 50 times more per capita than people in less developed countries. This means that developed countries also generate proportionately high quantities of waste material, which has serious health concerns.

Definition of Health Impact Assessment (HIA) by WHO: Health impact assessment is a combination of procedures, methods and tools by which a policy, program or project may be judged as to its potential effects on the health of a population, and the distribution of those effects within the population.

7.3.2 Climate and Health

Human civilizations have adapted mankind to live in a wide variety of climates. From the hot tropics to the cold arctic, in deserts, marshlands and in the high mountains. Both climate and weather have a powerful impact on human life and health issues.

Natural disasters created by extremes of weather (heavy rains, floods, hurricanes) which occur over a short period of time, can severely affect health of a community. Poor people are more vulnerable to the health impacts of climate variability than the rich. Of approximately 80,000 deaths which occur worldwide each year as a result of natural disasters about 95% are in poor countries. In weather-triggered disasters hundreds of people and animals die, homes are destroyed, crops and other resources are lost. Public health infrastructure, such as sewage disposal systems, waste management, hospitals and roads are damaged. The cyclone in Orissa in 1999 caused 10, 000 deaths. The total number of people affected was estimated at 10 to 15 million.

Human physiology can adapt to changes in weather, within certain limits. However, marked short-term fluctuations in weather lead to serious health issues. Heat waves cause heat-related illness and death (e.g., heat stroke). The elderly and persons with existing heart or respiratory diseases are more vulnerable. Heat waves in India in 1998 were associated with many deaths.

Climate plays an important role in vector-borne diseases transmitted by insects such as mosquitoes. These disease transmitters are sensitive to direct effects of climate such as temperature, rainfall patterns and wind. Climate affects their distribution and abundance through its effects on host plants and animals.

Malaria transmission is particularly sensitive to weather and climate. Unusual weather conditions, for example, a heavy downpour, can greatly increase the mosquito population and trigger an epidemic. In the desert and at highland fringes of malarias areas, malaria transmission is unstable and the human population lacks inherent protective immunity. Thus, when weather conditions (rainfall and temperature) favour transmission, serious epidemics occur in such areas. Fluctuations in malaria over the years have been linked to changes in rainfall associated with the El Niño cycle.

7.3.3 Infectious Diseases

Many infectious diseases have re-emerged with a vengeance. Loss of effective control over diseases such as malaria and tuberculosis, have led to a return of these diseases decades after being kept under stringent control.

Other diseases were not known to science earlier and seem to have suddenly hit our health and our lives during the last few decades. AIDS, due to the Human Immunodeficiency Virus (HIV) caused through sexual transmission and Severe Acute Respiratory Syndrome (SARS) are two such examples. While these cannot be directly related to environmental change, they affect the environment in which we live by forcing a change in lifestyles and behavior patterns. For example the SARS outbreak prevented people from several countries from travelling to other countries for months, severely affecting national economies, airline companies and the tourism industry.

Why have infectious diseases that were related to our environment that were under control suddenly made a comeback? Diseases such as tuberculosis have been effectively treated with anti-tubercular drugs for decades. These antibiotics are used to kill off the bacteria that cause the disease. However, nature's evolutionary processes are capable of permitting bacteria to mutate by creating new genetically modified strains. Those that change in a way so that they are not affected by the routinely used antibiotics begin to spread rapidly. This leads to a re-emergence of the disease. In the case of tuberculosis this has led to multi-drug resistant tuberculosis. This is frequently related to HIV which reduces an individual's immunity to bacteria such as mycobacterium tuberculosis that causes tuberculosis.

The newer broad-spectrum antibiotics, antiseptics, disinfectants, and vaccines once thought of as the complete answer to infectious diseases have thus failed to eradicate infectious diseases. Experts in fact now feel that these diseases will be the greatest killers in future and not diseases such as malignancy or heart disease.

While antibiotic resistance is a well-known phenomenon there are other reasons for the re-emergence of diseases. Overcrowding due to the formation of slums in the urban setting leads to several health hazards, including easier spread of respiratory diseases. Inadequate drinking water quality and poor disposal of human waste due to absence of a closed sewage system and poor garbage management are all urban health issues. This has led to a comeback of diseases such as cholera and an increased incidence of diarrhea and dysentery as well as infectious hepatitis (jaundice).

With increasing global warming disease patterns will continue to change. Tropical diseases spread by vectors such as the mosquito will undoubtedly spread malaria further away from the equator. Global warming will also change the distribution of dengue, yellow fever, encephalitis, etc. Warmer wetter climates could cause serious epidemics of diseases such as cholera. El Nino which causes periodic warming is likely to affect rodent populations. This could bring back diseases such as the plague.

CASE STUDY: Tuberculosis in India

Globalization is a worldwide process which includes the internationalization of communication, trade and economic organization. It involves parallel changes such as rapid social, economic and political adjustments. Whilst globalization has the potential to enhance the lives and living standards of certain population groups, for poor and marginalized populations in both the non-formal as well as formal economic sectors of developing countries, globalization enhances economic inequalities.

Tuberculosis (TB) kills approximately 2 million people each year. In India the disease has re-emerged and is now more difficult to treat. A global epidemic is spreading and becoming more lethal. The spread of HIV/AIDS and the emergence of multidrug-resistant tuberculosis are contributing to the increasing morbidity of this disease. In 1993, the World Health Organization (WHO) declared that tuberculosis is global emergency. It is estimated that between 2002 and 2020, approximately 1000 million people will be newly infected, over 150 million people will get sick, and 36 million will die of TB – if its control is not rapidly strengthened.

TB is a contagious disease that is spread through air. Only people who are sick with pulmonary TB are infectious. When infectious people cough, sneeze, talk or spit, they emit the tubercle bacilli into the air. When a healthy person inhales these, he gets infected by the disease. Symptoms include prolonged fever, coughing spells and weight loss.

It is estimated that, left untreated, each patient of active tuberculosis will infect on an average between 10 to 15 people every year. But people infected with TB will not necessarily get sick with the disease. The immune system can cause the TB bacilli, which is protected by a thick waxy coat, to remain dormant for years. When an individual's immune system is weakened, the chances of getting active TB are greater.

- Nearly 1% of the world's population is newly infected with TB each year.
- It is estimated that overall, one third of the world's population is likely to be infected with the tuberculosis bacillus at some point in time.
- Five to ten percent of people who are infected with TB (but who are not infected with HIV) become sick or infectious at some time during their life. (WHO, 2002).

Factors Contributing to the rise in tuberculosis

- TB kills about 2 million people each year (including persons infected with HIV).

- More than 8 million people become sick with TB each year, one person in the world every second.
- About 2 million TB cases per year occur in sub-Saharan Africa. This number is rising rapidly as a result of the HIV/AIDS epidemic. Around 3 million TB cases per year occur in South-east Asia.
- Over a quarter of a million TB cases per year occur in Eastern Europe.

There are 14 million TB patients in India, account for one third of the global cases of TB. Everyday 20,000 Indians contract TB and more than 1000 die due to this chronic illness. TB attacks working adults in the age group of 15 to 50 years.

HIV is accelerating the spread of TB

The link between HIV and TB affects a large number of people, each disease speeding the other's progress. HIV weakens the immune system. Someone who is HIV-positive and infected with TB is many times more likely to become seriously sick with TB rather than someone infected with TB who is HIV-negative. Tuberculosis is a leading cause of death among people who are HIV-positive, accounting for about 11% of AIDS deaths worldwide.

Poorly managed TB programmes are threatening to make TB incurable

Until 50 years ago, there were no drugs to cure tuberculosis. Now, strains that are resistant to one or more anti-TB drugs have emerged. Drug-resistant tuberculosis is caused by inconsistent or partial treatment, when patients do not take all their drugs regularly for the required period, when doctors or health workers prescribe inadequate treatment regimens or where the drug supply is unreliable. From a public health perspective, poorly supervised or incomplete treatment of TB is worse than no treatment at all. When people fail to complete standard treatment regimens, or are given the wrong treatment regimen, they may remain infectious. The bacilli in their lungs may develop resistance to anti-TB drugs. People they infect will have the same drug-resistant strain. While drug-resistant TB is treatable, it requires extensive chemotherapy that is often very expensive and is also more toxic to patients.

Malaria is a life-threatening parasitic disease transmitted by mosquitoes. The cause of malaria, a single celled parasite called plasmodium, was discovered in 1880. Later it was found that the parasite is transmitted from person to person through the bite of a female Anopheles mosquito, which requires blood for the growth of her eggs.

Today approximately 40% of the world's population, mostly those living in the world's poorest countries, risk getting malaria. The disease was once more widespread but it was successfully eliminated from many countries with temperate climates during the mid 20th century. Today malaria has returned and is found throughout the tropical and sub-tropical regions of the world and causes more than 300 million acute illnesses and at least one million deaths annually (WHO).

There are several types of human malaria. Falciparum malaria is the most dangerous type of infection and is most common in Africa south of the Sahara, where it accounts for extremely high mortality rates. There are also indications of the spread of *P. Falciparum* malaria in India and it has reappeared in areas where it had been eliminated.

The malaria parasite enters the human host when an infected Anopheles mosquito bites an individual. Inside the human host, the parasite undergoes a series of changes as part of its complex life-cycle. Its various stages allow plasmodia to evade the immune system, infect the liver and red blood cells, and finally develop into a form that is able to infect a mosquito again when it bites an infected person. Inside the mosquito, the parasite matures until it reaches the sexual stage where it can again infect a human host when the mosquito takes her next blood meal, 10 or more days later.

Malaria symptoms appear about 9 to 14 days after the mosquito bite, although this varies with different plasmodium species. Malaria produces high fever, headache, vomiting and body ache. If drugs are not available for treatment, or the parasites are resistant to them, the infection can progress rapidly to become life-threatening. Malaria can kill by infecting and destroying red blood cells (anemia) and by clogging the capillaries that carry blood to the brain (cerebral malaria) or other vital organs.

Malaria parasites are developing unacceptable levels of resistance to drugs. Besides this, many insecticides are no longer useful against mosquitoes transmitting the disease. Good environmental management by clearing pools of stagnant water during the monsoons is effective in reducing the number of mosquitoes.

Mosquito nets treated with insecticide reduce malaria transmission and child deaths. Prevention of malaria in pregnant women, through measures such as intermittent preventive treatment and the use of insecticide-treated nets (ITNs), results in improvement in maternal health, as well as infant health and survival. Prompt access to treatment with effective upto date medicines, such as artemisinin-based combination therapies (ACTs), saves lives. If countries can apply these and other measures on a wide scale and monitor them carefully, the burden of malaria on society will be significantly reduced.

7.3.4 Water-related Diseases

Water Supply, sanitation and hygiene development

Among the main problems are a lack of priority given to this sector, lack of financial resources, erratic water supply and sanitation services, poor hygiene related behaviour patterns, and inadequate sanitation in public places such as schools, hotels, hospitals, health centres, etc. One of the most important aspects is a lack of environmental education and awareness that these disease processes are related to poor environment management in various sectors.

Providing access to sufficient quantities of safe water, the provision of facilities for a sanitary disposal of excreta, and introducing sound hygiene related behaviour can reduce the morbidity and mortality caused by these risk factors.

Environmental Sanitation and Hygiene Development

About 2.4 billion people globally live under highly unsanitary conditions. Poor hygiene and behaviour pattern increase the exposure to risk of incidence and spread of infectious diseases. Water improperly stored in homes is frequently contaminated by inadequate management at the household level. This can be easily reduced through education and awareness of how waterborne diseases are transmitted.

Health and Water Resources Development

An important aspect related to water-related diseases (in particular: water-related vectorborne diseases) is attributable to the way water resources are developed and managed. In many parts of the world the adverse health impacts of dam construction, irrigation development and flood control is related to increased incidence of malaria, Japanese encephalitis, schistosomiasis, lymphatic filariasis and other conditions. Other health issues indirectly associated with water resources development include nutritional status, exposure to agricultural pesticides and their residues.

Waterborne Diseases

Arid areas with rapidly expanding populations are already facing a crisis over water. Conservation of water and better management is an urgent need. The demand and supply balance is a vital part of developing sustainable use of water. This is being termed the 'Blue Revolution' and needs governments, NGOs and people to work together towards a better water policy at international, national, state, regional and local levels. Locally good watershed management is a key to solving local rural problems. Present patterns of development are water hungry and water wasters. They do not address pollution and overuse. The linkages between managing water resources and health issues are have not been prioritized as a major source of environmental problems that require policy change, administrative capacity building and an increased financial support.

There are 4 major types of water related diseases:

1. **Waterborne diseases:** These are caused by dirty water contaminated by human and animal wastes, especially from urban sewage, or by chemical wastes from industry and agriculture. Some of these diseases, such as cholera and typhoid, cause serious epidemics. Diarrhea, dysentery, polio, meningitis, and hepatitis A and E, are caused due to improper drinking water. Excessive levels of nitrates because blood disorders when they pollute water sources. Pesticides entering drinking water in rural areas cause cancer, neurological diseases and infertility. Improving sanitation and providing treated drinking water reduces the incidence of these diseases.
2. **Water-based diseases:** Aquatic organisms that live a part of their life cycle in water and another part as a parasite in man, lead to several diseases. In India, guinea worm affects the feet. Round worms live in the small intestine, especially of children.
3. **Water related vector diseases:** Insects such as mosquitoes that breed in stagnant water spread diseases such as malaria and filariasis. Malaria that was effectively controlled in India has now come back as the mosquitoes have become resistant to insecticides. In addition, anti-malarial drugs are now unable to kill the parasites as they have become resistant to drugs. Change in climate is leading to the formation of new breeding sites. Other vector-borne diseases in India include dengue fever and filariasis. Dengue fever carries a high mortality. Filariasis leads to fever and chronic swelling over the legs.

   Eliminating mosquito breeding sites when pooling of water occurs in the monsoon, using fish to control mosquito larval populations, are ways to reduce these diseases without using toxic insecticides that have ill effects on human health.
4. **Water scarcity diseases:** In areas where water and sanitation are poor, there is a high incidence of diseases such as tuberculosis, leprosy, tetanus, etc., which occur when hands are not adequately washed.

   Arsenic in drinking water is a serious hazard to human health. It has attracted much attention since its recognition in the 1990s of its wide occurrence in well-water in Bangladesh. It occurs less frequently in most other countries. The main source of arsenic in drinking water is arsenic-rich rocks through which the water has filtered. It may also occur because of mining or industrial activity in some areas. WHO has worked with other UN organizations to produce a state-of-the-art review on arsenic in drinking water.

CASE STUDIES: Arsenic Poisoning – Bangladesh

More than half the population of Bangladesh is threatened by high levels of arsenic found in drinking water. This could eventually lead to an epidemic of cancers and other fatal diseases.

Rezaul Morol, a young Bangladeshi man, nearly died from arsenic poisoning caused by drinking arsenic-laden well-water for several years. The doctor advised Rezaul to stop drinking contaminated water and eat more protein-rich food such as fish. Since then Rezaul feels a lot better and is happy that his skin is healing.

Drinking water that is rich in arsenic leads to arsenic poisoning or arsenicsis. Excessive concentrations are known to occur in some areas. The health effects are generally delayed and the most effective preventive measure is supplying drinking water which is free of arsenic. Arsenic contamination of water is also due to industrial processes such as those involved in mining, metal refining, and timber treatment. Malnutrition may aggravate the effects of arsenic on blood vessels.

Water with high concentration of arsenic if used over 5 to 20 years, results in problems such as color changes on the skin, hard patches on the palms and soles, skin cancer, cancers of the bladder, kidney and lung, and diseases of the blood vessels of the legs and feet. It may also lead to diabetes, high blood pressure and reproductive disorders.

Natural arsenic contamination occurs in Argentina, Bangladesh, Chile, China, India, Mexico, Thailand and the United States. In China (in the Province of Taiwan) exposure to arsenic leads to gangrene, known as 'black foot disease'.

Long-term solutions for prevention of arsenicosis are based on providing safe drinking water:

- Deeper wells are often less likely to be contaminated.
- Testing of water for levels of arsenic and informing users.
- Monitoring by health workers. People need to be checked for early signs of arsenicosis - usually by observing skin problems in areas where arsenic in known to occur.
- Health education regarding harmful effects of arsenicosis and how to avoid them.

Diarrhea

Though several types of diarrhea which give rise to loose motions and dehydration occur all over the world, this is especially frequently observed in developing countries. It causes 4% of all deaths. In another 5% it leads to loss of health. It is caused by gastrointestinal infections which kill around 2.2 million people globally each year. Most of these are children in developing countries. The use of contaminated water is an important cause of this group of conditions. Cholera and dysentery cause severe, sometimes life threatening and epidemic forms of these diseases.

Effects on health: Diarrhoea is the frequent passage of loose or liquid stools. It is a symptom of various gastrointestinal infections. Depending on the type of infection, diarrhea may be watery (for .example in cholera caused by vibrio cholera) or passed with blood and mucous (in dysentery caused by an amoeba, E Histolitica).

Depending on the type of infection, it may last a few days, or several weeks. Severe diarrhea can become life threatening due to loss of excessive fluid and electrolytes such as sodium and potassium in watery diarrhea. This is particularly fatal in infants and young children. It is also dangerous in malnourished individuals and people with poor immunity.

The impact of repeated diarrhea on nutritional status is linked in a vicious cycle in children. Chemical or non-infectious intestinal conditions can also result in diarrhea.

Causes of diarrhea: Diarrhoea is caused by several bacterial, viral and parasitic organisms. They are mostly spread by contaminated water. It is more common when there is a shortage of clean water for drinking, cooking and cleaning. Basic hygiene is important in its prevention.

Water contaminated with human feces surrounding a rural water source, or from municipal sewage, septic tanks and latrines in urban centres, are important factors in the spread of these diseases. Feces of domestic animals also contain microorganisms that can cause diarrhea through water.

Diarrhoea is spread from one individual to another due to poor personal hygiene. Food is a major cause of diarrhea when it is prepared or stored in unhygienic conditions. Water can contaminate food such as vegetables during irrigation. Fish and seafood from polluted water is a cause of severe diarrhea.

The infectious agents that cause diarrhea are present in our environment. In developed countries where good sanitation is available, most people get enough safe drinking water. Good personal and domestic hygiene prevents this disease which is predominantly seen in the developing world. About 1 billion people do not have access to clean water sources and 2.4 billion have no basic sanitation (WHO website). In Southeast Asia, diarrhea is responsible for 8.5% of all deaths. In 1998, diarrhea was estimated to have killed 2.2 million people, most of whom were under 5 years of age (WHO,2000).

Interventions: Key measures to reduce the number of cases of diarrhoea include:

- Access to safe drinking water.
- Improved sanitation.
- Good personal and food hygiene.
- Health education about how these infections spread.

Key measures to treat diarrhea include:

- Giving more fluids than usual, (oral rehydration) with salt and sugar, to prevent dehydration.
- Continue feeding.
- Consulting a health worker if there are signs of dehydration or other problems.

In rural India, during the last decade public education through posters and other types of communication strategies has decreased infant mortality due to diarrhea in several states. Posters depicting a child with diarrhoea being given water, salt and sugar solution to reduce death from dehydration has gone along way in reducing a serious condition requiring hospitalization and intravenous fluids, as well as mortality.

7.3.5 Risks Due to Chemicals in Food

Food contaminated by chemicals is a major worldwide public health concern. Contamination may occur through environmental pollution of the air, water and soil. Toxic metals, PCBs and dioxins, or the intentional use of various chemicals, such as pesticides, animal drugs and other agrochemicals have serious consequences on human health. Food additives and contaminants used during food manufacture and processing adversely affects health.

Diseases spread by food: Some food-borne diseases though well recognized, have recently become more common. For example, outbreaks of salmonellosis which have been reported for decades, has increased within the last 25 years. In the Western hemisphere and in Europe, Salmonella serotype Enteritidis (SE) has become a predominant strain. Investigations of SE outbreaks indicate that its emergence is largely related to consumption of poultry or eggs.

While cholera has devastated much of Asia and Africa for years, its reintroduction for the first time in almost a century on the South American continent in 1991 is an example of a well recognized infectious disease re-emerging in a region after decades. While cholera is often water-borne, many foods also transmit infection. In Latin America, ice and raw or under processed seafood are important causes for cholera transmission.

Infection with a specific type of *Escherichia coli* (*E. coli*) was first described in 1982. Subsequently, it has emerged rapidly as a major cause of bloody diarrhea and acute renal failure. The infection is sometimes fatal, particularly in children. Outbreaks of infection, generally associated with beef, have been reported in Australia, Canada, Japan, United States, in various European countries, and in southern Africa. Outbreaks have also implicated alfalfa sprouts, unpasteurized fruit juice, lettuce, game meat (meat of wild animals) and cheese curd.

In 1996, an outbreak of *Escherichia coli* in Japan affected over 6300 school children and resulted in 2 deaths.

Listeria monocytogenes (Lm): The role of food in the transmission of this condition has been recognized recently. In pregnant women, infections with Lm causes abortion and stillbirth. In infants and persons with a poor immune system it may lead to septicemia (blood poisoning) and meningitis. The disease is most often associated with consumption of foods such as soft cheese and processed meat products that are kept refrigerated for a long time, because Lm can grow at low temperatures. Outbreaks of listeriosis have been reported from many countries, including Australia, Switzerland, France and the United States. Two recent outbreaks of Listeria monocytogenes in France in 2000 and in the USA in 1999 were caused by contaminated pork tongue and hot dogs respectively.

Food-borne treaties (worms) are increasing in South-east Asia and Latin America. This is related to a combination of intensive aquaculture production in unsanitary conditions, and consumption of raw or lightly processed freshwater fish and fishery products. Food-borne diseases can cause acute liver disease, and may lead to liver cancer. It is estimated that 40 million people are affected worldwide.

Bovine Spongiform Encephalopathy (BSE) is a fatal, transmittable, neurodegenerative disease of cattle. It was first discovered in the United Kingdom in 1985. The cause of the disease was traced to an agent in sheep, which contaminated recycled bovine carcasses used to make meat and bone meal additives for cattle feed. Recycling of the BSE agent developed into a common source epidemic of more than 180,000 diseased animals in the UK alone. The agent affects the brain and spinal cord of cattle which produces sponge-like changes visible under a microscope. About 19 countries have reported BSE cases and the disease is no longer confined to the European Community. A case of BSE has been reported in a cattle herd in Japan.

In human populations, exposure to the BSE agent (probably in contaminated bovine-based food products) has been strongly linked to the appearance in 1996 of a new transmissible spongiform encephalopathy of humans called variant Creutzfeldt-Jakob Disease (vCJD). By January 2002, 119 people developed vCJD, most from the UK but five cases have been reported from France.

7.3.6 Cancer and Environment

Cancer is caused by the uncontrolled growth and spread of abnormal cells that may affect almost any tissue of the body. Lung, colon, rectal and stomach cancer are among the five most common cancers in the world for both men and women. Among men, lung and stomach cancer are the most common cancers worldwide. For women, the most common cancers are breast and cervical cancer. In India, oral and pharyngeal cancers form the most common type of cancer which is related to tobacco chewing.

More than 10 million people are diagnosed with cancer in the world every year. It is estimated that there will be 15 million new cases every year by 2020. Cancer causes 6 million deaths every year – or 12% of deaths worldwide.

The causes of several cancers are known. Thus prevention of at least one-third of all cancers is possible. Cancer is preventable by stopping smoking, providing healthy food and avoiding exposure to cancer-causing agents (carcinogens). Early detection and effective treatment is possible for a further one-third of cases. Most of the common cancers are curable by a combination of surgery, chemotherapy (drugs) or radiotherapy (X-rays). The chance of cure increases if cancer is detected early.

Cancer control is based on the prevention and control of cancer by:

- Promotion and strengthening of comprehensive national cancer control programs.
- Building international networks and partnerships for cancer control.
- Promotion of organized, evidence-based interventions for early detection of cervical and breast cancer.

- Development of guidelines on disease and program management.
- Advocacy for a rational approach to effective treatments for potentially curable tumors.
- Support for low-cost approaches to respond to global needs for pain relief and palliative care.

Prevention of cancer: Tobacco smoking is the single largest preventable cause of cancer in the world. It causes 80 to 90% of all lung cancer deaths. Another 30% of all cancer deaths, especially in developing countries include deaths from cancer of the oral cavity, larynx, esophagus and stomach which are related to tobacco chewing. Preventive measures include bans on tobacco advertising and sponsorship, increased tax on tobacco products, and educational programmes which are undertaken to reduce tobacco consumption.

Dietary modification is an important approach to cancer control. Overweight individuals and obesity are known to be associated with cancer of the esophagus, colon, rectum, breast, uterus and kidney. Fruit and vegetables may have a protective effect against many cancers. Excess consumption of red and preserved meat may be associated with an increased risk of colorectal cancer.

Infectious agents are linked with 22% of cancer deaths in developing countries and 6% in industrialized countries. Viral hepatitis B and C cause cancer of the liver. Human papilloma virus infection causes cancer of the cervix. The bacterium Helicobacter pylori increases the risk of stomach cancer. In some countries the parasitic infection schistosomiasis increases the risk of bladder cancer. Liver fluke increases the risk of cancer of the bile ducts. Preventive measures include vaccination and prevention of infection. Excessive solar ultraviolet radiation increases the risk of all types of cancer of the skin. Avoiding excessive exposure to the sun, use of sunscreens and protective clothing are effective preventive measures. Asbestos is known to cause lung cancer. Aniline dyes have been linked to bladder cancer. Benzene can lead to leukaemia (blood cancer). The prevention of certain occupational and environmental exposure to several chemicals is an important element in preventing cancer.

7.4 HUMAN RIGHTS

Several environmental issues are closely linked to human rights. These include the equitable distribution of environmental resources, the utilization of resources and Intellectual Property Rights (IPRs), conflicts between people and wildlife especially around PAs, resettlement issues around development projects such as dams and mines, and access to health to prevent environment related diseases.

7.4.1 Equity

One of the primary concerns in environmental issues is how wealth, resources and energy must be distributed in a community. We can think of the global community, regional community issues, national concerns and those related to a family or at the individual level. While economic disparities remain a fact of life, we as citizens of a community must appreciate that a widening gap between the rich and the poor, between men and women, or between the present and future generations must be minimized if social justice is to be achieved. Today the difference between the economically developed world and the developing countries is unacceptably high. The access to a better lifestyle for men as against women is inherent in many cultures. Last but not the least, we in the present generation cannot greedily use up all our resources leaving future generations increasingly impoverished. Rights to land, water, food, and housing are all a part of our environment that we all share. However, while some live unsustainable lifestyles with consumption patterns that the resource base cannot support, many others live well below the poverty line. Even in a developing country such as ours, there are enormous economic inequalities. This requires an ethic in which an equitable distribution becomes a part of everyone's thinking. The people who live in the countries of the North and the rich from the countries in the South will have to take steps to reduce their resource use and the waste they generate. Both the better off sectors of society and the less fortunate need to develop their own strategies of sustainable living and communities at each level must bring about more equitable patterns of wealth.

The right to the use of natural resources that the environment holds is an essential component of human rights. It is related to disparities in the amount of resources available to different sectors of society. People who live in wilderness communities are referred to as ecosystem people. They collect food, fuelwood, and non-wood products, fish in aquatic ecosystems, or hunt for food in forests and grasslands. When land use patterns change from natural ecosystem to more intensively used farmland and pasture land the rights of these indigenous people are usually sacrificed. Take the case of subsidies given to the pulp and paper industry for bamboo which makes it several times cheaper for the industry than for a rural individual who uses it to build his home. This infringes on the human right to collect resources they have traditionally used free of cost. Another issue is the rights of small traditional fishermen who have to contend against mechanised trawlers that impoverish their catch and over-harvest fish in the marine environment. These people's right to a livelihood conflicts with the powerful economic interests of large-scale organized fisheries.

There are serious conflicts between the rights of rural communities for even basic resources such as water, and industrial development which requires large amounts of water for sustaining its productivity. The right to land or common property resources of tribal people is infringed upon by large development projects such as dams, mining and protected areas. Movements to protect the rights of indigenous peoples are growing worldwide. Reversing actions that have already been taken decades ago is a complex problem that has no simple solutions. In many cases a just trade-off is at best achieved through careful and sensitively managed negotiations. This needs a deep appreciation of local environmental concerns as well as sensitivity to the rights of local people.

7.4.2 Nutrition, Health and Human Rights

There are links between environment, nutrition and health which must be seen from a human-rights perspective. Proper nutrition and health are fundamental human rights. The right to life is a Fundamental Right in our constitution. As a deteriorating environment shortens life spans, this in effect has an impact on our fundamental constitutional right.

Nutrition affects and defines the health status of all people, rich and poor. It is linked to the way we grow, develop, work, play, resist infection and reach our aspirations as individuals, communities and societies. Malnutrition makes people more vulnerable to disease and premature death. Poverty is a major cause as well as a consequence of ill-health. Poverty, hunger, malnutrition and poorly managed environments together affect health and weaken the socio-economic development of a country. Nearly 30% of humanity, especially those in developing countries – infants, children, adolescents, adults, and older persons are affected by this problem. A human rights approach is needed to appreciate and support millions of people left behind in the 20th century's health revolution. We must ensure that our environmental values and our vision are linked to human rights and create laws to support those that need a better environment, better health and a better lifestyle.

Health and sustainable human development are equity issues. In our globalized 21st century, equity must begin at the bottom, hand in hand with a healthy environment, improved nutrition, and sustainable lifestyles. Putting first things first, we must also realize that resources allocated to preventing and eliminating disease will be effective only if the underlying causes such as malnutrition and environmental concerns, as well as their consequences, are successfully addressed.

7.4.3 Intellectual Property Rights and Community Biodiversity Registers

Traditional people, especially tribals living in forests, have used local plants and animals for generations. This storehouse of knowledge leads to many new 'discoveries' for modern pharmaceutical products. The revenue generated from such 'finds' goes to the pharmaceutical industry that has done the research and patented the product. This leaves the original tribal user with nothing while the industry could earn billions of rupees. To protect the rights of indigenous people who have used these products, a possible tool is to create a Community Biodiversity Register of local products and their uses so that its exploitation by the pharmaceutical industry would have to pay a royalty to the local community. This however has still not been generally accepted. Mechanisms have to be worked out so that the local traditional user's rights are protected.

Traditional medicine refers to health practices, approaches, knowledge and beliefs that incorporate plant, animal and mineral based medicines, frequently of local or regional origin. It may be linked to spiritual therapies, manual techniques and exercises. These may be used singly or in combination to treat, diagnose and prevent illnesses or maintain well-being. Traditional medicine is often handed down through the generations or may be known to a special caste or tribal group.

Traditional medicine has maintained its popularity in all regions of the developing world and its use is rapidly spreading in industrialized countries. In India, some of our primary health care needs are taken care of entirely by traditional medicine, while in Africa; up to 80% of the population uses it for primary health care. In industrialized countries, adaptations of traditional medicine are termed "Complementary"or "Alternative" Medicine (CAM).

While there are advantages to traditional medicine as it is cheap and locally available, there are diseases which it cannot treat effectively. This is a risk, as patients who use these alternative medicinal practices may rely on an ineffective measure. The consequences could be a serious delay in diagnosis and effective treatment of a treatable condition. There is a need to carefully research the claims of traditional practices to ensure that they are effective.

In addition to patient safety issues, there is the risk that a growing herbal market and its great commercial benefit poses a threat to biodiversity through the over-harvesting of the raw material for herbal medicines and other natural health care products. This has been observed in the case of several Himalayan plants. If extraction from the wild is not controlled, this can lead to the extinction of endangered plant species and the destruction of natural habitats of several species.

Another related issue is that at present, the requirements for protection provided under international standards for patent laws and by most national conventional patent laws are inadequate to protect traditional knowledge and biodiversity.

There are tried and tested scientific methods and products that have their origins in different traditional medicinal methods. Twenty-five per cent of modern medicines are made from plants first used traditionally. Yoga is known to reduce asthma attacks. Traditional medicine has been found to be effective against several infectious diseases.

A US company was granted a patent for discovering extracts of arhar (pigeon pea or Cajanuscajan) in the treatment of diabetes, hypoglycemia, obesity and blockage of arteries. The use of pigeon pea extracts in India is well known. CSIR has challenged this patent as it infringes on India's traditional knowledge, although challenging the patent is difficult, as India's scientific documentation of its traditional knowledge is quite poor.

Over one-third of the population in developing countries lack access to essential allopathic medicines. The provision of safe and effective TM/ CAM therapies could become a tool to increase access to health care.

7.5 VALUE EDUCATION

Value education in the context of our environment is expected to bring about a new sustainable way of life. Education both through formal and non-formal processes must thus address understanding environmental values, valuing nature and cultures, social justice, human heritage, equitable use of resources, managing common property resources and appreciating the cause of ecological degradation.

Essentially, environmental values cannot be taught. They are inculcated through a complex process of appreciating our environmental assets and experiencing the problems caused due to our destruction of our environment. The problems that are created by technology and economic growth are a result of our improper thinking on what 'development' means. Since we still put a high value only on economic growth, we have no concern for aspects such as sustainability or equitable use of resources. This mindset must change before concepts such as sustainable development can be acted upon.

Unsustainable development is a part of economic growth of the powerful while it makes the poor poorer. Consumerism is one aspect of this process favoured by the rich. As consumption of resources has till recently been an index of development, consumerism has thrived. It is only recently that the world has come to realize that there are other more important environmental values that are essential to bring about a better way of life.

Values in environment education must bring in several new concepts. Why and how can we use less resources and energy? Why do we need to keep our surroundings clean? Why should we use less fertilizers and pesticides in farms? Why is it important for us to save water and keep our water sources clean? Or separate our garbage into degradable and non-degradable types before disposal? All these issues are linked to the quality of human life and go beyond simple economic growth. They deal with a love and respect for nature. These are the values that will bring about a better humanity, one in which we can live healthy, productive and happy lives in harmony with nature.

What are values?

Values deal with one's own principles and standards from which we judge what is right and wrong behaviour.

7.5.1 Environmental Values

Every human being has a great variety of feelings for different aspects of his or her surroundings. The Western, modern approach values the resources of nature for their utilitarian importance alone. However, true environmental values go beyond valuing a river for its water, a forest for its timber and non-wood forest products, or the sea for its fish. Environmental values are inherent in feelings that bring about sensitivity for preserving nature as a whole. This is a more spiritual,

Eastern traditional value. There are several writings and sayings in Indian thought that support the concept of the oneness of all creation, of respecting and valuing all the different components of nature. Our environmental values must translate to pro-conservation actions in all our day-to-day activities. Most of our actions have adverse environmental impacts unless we consciously avoid them. The sentiment that attempts to reverse these trends is enshrined in our environmental values. Values lead to a process of decision making which leads to action. For value education in relation to the environment, this process is learnt through an understanding and appreciation of nature's oneness and the importance of its conservation.

Humans have an inborn desire to explore nature. Wanting to unravel its mysteries is a part of human nature. However, modern society and educational processes have invariably suppressed these innate sentiments. Once exposed to the wonders of the wilderness, people tend to bond closely to nature. They begin to appreciate its complexity and fragility and this awakens a new desire to want to protect our natural heritage. This feeling for Nature is a part of our Constitution, which strongly emphasizes this value. Concepts of what constitutes right and wrong behavior changes with time. Values are not constant. It was once considered 'sport' to shoot animals. It was considered a royal, brave and much desirable activity to kill a tiger. In today's context, with wildlife reduced to a tiny fraction of what there was in the past, it is now looked down upon as a crime against biodiversity conservation. Thus, the value system has been altered with time. Similarly with the large tracts of forest that existed in the past, cutting a few trees was not a significant criminal act. Today this constitutes a major concern. We need a strong new environmental value system in which felling trees is considered unwise behavior. With the small human numbers in the past, throwing away a little household degradable garbage could not have been considered wrong. But with enormous numbers of people throwing away large quantities of non-degradable waste, it is indeed extremely damaging to the environment and our value system must prevent this through a strong environmental value education system.

Appreciating the negative effects of our actions on the environment must become a part of our day-to-day thinking. Our current value systems extol economic and technical progress as being what we need in our developing country.

Environmental values based on the Constitution of India

Article 48A: "The state shall endeavour to protect and improve the environment and to safeguard the forests and wildlife in the country."

Article 51A (g): The constitution expects that each citizen of the country must "protect and improve the natural environment, including forests, lakes, rivers and wildlife, and to have compassion for all living creatures."

While we do need economic development, our value system must change to one that makes people everywhere support a sustainable form of development so that we do not have to bear the cost of environmental degradation.

Environmental problems created by development are due neither to the need for economic development, nor to the technology that produces pollution, but rather to a lack of awareness of the consequences of unlimited and unrestrained anti-environmental behaviour. Looked at in this way, it deals with concepts of what is appropriate behaviour in relation to our surroundings and to other species on the earth. How we live our lives in fact shapes our environment. This is what the environmental values are about.

Each action by an individual must be linked to its environmental consequences in his/her mind so that a value is created that leads to strengthening pro-environmental behaviour and preventing anti-environmental actions. This cannot happen unless new educational processes are created that provide a meaning to what is taught at school and college level. Every small child while growing up asks questions like 'What does this mean?' They want an explanation for things happening around them that can help them make decisions and through this process develop values. It is this innate curiosity that leads to a personalized set of values in later life. Providing appropriate 'meanings' for such questions related to our own environment brings in a set of values that most people in society begin to accept as a norm. Thus, pro-environmental actions begin to move from the domain of individuals to that of a community.

At the community level, this occurs only when a critical number of people become environmentally conscious so that they constitute a pro-environment lobby force that makes governments and other people accept good environmental behaviour as an important part of development.

Strategies for sustainable living

- I will work towards the protection of our environment and the preservation of our wild species.
- I will work towards this with other like-minded individuals.
- I will consciously avoid committing acts that damage our environment and will publicly assert my dislikes for acts against the environment.
- I will not permit others to cause harm to the wilderness and our wild species.
- I will use resources carefully by reducing, reusing and recycling whatever I use such as water, paper, plastic, and metal and glass articles. I will not carelessly throwaway items that are made of our precious natural resources. I will use energy carefully and close off electrical appliances when not in use.
- I will not waste energy by using a fuel based vehicle when I can walk or cycle.
- I will visit our wondrous wild places with clean air, water, soil, and all their plants and animals, and become party to their conservation.
- I will not permit any individual or government action spoil our environment or damage wilderness without protest.
- I will always care for Mother Earth.
- I will try not to damage her knowingly or unknowingly.

What professions require making value judgments that greatly influence our environment? Evidentially, nearly every profession can and does influence our environment, but some do so more than others. Policy-makers, administrators, land use planners, media, architects, medical personnel, health care workers, agriculturalists, agricultural experts, irrigation planners, mining experts, foresters, forest planners, industrialists and, most importantly, teachers at school and college level, are all closely related to pro-environmental outcomes.

Environmental values have linkages to varied environmental concerns. While we value resources that we use as food, water and other products, there are also environmental services that we must appreciate. These include nature's mechanisms in cleaning up air by removing carbon dioxide and adding oxygen by plant life, recycling water through the water cycle of nature, maintaining climate regimes, etc.

But there are other aesthetic, ethical values that are equally important aspects of our environment that we do not appreciate consciously. While every species is of importance in the web of life, there are some which man has come to admire for their beauty alone. The tiger's magnificence, the whale and elephant's giant size, the intelligence of our cousins the primates, the graceful flight of a flock of cranes, are parts of nature that we cannot help but admire. The lush splendour of an evergreen forest, the great power of the ocean's waves, and the tranquillity of the Himalayan mountains are things that each of us values even if we do not experience it ourselves. We value its being there on the earth for us. This is called its 'existence value'. The list of wondrous aspects of nature's intricate connections is indeed awe-inspiring. This is also a part of our environment that we must value for its own sake. This is the oneness of nature.

We must equally look at our environment beyond the wild sphere. There is incredible beauty in some man-modified landscapes, the coloured patterns of farmland or the greens of a tea or coffee plantation in the hills.

Urban gardens and open space are also valuable and thus must be of prime concern to urban planners. These green spaces act as not only the lungs of a city, but also provide much needed psychological support. The mental peace and relaxation provided by such areas needs to be valued, although it is difficult to put a price tag on these values. Nevertheless, these centres of peace and tranquillity give urban dwellers an opportunity to balance their highly man-modified environments with the splash of green of a garden space.

Environmental values must also stress on the importance of preserving ancient structures. The characteristic architecture, sculpture, artworks and crafts of ancient cultures is an invaluable environmental asset. It tells us where we have come from, where we are now, and perhaps where we should go. Architectural heritage goes beyond preserving old buildings, to conserving whole traditional landscapes in rural areas and streetscapes in urban settings. Unless we learn to value these landscapes, they will disappear and our heritage will be lost.

As environmentally conscious individuals we need to develop a sense of values that are linked with a better and more sustainable way of life for all people. There are several positive as well as negative aspects of behaviour that are linked to our environment. The positive feelings that support environment include a value for nature, cultures, heritage, and equity. We also need to become more sensitive to aspects that have negative impacts on the environment. These include our attitude towards degradation of the environment, loss of species, pollution, poverty, corruption in environmental management, the rights of future generations and animal rights.

Several great philosophers have thoughts that have been based on, or embedded, in pro-environmental behaviour. Mahatma Gandhi and Rabindranath Tagore are among the many internationally well known scholars whose thoughts have included values that are related to environmental consciousness. We need to appreciate these values to bring about a better way of life on the earth for all people and all living creatures.

7.5.2 Valuing Nature

The most fundamental environmental sentiment is to value nature herself. Appreciating her magnificence and treasuring life it leads to positive feelings that are a manifestation of pro-environmental consciousness. The oneness of our lives with the rest of nature and a feeling that we are only a miniscule part of nature's complex web of life becomes apparent, when we begin to appreciate the wonders of nature's diversity. We must appreciate that we belong to a global community that includes another 1.8 million known living forms. Nothing makes us more conscious of this wondrous aspect of our earth's diversity than a walk through the wilderness, feeling and exploring its beauty and experiencing its infinite variety. The tiny creatures that live complex lives and the towering trees are all a part of this phenomenon we call 'life'. Today, man does not even know if other complex forms of life exist outside our own solar system in distant space. We may be alone in space or may be accompanied by other, completely different, living forms. But for now we only know for sure that the earth's life forms are unique. We thus have a great responsibility to protect life in all its glorious forms and must therefore respect the wilderness with all its living creatures, where man's own hand has not created changes that have led to perturbing natural habitats. We need to develop a sense of values that lead us to protect what is left of the wilderness by creating effective national parks and wildlife sanctuaries. However this cannot be done to the detriment of the millions of tribal or indigenous people who live in wilderness ecosystems. There are thus conflicting values that need to be balanced carefully. On the one hand we need to protect natural ecosystems, while on the other; we must protect the rights of local people.

Deep ecology

In the 1970s a new thinking on environmental concerns began to emerge, protecting nature and the wilderness for its own sake, which is now referred to as 'Deep Ecology'. It was fostered by the thinking of Arne Names, a Norwegian professor of Philosophy and a great believer in Gandhian thinking and Buddhism. It recognizes the intrinsic value of all living beings and looks upon mankind as a small segment of a great living community of life forms. It teaches that the well-being and flourishing of human and non-human life on the earth have values in themselves and that these values are independent of the usefulness of the non-human world for human purposes.

Yet apart from valuing the diversity of life itself, we must also learn to value and respect diverse human cultures. Many of the tribal cultures of our country are vanishing because those with more dominant and economically advanced ways of life do not respect their lifestyles that are in fact closer to nature and frequently more sustainable. We believe that our modern technology-based lifestyles are the sole way for society to progress. Yet this is only a single dimension of life that is based on economic growth.

While currently the environmental movement focuses on issues that are concerned with the management of the natural environment for the 'benefit' of man, Deep ecology promotes an approach that is expected to bring about a more appropriate ecological balance on the earth and is akin to a spiritual approach to nature. This has great long-term implications not only for humans but for the whole of nature.

For example, some environmentalists emphasize the need to preserve wilderness for its aesthetic and utilitarian functions. Wilderness is preserved because it is scenic and serves the purpose of tourism for nature lovers, and has recreational and economic value. Other environmentalists stress that the goal is for protecting the useful ecological functions of the wilderness, its services and goods that we use. Deep ecologists on the other hand stress that wilderness preservation is a means to achieve the conservation and protection of biological diversity. Thus, it is not enough to protect bits of what is left of the wilderness but to make at tempts to restore degraded areas to their former natural ecological state. In a country such as India, with its enormous population coupled with poverty on the one hand and the need for economic industrial growth on the other, this will be extremely difficult to achieve.

Another new approach is that of 'Gaia', the hypothesis that the earth is itself like one giant form of throbbing life consisting of all the unquantifiable numbers of individuals of its millions of known and unknown species.

7.5.3 Valuing Cultures

Every culture has a right to exist. Tribal people are frequently most closely linked with nature and we have no right to foist on them our own modern way of life. The dilemma is how to provide them with modern health care and education that gives them an opportunity to achieve a better economic status without disrupting their culture and way of life. This will happen only if we value their culture and respect their way of life.

7.5.4 Social Justice

As the divide widens between those people who have access to resources and wealth, and those who live near or below the poverty line, it is the duty of those who are better off to protect the rights of the poor who do not have the means to fight for their rights. If this is not respected the poor will eventually rebel, anarchy and terrorism will spread and the people who are impoverished will eventually form a desperate seething revolution to better their own lot. The developing world would face a crisis earlier than the developed countries unless the rights of poor people that are fundamental to life are protected.

Modern civilization is a homogeneous culture, based until recently on a belief that modern science holds the answer to everything. We are now beginning to appreciate that many ancient and even present-day sequestrated cultures have a wisdom and knowledge of their own environments that is based on a deep sense of respect for nature. Tribal cultures have over many generations used indigenous medicines which are proving to be effective against diseases. They have produced unique art forms such as painting, sculpture, and crafts that are beautiful and can enrich living experiences for everyone. They have their own poetry, songs, dance and drama all art forms that are unfortunately being rapidly lost as we introduce a different set of modern values to them through television and other mass media. The world will be culturally impoverished if we allow these indigenous people to lose their traditional knowledge which includes sustainable use of water, land and resources with a low impact on biodiversity. They will soon lose the beauty within their homes that is based on the things they make from nature. The art of the potter will be lost forever to the indestructible plastic pot. The bamboo basket weaver who makes a thing of beauty that is so user friendly and aesthetically appealing, will give place to yet another plastic box. Much that is beautiful and hand-crafted will disappear if we do not value these diverse aspects of human cultures.

7.5.5 Human Heritage

The earth itself is a heritage left to us by our ancestors for not only our own use but for the generations to come. There is much that is beautiful on our earth – the undisturbed wilderness, a traditional rural landscape, the architecture of a traditional village or town, and the value of a historical monument or place of worship. These are all parts of human heritage.

Heritage preservation is now a growing environmental concern because much of this heritage has been undervalued during the last several decades and is vanishing at an astonishing pace. While we admire and value the Ajanta and Ellora Caves, the temples of the 10th to 15th centuries that led to different and diverse styles of architecture and sculpture, the Moghul styles that led to structures such as the Taj Mahal, or the unique environment-friendly colonial buildings, we have done little to actively preserve them. As environmentally conscious individuals we need to lobby for the protection of the wilderness and our glorious architectural heritage.

7.5.6 Equitable Use of Resources

An unfair distribution of wealth and resources, based on a world that is essentially only for the rich, will bring about a disaster of unprecedented proportions. Equitable use of resources is now seen as an essential aspect of human well-being and must become a shared point of view among all socially and environmentally conscious individuals. This includes an appreciation of the fact that economically advanced countries and the rich in even poor nations consume resources at much greater levels than the much larger poorer sectors of humanity in the developing world. In spite of the great number of people in the more populous developing countries, the smaller number of people in developed countries use more resources and energy than those in the developing world. This is equally true of the small number of rich people in poor countries whose per capita use of energy and resources, and the generation of waste based on the one time use of disposable products, leads to great pressures on the environment. The poor while polluting the environment have no way to prevent it. The rich damage the environment through carelessness that proves only that they have no appreciation for environmental safety. As we begin to appreciate that we need more sustainable lifestyles we also begin to realize that this cannot be brought about without a more equitable use of resources.

7.5.7 Common Property Resources

Our environment has a major component that does not belong to individuals. There are several commonly owned resources that all of us use as a community. The water that nature recycles, the air that we all breathe, the forests and grasslands which maintain our climate and soil, are all common property resources. When government took over the control of community forests in British times, the local people who until then had controlled their use through a set of norms that were based on equitable use began to overexploit resources on which they now had no personal stake. Bringing back such traditional management systems is extremely difficult. However, in the recent past managing local forests through village level forest protection committees has shown that if people know that they can benefit from the forests, they will begin to protect them. This essentially means sharing the power to control forests between the forest department and local people.

7.5.8 Ecological Degradation

In many situations valuable ecological assets are turned into serious environmental problems. This is because we as a society do not strongly resist forces that bring about ecological degradation. These consist of sectors of society that use a 'get-rich-quick' approach to development. While ecological degradation has frequently been blamed on the needs of fuelwood and fodder of growing numbers of rural people, the rich, urbanized, industrial sector is responsible for greater ecological damage. Changes in land use from natural ecosystems to more intensive utilization such as turning forests into monoculture forestry plantations, or tea and coffee estates, or marginal lands into intensive agricultural patterns such as sugarcane fields or changes into urban or industrial land carry an ecological price. Wetlands, for example, provide usable resources and a variety of services not easily valued in economic terms, and when destroyed to provide additional farmland, in many cases produce lower returns. A natural forest provides valuable non-wood forest products whose economic returns far outweigh that provided by felling the forest for timber. These values must form a part of a new conservation ethic. We cannot permit unsustainable development to run onwards at a pace in which our lives will be overtaken by a development strategy that must eventually fail as earth's resources are consumed and ecosystem rendered irreparable.

7.6 HIV/AIDS

The Human Immunodeficiency Virus (HIV) causes Acquired Immuno deficiency Syndrome (AIDS) through contact with tissue fluids of infected individuals, especially through sexual contact. As it reduces an individual's resistance to disease, it causes infected individuals to suffer from a large number of environment related diseases and reduces the ability of infected individuals to go about their normal lives. It affects their income generation and/or their ability to utilize natural resources. As more and more people are affected, this disease will also have impacts on our natural resource base, as utilization patterns change to unsustainable levels. The inability of these patients to have the strength to access natural resources also affects the outcome of the disease process, as their overall health and well-being is likely to worsen the course of the disease when their nutritional status suffers.

In sub-Saharan Africa where the infection has become highly prevalent, it is leading to great suffering and worsening poverty. The capacity of these patients to work for their usual sources of income generation is lost. An increasing proportion of the poor are affected. It is evident that it is going to be increasingly difficult to manage environments sustainably, as natural resources on which the poor debilitated patients depend continue to be degraded. Incomes lost due to the stigma of HIV/AIDS must be met by the sufferers by overexploiting their resource base. People affected by the disease inevitably try to get whatever they can from their natural resource base as they are not in position to think of the future. In Africa, this has led to degradation of the ecosystem and an increase of pressures from other impacts such as overuse of medicinal plants and poaching for wildlife. In South Africa, for example, people have a mistaken belief that turtle eggs can cure HIV/ AIDS, thus leading to the eggs being over-harvested. As males die of the disease, work on agricultural land has to be taken over by already overworked women and their children, affecting land management and productivity. Providing balanced diets and nutritional support for these poverty stricken patients can be partially addressed by better natural resource management such as forestation, access to clean water and wholesome food.

HIV/AIDS seriously affects the patient's working environment. It creates fear in the minds of co-workers. It must be clearly understood that AIDS is not spread by casual contact during work. Patients have a right to continue to work as before along with unaffected individuals. As patients are unable to continue their original hard labour related work, it is essential that alternative sources of work must be created for them.

Educators and extension information, in the formal and non-formal educational sectors, must address the issues related to the linkages between natural resource management and this disease, as well as the need to remove the social stigma attached to it.

HIV/ AIDS has a serious impact on the socio- economic fabric of society. By 2002, India had an estimated 3.97 million infected individuals. There is a great need to organize AIDS education on prevention and management of the disease. This needs to be done through the formal educational sector and by using non-formal methods. Education is also important to reduce the stigma and discrimination against these patients. In India, women who are not socially empowered are at a great disadvantage as they are powerless to demand safe sex from their partners. Women also have an added burden of caring for HIV infected husbands. This produces enormous economic stresses on their family. HIV in India is rapidly moving from a primarily urban sector disease to rural communities.

Research in Nepal has shown a linkage between rural poverty, deforestation and a shift of population.

To urban areas resulting in a rising number of AIDS patients. Prior to 1992, it was mainly seen in males who migrated to urban centres.In more recent times, a growing number of women are moving to Indian cities as sex workers. Women engaged in prostitution find it difficult to make partners take protective measures, such as the use of condoms. A large proportion became victims of the disease.

Blood transfusion from an infected person can also lead to HIV/AIDS in the recipient, as well as drug abuse by sharing needles with an infected person. In sexually transmitted AIDS, the use of condoms during intercourse is a key to preventing the disease. Behavioural change, where the number of individuals who have multiple partners, towards strictly single partners, reduces the risk of HIV/AIDS and thus reduces incidence of the disease in society. However, the most important measure to prevent AIDS is the proper use of condoms that forms a barrier to the spread of the virus during intercourse.

7.7 WOMAN AND CHILD WELFARE

There are several environmental factors that are closely linked to the welfare of women and children. Each year, close to eleven million children worldwide are estimated to have died from the effects of disease and inadequate nutrition. Most of these deaths are in the developing world. In some countries, more than one in five children die before they are 5 years old. Seven out of 10 of childhood deaths in developing countries can be attributed to five main causes, or a combination of them. These are pneumonia, diarrhea, measles, malaria and malnutrition. Around the world, three out of every four children suffer from at least one of these conditions.

The diagnosis of common childhood disease conditions

Presenting complaint possible cause or associated condition

Cough and/or pneumonia

Fast breathing severe anemia

P. Falciparum malaria

Lethargy or cerebral malaria, unconsciousness meningitis

Severe dehydration

Very severe pneumonia

Measles, rash, pneumonia

Diarrhea

Ear infection

"Very sick" young infant, pneumonia meningitis sepsis

Respiratory conditions: Most respiratory diseases are caused by or are worsened by polluted air. Crowded ill-ventilated homes and living in smokey households with open fires can trigger respiratory conditions especially in children.

CASE STUDY: Chula Issue

The World Health Organization estimates that 1.6 billion early deaths occur annually from cooking stove pollution. 400,000 to 550,000 children under five and women die each year in India due to indoor smoke. Chula smoke is the third highest cause of disease and death after dirty water and lack of sanitation. Hence by providing access to clean water, sanitation, food and ventilated homes, over half the diseases and premature deaths could be avoided in India.

Pneumonia: Acute respiratory infections (ARI), most frequently pneumonia, is a major cause of death in children under five, killing over two million children annually. Upto 40% of children seen in health centres suffer from respiratory conditions and many deaths attributed to other causes are, in fact, "hidden" ARI deaths. Children may die very quickly from the infection and thus need treatment urgently. Most patients of pneumonia can be treated with oral antibiotics. Correct management could save over 1 million lives per year globally.

Gastrointestinal conditions: Contaminated water and food causes widespread ill health especially in children.

Diarrhea: Diarrhea is caused by a wide variety of infections. Urgent diagnosis and treatment of diarrhea is a priority for saving a child's life. Treating malnutrition that often accompanies diarrhea can further reduce mortality. Increasing vigilance to detect other diseases that can occur concurrently with diarrhea, such as measles or malaria, is an important measure.

Two million children die each year in developing countries from diarrheal diseases, the second most serious killer of children under five worldwide. In most cases diarrhea is preventable and children can be saved by early treatment. Correct management of diarrhea could save the lives of up to 90% of children who currently die by promoting rapid and effective treatment through standardized management, including antibiotics and simple measures such as oral rehydration using clean boiled water with salt and sugar. In severe cases intravenous fluids must be started. Improved hygiene and management of the home and surroundings is the most important preventative measure, as well as improved nutrition. Increased breastfeeding and measles vaccination have also been observed to have reduced the number of cases of diarrhea.

Measles: Measles is a rash that appears with fever and body ache in children and is caused by a virus. It infects over 40 million children and kills over 8,00,000 children under the age of five. Prevention includes wider immunization coverage, rapid referral of serious cases, prompt recognition of conditions that occur in association with measles, and improved nutrition, including breastfeeding, and vitamin A supplementation. Measles is prevented by a vaccine. Young children with measles often develop other diseases such as acute respiratory infections, diarrhea and malnutrition that are all linked to poor environmental conditions in their surroundings. Children who survive an attack of measles are more vulnerable to other dangerous infections for several months. Effective prevention and treatment could save 7,00,000 lives per year.

Malaria: This condition is closely linked to pooling and stagnation of water in tropical environments. Malaria is a widespread tropical disease which is caused by a parasite transmitted to humans by mosquitoes. It has proved difficult to control because mosquitoes have become resistant to insecticides used against them and because the parasite has developed resistance in some areas to the cheap and effective drugs that used to provide good protection in the past. However, alternative newer drug therapies have been developed for use in areas where resistant parasites are found. In India the disease was nearly wiped out a few decades ago but has now re-emerged in many parts of the country. Correct management could save 5,00,000 lives per year. Approximately 7,00,000 children die of malaria globally each year, most of them in sub-Saharan Africa. Young children are particularly vulnerable because they have not developed the partial immunity that results from surviving repeated infections.

Deaths from malaria can be reduced by several measures, including encouraging parents to seek prompt care, accurate assessment of the condition of the child, prompt treatment with appropriate anti-malarial drugs, recognition and treatment of other co-existing conditions, such as malnutrition and anemia, and prevention by using mosquito proof bed nets. Because fever may be the only sign of malaria, it can be difficult to distinguish it from other potentially life-threatening conditions.

Poverty-environment-malnutrition: There is a close association between poverty, a degraded environment, and malnutrition. This is further aggravated by a lack of awareness on how children become malnourished.

Malnutrition: Although malnutrition is rarely listed as the direct cause of death, it contributes to about half of all childhood deaths. Lack of access to food, poor feeding practices and infection, or a combination of the two, are major factors in mortality.

Infection, particularly frequent or persistent diarrhea, pneumonia, measles and malaria, undermines nutritional status. Poor feeding practices — inadequate breastfeeding, providing the wrong foods, giving food in insufficient quantities, contribute to malnutrition. Malnourished children are more vulnerable to disease.

Promoting breastfeeding, improving feeding practices, and providing micronutrient supplements routinely for children who need them are measures that reduce mortality.

The nutritional status and feeding practices of every child less than two years of age, and those with a low weight for their age must be intensively managed. Counselling of parents on the correct foods for each age group and helping them to overcome various feeding problems is an essential health care measure.

Children between 6 months and 2 years of age are at increased risk of malnutrition when there is a transition between breastfeeding and sharing fully in the family diet. Changing family habits and the kinds of food offered to children is an important measure. Talking to mothers individually about home care and their child's feeding, with relatively simple changes to better feeding practices, such as helping them to eat rather than leaving them to fend for themselves, can ensure that a child gets enough to eat.

A minor increase in breastfeeding could prevent up to 10% of all deaths of children under five:

When mothers breastfeed exclusively during at least the first four months and, if possible, six months of life, there is a decrease in episodes of diarrhea and, to a lesser extent, respiratory infections. Even small amounts of water-based drinks decreases breast milk intake and lead to lowered weight gain. This increases the risk of diarrhea. Continuing to breastfeed up to two years of age, in addition to giving complementary foods, maintains good nutritional status and helps prevent diarrhea.

Encouraging maximum support to mothers to establish optimal breastfeeding from birth, equipping health workers with counselling skills, and providing individual counselling and support for breastfeeding mothers are measures that reduce malnutrition. Mothers often give their babies other food and fluids before six months because they doubt their breast milk supply is adequate. A one-on-one counselling with mothers on breastfeeding techniques and its benefits helps reduce incidence of malnutrition.

There are strong connections between the status of the environment and the welfare of women and children in India. Women, especially in lower income group families, both in the rural and urban sector, work longer hours than men. Their work pattern differs and is more prone to health hazards. The daily collection of water, fuelwood and fodder is an arduous task for rural women. In urban areas, where lower economic group women live in crowded smoke filled shantys in unhygienic slums, they spend long hours indoors, which is a cause of respiratory diseases. In urban centres, a number of women eke out a living by garbage picking. They separate plastics, metal and other recyclable material from the waste produced by the more affluent groups of society. During this process, they can get several infections. Thus they are providing an environmental service of great value, but earn a pittance from this work.

Women are often the last to get enough nutrition as their role in traditional society is to cook the family meal and feed their husband and children. This leads to malnutrition and anemia due to inadequate nutrition.

The sorry plight of women includes the fact that the girl child is given less attention and educational facilities as compared to boys in India. Thus they are unable to compete with men in later life. This social-environmental divide is a major concern that needs to be corrected throughout the country.

7.8 ROLE OF INFORMATION TECHNOLOGY IN ENVIRONMENT AND HUMAN HEALTH

The understanding of environmental concerns and issues related to human health has exploded during the last few years due to the sudden growth of Information Technology. The computer age has turned the world around due to the incredible rapidity with which IT spreads knowledge. IT can do several tasks extremely rapidly, accurately and spread the information through the world's networks of millions of computer systems. A few examples of the use of computer technology that aid environmental studies include software such as using geographical information systems (GIS). GIS is a tool to map land use patterns and document change by studying digitized top sheets and/or satellite imagery. Once this is done, an expert can ask a variety of questions which the software can answer by producing maps which helps in land use planning.

CASE STUDY: Karnataka's GIS Scheme

Karnataka's GIS scheme, Bhomi has revolutionized the way farmers access their land records. Farmers can now get a copy of the records of rights, tenancy and crops from a computerized information kiosk without harassment and bribes. Karnataka has computerized 20 million records of land ownership of 6.7 million farmers in the state.

The Internet with its thousands of websites has made it extremely simple to get the appropriate environmental information for any study or environmental management planning. This not only assists scientists and students but is a powerful tool to help increase public awareness about environmental issues.

Specialized software can analyze data for epidemiological studies, population dynamics and a variety of key environmental concerns.

The relationship between the environment and health has been established due to the growing utilization of computer technology. This looks at infection rates, morbidity or mortality and the etiology (causative factors) of a disease. As knowledge expands, computers will become increasingly efficient. They will be faster, have greater memories and even perhaps begin to think for themselves.

Learning Outcome

The main learning outcomes from this chapter should be advance level understanding of interaction between human population and environment. How the relation between two evolved in due course of time and what lessons can be learned for future sustainable development. The chapter provided in detail study material for; population growth variation among nations; Population explosion — Family Welfare Programme; Environmental and human health; Human Rights; Value Education; HIV/AIDS; Women and Child Welfare; Role of Information Technology in Environment and human health; and Case Studies.

Readers are also advised to carry out following exercise:

Exercise

1. Write a 1000 words essay on "Impact of environmental degradation on human health"
2. Give examples of human created environmental problem and possible measure to tackle the problem.
3. Write a critical review on threats faced by environment we are living in by tremendous increase in population.

Chapter 8

Waste Pollution and Management

8.1 INTRODUCTION

Waste management is the collection, transport, processing, recycling or disposal, managing and monitoring of waste materials. The term usually relates to materials produced by human activity, and is generally undertaken to reduce their effect on health, the environment or aesthetics. Waste management is also carried out to recover resources from it. Waste management can involve solid, liquid, gaseous or radioactive substances, with different methods and fields of expertise for each.

Figures 8.1 and 8.2 waste disposal programme and waste to energy. Solid wastes are generated due to human as well as industrial activities. These wastes are to be collected transported to the site and treated properly. This is very essential as the environments are to be kept clean. Innovation in solid waste industry makes it possible for us to use waste as a source of renewable and sustainable energy. Waste-to-energy facilities burn garbage in power plants equipped with modern pollution control equipment to clean emissions. About 10 per cent of America's garbage is disposed in these plants. According to the Integrated Waste Services Association, as of 2008, America's solid waste industry operated 87 waste-to-energy facilities that generated more than 2700 megawatts of electricity, enough to power 1.7 million homes.

8.2 SOLID WASTE MANAGEMENT: CAUSES AND CONTROL MEASURES

Waste management practices differ for developed and developing nations, for urban and rural areas, and for residential and industrial producers. Management for non-hazardous waste residential and institutional waste in metropolitan areas is usually the responsibility of local government authorities, while management for non-hazardous commercial and industrial waste is usually the responsibility of the generator.

Human and animal activities generate different kinds of wastes. These wastes are generally in solid form, and may cause pollution of land, water, and air unless treated and disposed of. The process of collection, transportation, treatment, and disposal can be grouped under solid waste management.

The main sources for solid wastes are domestic, commercial, industrial, municipal, and agricultural wastes. It will be good to consider some of the example of poor domestic waste management is Uganda. There are several causes, however some of them of discussed here. Lack

Fig. 8.1 Waste-disposal.

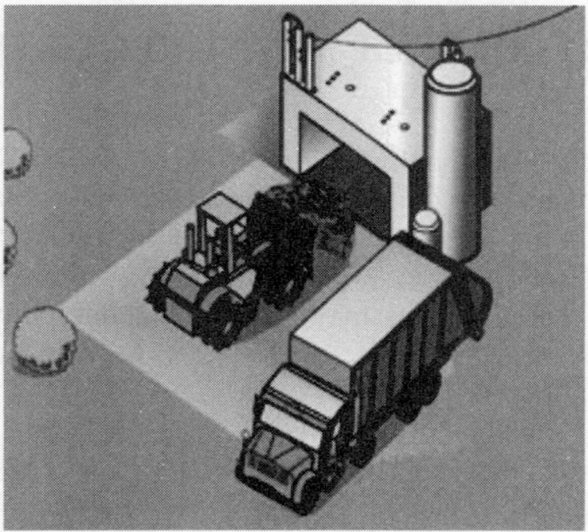

Fig. 8.2 Waste to energy.

of dumping sites where to deposit the solid waste. This is because the issue of waste management is new in the country. It wasn't considered to be a problem before.

Ignorance of the masses about the need to dispose of these wastes well and how to dispose of them (the wastes). There is lack of enough literacy programmes on waste management which leaves most of the people backward on waste management. This is because of poor or no sensitization of the masses by the government and other organizations of Uganda.

Inefficient collection methods which are mainly due to lack of funds to provide the necessary machinery. In Uganda, machinery like the trucks that carry the waste from the various areas have poor covering systems such that even the waste goes on leaking on the road while being transported, and even there are few places with proper garbage containers or at times the containers are over flooded when there are rain showers.

Poor government attitude towards waste management. From a citizen's point of view, it is realized that very little money from the government is directed towards waste management, with most of it going towards industrialization. This leads to poor purchase of collecting equipment.

Another cause is the poverty that exists in Uganda. This undoubtedly leads to masses buying cheap non biodegradable containers which are not easy to dispose of, and also substitutes like paper bags are not easily available to the rural poor areas.

Also the low expense of these solid wastes especially polythene bags which are very cheap as compared to other containers makes them very common, which makes their proper disposal very difficult.

Lack of trained manpower/personnel to deal with garbage collecting machinery and to ensure the proper disposal of the solid waste for example door to door collectors in most advanced countries.

Lack of recycling facilitates in most parts of the country for instance most of the polythene bags used are not recycled by the manufacturing industries. People don't work together as a community for the betterment for the environment. Voluntary organizations such as scouts put little emphasis in such activities.

8.3 EFFECTS AND CONTROL MEASURES OF URBAN AND INDUSTRIAL WASTES

Improper waste management not only affects our day-to-day life but also affects our society as a whole. Also its implications become quite visible on related stakeholder with time. Some of the major effects of wastes pollution and possible control measure are discussed below.

(a) **Health Hazard:** If solid wastes are not collected and allowed to accumulate, they may create unsanitary conditions. This may lead to epidemic outbreaks. Many diseases like cholera, diarrhea, dysentery, plague, jaundice, or gastrointestinal diseases may spread and cause loss of human lives.

In addition, improper handling of the solid wastes is a health hazard for the workers who come in direct contact with the waste.

(b) **Environmental Impact:** If the solid wastes are not treated properly, decomposition and putrefaction may take place, causing land and water pollution when the waste products percolate down into the underground water resources. The organic solid waste during decomposition may generate obnoxious odours. Stray dogs and birds may sometimes invade garbage heaps and may spread it over the neighbourhood causing unhygienic and unhealthy surroundings.

The main purpose of solid waste management is to minimize the adverse effects on the environment. The steps involved to control these effects are:

(a) **Collection of solid wastes:** Collection of wastes includes gathering the waste, transporting it to a centralized location, and then moving it to the site of disposal.

The collected waste is then separated into hazardous and non-hazardous materials. There are many waste separation technologies available such as air stripping, stream stripping, carbon absorption, and precipitation.

(b) Disposal of solid wastes: Before the final disposal of the solid wastes, it is processed to recover the usable resources and to improve the efficiency of the solid waste disposal system. The main processing technologies are compaction, incineration, and manual separation. The appropriate solid waste disposal method has to be selected, keeping in view the following objectives:
- should be economically viable
- should not create a health hazard
- should not cause adverse environmental effects
- should not result in unpleasant sight, odor, and noise

(c) Utilization of wastes: The solid wastes can be properly utilized to reap the benefits such as:
- conservation of natural resources
- economic development
- generate many useful products
- employment opportunities
- control of air pollution

8.4 INTEGRATED WASTE MANAGEMENT

Integrated waste management using LCA (Life Cycle Analysis) attempts to offer the most benign options for waste management. For mixed MSW (Municipal Solid Waste) a number of broad studies have indicated that waste administration, then source separation and collection followed by reuse and recycling of the non-organic fraction and energy and compost/fertilizer production of the organic waste fraction via anaerobic digestion to be the favoured path. There are several methods for integrated waste management and some of them are discussed as follows.

(a) Landfill: Disposing of waste in a landfill involves burying the waste, and this remains a common practice in most countries. Landfills were often established in abandoned or unused quarries, mining voids or borrow pits. A properly designed and well-managed landfill can be a hygienic and relatively inexpensive method of disposing of waste materials. Older, poorly designed or poorly managed landfills can create a number of adverse environmental impacts such as wind-blown litter, attraction of vermin, and generation of liquid leachate. Another common by product of landfills is gas (mostly composed of methane and carbon dioxide), which is produced as organic waste breaks down anaerobically. This gas can create odour problems, kill surface vegetation, and is a greenhouse gas.

Design characteristics of a modern landfill include methods to contain leachate such as clay or plastic lining material. Deposited waste is normally compacted to increase its density and stability, and covered to prevent attracting vermin (such as mice or rats). Many landfills also have landfill gas extraction systems installed to extract the landfill gas. Gas is pumped out of the landfill using perforated pipes and flared off or burnt in a gas engine to generate electricity.

Fig. 8.3 Landfill site.

Fig. 8.4 Incineration plant in Vienna.

Landfills are engineered areas where waste is placed into the land. Landfills usually have liner systems and other safeguards to prevent polluting the groundwater. The placement of solid waste in landfills is the probably the oldest and definitely the most prevalent form of ultimate garbage disposal. From the outset, it must be recognized that many "landfills" are nothing more than open, sometimes controlled, dumps. The difference between landfills and dumps is the level of engineering, planning, and administration involved. Open dumps are characterized by the lack of engineering measures, no leachate management, no consideration of landfill gas management, and few, if any, operational measures such as registration of users, control of the number of "tipping fronts" or compaction of waste. In an examination of landfills throughout the developing world in 1997-1998, Johannessen (1999) found varying amounts of planning and engineering in MSW dumping; among the various regions visited, African nations (with the exception of South Africa) had the fewest engineered landfills, with most nations practising open dumping for waste disposal; waste managers in Asian and Latin American nations were more likely to be aware of environmental effects of improper landfill design and were much more likely to design and implement some control measures, however limited in scope. 'Sanitary' landfills, on the other hand, are sites where waste is allowed to decompose into biologically and chemically inert materials in a setting isolated from the environment. Cointreau (1982) outlined four features that must be present in order for a landfill to be considered sanitary:

- Full or partial hydrogeological isolation through the use of liners to prevent leachate infiltration into the soil and groundwater; collection and treatment infrastructure should be used where leachate is expected to be generated.
- Formal engineering preparations with an examination of geological and hydrological features and related environmental impact analysis, waste tipping plan and final site restoration plan.
- Permanent control, with trained and equipped staff to supervise construction etc.
- Planned waste emplacement and covering, with waste and soil placed in compacted layers as well as daily and final soil cover to reduce water infiltration and reduce odours and pests. Other practical and social considerations must be addressed when planning landfills, especially in the context of developing nations and their problems as outlined above. One of the most important is the siting of landfills in proximity to urban areas. Naturally, there are few people who would be excited by having a landfill in their backyard; however it is important to realize that landfills must be located within a reasonable distance to population concentrations along a good road system. If they are located too far from collection points and transfer stations, waste transport could become prohibitively expensive due to the distance the waste is transported. In the poorest of nations, development agencies may finance all or part of landfill construction. However, if it is sited far from the urban area, without regard to the ability of the government to transport waste with its limited finances, scattered unregulated dumping will become financially attractive once again.

(b) **Incineration:** Incineration is a disposal method in which solid organic wastes are subjected to combustion so as to convert them into residue and gaseous products. This method is useful for disposal of residue of both solid waste management and solid residue from waste water management. This process reduces the volumes of solid waste to 20 to 30 per cent of the original volume. Incineration and other high temperature waste treatment systems are sometimes described as "thermal treatment". Incinerators convert waste materials into heat, gas, steam and ash.

Incineration is carried out both on a small scale by individuals and on a large scale by industry. It is used to dispose of solid, liquid and gaseous waste. It is recognized as a practical method of disposing of certain hazardous waste materials (such as biological medical waste). Incineration is a controversial method of waste disposal, due to issues such as emission of gaseous pollutants.

Incineration is common in countries such as Japan where land is scarce, as these facilities generally do not require as much area as landfills. Waste-to-energy (WtE) or energy-from-waste (EfW) are broad terms for facilities that burn waste in a furnace or boiler to generate heat, steam or electricity. Combustion in an incinerator is not always perfect and there have been concerns about pollutants in gaseous emissions from incinerator stacks. Particular concern has focused on some very persistent organics such as dioxins, furans, PAHs which may be created which may have serious environmental consequences.

(c) **Recycling:** Recycling refers to the collection and reuse of waste materials such as empty beverage containers. The materials from which the items are made can be reprocessed into new products. Material for recycling may be collected separately from general waste using dedicated bins and collection vehicles, or sorted directly from mixed waste streams.

The most common consumer products recycled include aluminium such as beverage cans, copper such as wire, steel food and aerosol cans, old steel furnishings or equipment, polyethylene and PET bottles, glass bottles and jars, paperboard cartons, newspapers, magazines and light paper, and corrugated fiberboard boxes.

PVC, LDPE, PP, and PS are usually composed of a single type of material, making them relatively easy to recycle into new products. The recycling of complex products (such as computers and electronic equipment) is more difficult, due to the additional dismantling and separation required.

The type of recycling material accepted varies by city and country. Each city and country have different recycling programmes in place that can handle the various types of recyclable materials.

(d) **Biological reprocessing:** Waste materials that are organic in nature, such as plant material, food scraps, and paper products, can be recycled using biological composting and digestion processes to decompose the organic matter. The resulting organic material is then recycled as mulch or compost for agricultural or landscaping purposes. In addition, waste gas from the process (such as methane) can be captured and used for generating electricity and heat (CHP/cogeneration) maximising efficiencies. The intention of biological processing in waste management is to control and accelerate the natural process of decomposition of organic matter.

Fig. 8.5 An active compost heap.

There is a large variety of composting and digestion methods and technologies varying in complexity from simple home compost heaps, to small town scale batch digesters, industrial-scale enclosed-vessel digestion of mixed domestic waste. Methods of biological decomposition are differentiated as being aerobic or anaerobic methods, though hybrids of the two methods also exist.

Anaerobic digestion of the organic fraction of MSW Municipal Solid Waste has been found to be in a number of LCA analysis studies to be more environmentally effective, than landfill, incineration or pyrolysis. The resulting biogas (methane) though must be used for cogeneration (electricity and heat preferably on or close to the site of production) and can be used with a little upgrading in gas combustion engines or turbines. With further upgrading to synthetic natural gas it can be injected into the natural gas network or further refined to hydrogen for use in stationary cogeneration fuel cells. Its use in fuel cells eliminates the pollution from products of combustion.

An example of waste management through composting is the Green Bin Program in Toronto, Canada, where source separated organics (such as kitchen scraps and plant cuttings) are collected in a dedicated container and then composted.

(e) **Waste-to-energy:** The energy content of waste products can be harnessed directly by using them as a direct combustion fuel, or indirectly by processing them into another type of fuel. Recycling through thermal treatment ranges from using waste as a fuel source for cooking or heating, to anaerobic digestion and the use of the gas fuel, to fuel for boilers to generate steam and electricity in a turbine. Pyrolysis and gasification are two related forms of thermal treatment where waste materials are heated to high temperatures with limited oxygen availability. The process usually occurs in a sealed vessel under high pressure. Pyrolysis of solid waste converts the material into solid, liquid and gas products. The liquid and gas can be burnt to produce energy or refined into other chemical products (chemical refinery). The solid residue (char) can be further refined into products such as activated carbon. Gasification and advanced plasma are gasification are used to convert organic materials directly into a synthetic gas (syngas) composed of carbon monoxide and hydrogen. The gas is then burnt to produce electricity and steam. An alternative to pyrolisis is high temperature and pressure supercritical water decomposition (hydrothermal monophasic oxidation).

(f) **Waste minimization:** An important method of waste management is the prevention of waste material being created, also known as waste reduction. Methods of avoidance include reuse of second-hand products, repairing broken items instead of buying new, designing products to be refillable or reusable (such as cotton instead of plastic bags), encouraging consumers to avoid using disposable products (such as disposable cutlery), removing any food/liquid remains from cans, packaging, and products that use less material to achieve the same purpose (for example, light-weigh of beverage cans).

(g) **Waste collection vehicle**, **dustbin**, and **waste sorting:** Waste collection methods vary widely among different countries and regions. Domestic waste collection services are often provided by local government authorities, or by private companies in the industry. Some areas, especially those in less developed countries, do not have a formal waste-collection system.

Examples of waste handling systems include:

In Australia, curbside collection is the method of disposal of waste. Every urban domestic household is provided with three bins: one for recyclables, another for general waste and another for garden materials — these bins are provided by the municipality if requested. Also, many households have compost bins; but this is not provided by the municipality. To encourage recycling, municipalities provide large recycle bins, which are larger than general waste bins. Municipal, commercial and industrial, construction and demolition waste is dumped at landfills and some is recycled. Household waste is segregated: recyclables sorted and made into new products, and general waste is dumped in landfill areas. This suggests that Australians are in favour of reduced or no landfilling and the recycling of waste. Of the total waste produced in 2002–03, '30% of municipal waste, 45% of commercial and industrial waste and 57% of construction and demolition waste' was recycled. Energy is produced from waste as well: some landfill gas is captured for fuel or electricity generation. Households and industries are not charged for the volume of waste they produce.

In Europe and a few other places around the world, a few communities use a proprietary collection system known as Envac, which conveys refuse via underground conduits using a vacuum system. Other vacuum-based solutions include the Metro Taifun single-line and ring-line systems.

In Canadian urban centres curbside collection is the most common method of disposal, whereby the city collects waste and/or recyclables and/or organics on a scheduled basis. In rural areas people often dispose of their waste by hauling it to a transfer station. Waste collected is then transported to a regional landfill.

In Taipei, the city government charges its households and industries for the volume of rubbish they produce. Waste will only be collected by the city council if waste is disposed of in government issued rubbish bags. This policy has successfully reduced the amount of waste the city produces and increased the recycling rate.

1In Israel, the Arrow Ecology company has developed the ArrowBio system, which takes trash directly from collection trucks and separates organic and inorganic materials through gravitational settling, screening, and hydro-mechanical shredding. The system is capable of sorting huge volumes of solid waste, salvaging recyclables, and turning the rest into biogas and rich agricultural compost. The system is used in California, Australia, Greece, Mexico, the United Kingdom and in Israel. For example, an Arrow Bioplant that has been operational at the Hiriya landfill site since December 2003 serves the Tel Aviv area, and processes up to 150 tonnes of garbage a day.

8.5 TECHNOLOGIES

Traditionally, the waste management industry has been slow to adopt new technologies such as RFID (Radio Frequency Identification) tags, GPS and integrated software packages which enable better quality data to be collected without the use of estimation or manual data entry.

Technologies like RFID tags are now used to collect data on presentation rates for curb-side pick-ups which is useful when examining the usage of recycling bins or similar.

Benefits of GPS tracking is particularly evident when considering the efficiency of ad hoc pick-ups (like skip bins or dumpsters) where the collection is done on a consumer request basis.

Integrated software packages are useful in aggregating this data for use in optimisation of operations for waste collection operations.

Rear vision cameras are commonly used for OH&S reasons and video recording devices are becoming more widely used, particularly concerning residential services and contaminations of the waste stream.

8.6 WASTE MANAGEMENT CONCEPTS

There are a number of concepts about waste management which vary in their usage between countries or regions. Some of the most general, widely used concepts include:

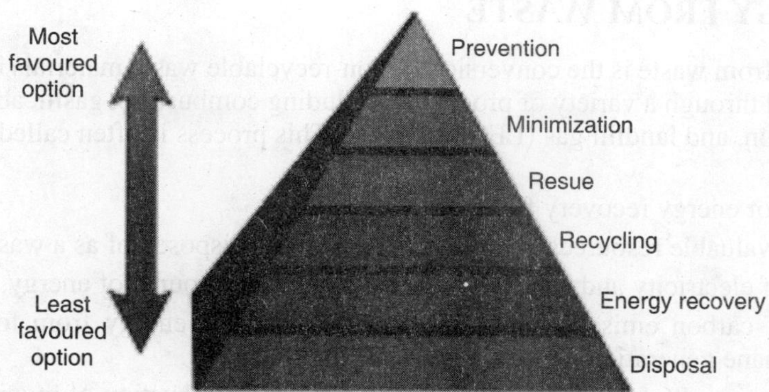

Fig. 8.6 Diagram of the waste hierarchy.

(a) **Waste hierarchy:** The waste hierarchy refers to the "3 Rs" — reduce, reuse and recycle, which classify waste management strategies according to their desirability in terms of waste minimization. The waste hierarchy remains the cornerstone of most waste minimization strategies. The aim of the waste hierarchy is to extract the maximum practical benefits from products and to generate the minimum amount of waste.

(b) **Extended producer responsibility:** Extended producer responsibility (EPR) is a strategy designed to promote the integration of all costs associated with products throughout their life cycle (including end-of-life disposal costs) into the market price of the product. Extended producer responsibility is meant to impose accountability over the entire lifecycle of products and packaging introduced to the market. This means that firms which manufacture, import and/or sell products are required to be responsible for the products after their useful life as well as during manufacture.

(c) **Polluter pays principle:** The polluter pays principle is a principle where the polluting party pays for the impact caused to the environment. With respect to waste management, this generally refers to the requirement for a waste generator to pay for appropriate disposal of the waste.

(d) **Education and awareness:** Education and awareness in the area of waste and waste management is increasingly important from a global perspective of resource management. The Talloires Declaration is a declaration for sustainability concerned about the unprecedented scale and speed of environmental pollution and degradation, and the depletion of natural resources. Local, regional, and global air pollution; accumulation and distribution of toxic wastes; destruction and depletion of forests, soil, and water; depletion of the ozone layer and emission of "green house" gases threaten the survival of humans and thousands of other living species, the integrity of the earth and its biodiversity, the security of nations, and the heritage of future generations. Several universities have implemented the Talloires Declaration by establishing environmental management and waste management programmes, e.g. the waste management university project. University and vocational education are promoted by various organizations, e.g., WAMITAB and Chartered Institution of Wastes Management. Many supermarkets encourage customers to use their reverse vending machines to deposit used purchased containers and receive a refund from the recycling fees. Brands that manufacture such machines include Tomra and Envipco.

8.7 ENERGY FROM WASTE

Energy recovery from waste is the conversion of non-recyclable waste materials into useable heat, electricity, or fuel through a variety of processes, including combustion, gasification, pyrolization, anaerobic digestion, and landfill gas (LFG) recovery. This process is often called waste-to-energy (WTE).

The benefits of energy recovery from waste include:

Recovery of valuable resources that might otherwise be disposed of as a waste.

Generation of electricity and heat from waste, a renewable source of energy.

Reduction of carbon emissions by offsetting the need for energy from fossil sources and reduction of methane generation from landfills.

Currently there are 86 facilities in the United States for combustion of municipal solid waste (MSW), with energy recovery. These facilities are located in 25 states, mainly in the Northeast. No new plants have been built in the US since 1995, but some plants have expanded to handle additional waste and create more energy. The 86 facilities have the capacity to produce 2,720 megawatts of power per year by processing more than 28 million tonnes of waste per year. According to Municipal Solid Waste in the US: Facts and Figures, in 2008 we combusted about 32 million tons of MSW (about 13 per cent) for energy recovery. After energy is recovered, approximately ten percent of the volume remains as ash. This ash is generally sent to a landfill. Please visit EPA's Landfill Methane Outreach Program (LMOP) for additional information on how energy is recovered from landfills.

(a) **About LMOP:** The U.S. EPA's Landfill Methane Outreach Program (LMOP) is a voluntary assistance program that helps to reduce methane emissions from landfills by encouraging the recovery and use of landfill gas (LFG) as an energy resource. LMOP forms partnerships with communities, landfill owners, utilities, power marketers, states, project developers, tribes, and non-profit organizations to overcome barriers to project development by helping them assess project feasibility, find financing, and market the benefits of project development to the community. EPA launched LMOP to encourage productive use of this resource as part of the United States' commitment to reduce greenhouse gas emissions under the United Nations Framework Convention on Climate Change.

(b) **Methane Emissions From Landfills:** Municipal solid waste (MSW) landfills are the third largest source of human-related methane emissions in the United States, accounting for approximately 17 per cent of these emissions in 2009. At the same time, methane emissions from landfills represent a lost opportunity to capture and use a significant energy resource. LFG is created as solid waste decomposes in a landfill. This gas consists of about 50 per cent methane (the primary component of natural gas), about 50 per cent carbon dioxide (CO_2), and a small amount of non-methane organic compounds. To learn more about methane emissions from landfills in the United States, visit EPA's methane emissions and sources page. For more information on methane emissions from landfills internationally, visit EPA's International Analyses site. Additionally, EPA issued a rule (40 CFR Part 98) in October 2009 that requires the reporting of greenhouse gases. The rule requires reporting of greenhouse gas emissions from large sources and suppliers in the United States, and is intended to collect accurate and timely emissions data to inform future policy decisions. More information on the rule may be found on the Greenhouse Gas Reporting Program page.

Instead of escaping into the air, landfill gas (LFG) can be captured, converted, and used as an energy source. Using LFG helps to reduce odours and other hazards associated with LFG emissions, and it helps prevent methane from migrating into the atmosphere and contributing to local smog and global climate change.

LFG is extracted from landfills using a series of wells and a blower/flare (or vacuum) system. This system directs the collected gas to a central point where it can be processed and treated depending upon the ultimate use for the gas. From this point, the gas can be flared, used to generate electricity, replace fossil fuels in industrial and manufacturing operations, or upgraded to pipeline–quality gas where the gas may be used directly or processed into an alternative vehicle fuel.

There are various components of an LFG collection system, e.g., LFG well, LFG wellhead, LFG wells and collection piping, LFG system, LFG treatment/blower/flare station and LFG flare. There are several options for converting LFG to energy. Below are descriptions of some of the typical project types.

(i) Electricity Generation: The generation of electricity from LFG makes up about two-thirds of the currently operational projects in the United States. Electricity for on-site use or sale to the grid can be generated using a variety of different technologies, including internal combustion engines, turbines, microturbines, and fuel cells. The vast majority of projects use internal combustion (reciprocating) engines or turbines, with micro turbine technology being used at smaller landfills and in niche applications. Technologies such as Stirling and organic Rankine cycle engines and fuel cells are still in development.

(ii) Direct Use: Directly using LFG to offset the use of another fuel (e.g., natural gas, coal, fuel oil) is occurring in about one-third of the currently operational projects. This direct use of LFG can be in a boiler, dryer, kiln, greenhouse, or other thermal application. LFG can also be used directly to evaporate leachate. Innovative direct uses include firing pottery and glass-blowing kilns; powering and heating greenhouses and an ice rink; and heating water for an aquaculture (fish farming) operation. Current industries using LFG include auto manufacturing, chemical production, food processing, pharmaceuticals, cement and brick manufacturing, wastewater treatment, consumer electronics and products, paper and steel production, and prisons and hospitals, just to name a few.

(iii) Cogeneration: Cogeneration (also known as combined heat and power or CHP) projects using LFG generate both electricity and thermal energy, usually in the form of steam or hot water. Several cogeneration projects have been installed at industrial operations, using both engines and turbines. The efficiency gains of capturing the thermal energy in addition to electricity generation can make these projects very attractive.

Graphical representation of the planned SC Johnson cogeneration project in Racine, Wisconsin

(iv) Alternate Fuels: Production of alternate fuels from LFG is an emerging area. LFG has been successfully delivered to the natural gas pipeline system as both a high–Btu and medium–Btu fuel. LFG has also been converted to vehicle fuel in the form of compressed natural gas and liquefied natural gas. Projects to convert LFG to methanol are in the planning stages.

(v) Benefits of Landfill Gas Energy Using LFG for energy is a win-win opportunity. LFG utilization projects involve citizens, non-profit organizations, local governments, and industry in sustainable community planning and create partnerships. These projects go hand-in-hand with community and corporate commitments to cleaner air, renewable energy, economic development, improved public welfare and safety, and reduction in greenhouse (global warming) gases. By linking communities with innovative ways to deal with their LFG, LMOP contributes to the creation of livable communities that enjoy increased environmental protection, better waste management, and responsible community planning.

It directly reduces greenhouse gas emissions. MSW landfills are the third largest human-generated source of methane emissions in the United States, releasing an estimated 27.5 million metric tonnes of carbon equivalent to the atmosphere in 2009 alone. Given that all landfills generate methane, it makes sense to use the gas for the beneficial purpose of energy generation rather than emitting it to the atmosphere. Methane is a very potent greenhouse gas that is a key contributor to global climate change (over 21 times stronger than CO_2). Methane also has a short (10-year) atmospheric life. Because methane is both potent and short-lived, reducing methane emissions from MSW landfills is one of the best ways to achieve a near-term beneficial impact in mitigating global climate change.

It is estimated that an LFG energy project will capture roughly 60 to 90 per cent of the methane emitted from the landfill, depending on system design and effectiveness. The captured methane is destroyed (converted to water and the much less potent CO_2) when the gas is burnt to produce electricity. The LFGE Benefits Calculator can be used to estimate greenhouse gas reductions from LFG recovery projects.

CO_2 emissions from MSW landfills are not considered to contribute to global climate change because carbon was contained in recently living biomass. The same CO_2 would be emitted as a result of the natural decomposition of the organic waste materials outside the landfill environment. It indirectly reduces air pollution by offsetting the use of non-renewable resources.Producing energy from LFG avoids the need to use non-renewable resources such as coal, oil, or natural gas to produce the same amount of energy. This can avoid gas end-user and power plant emissions of CO_2 and criteria pollutants such as sulfur dioxide (which is a major contributor to acid rain), particulate matter (a respiratory health concern), nitrogen oxides (NO_X), and trace hazardous air pollutants.

Note that—like all combustion devices—LFG electricity generation devices emit some NO_X, which can contribute to local ozone and smog formation. Depending on the fuels and technologies used by the power plant and the landfill project, the NO_X emission reductions from the power plant may not completely offset the NO_X emitted from the LFG electricity project. Overall, however, LFG electricity generation projects significantly improve the environment, because of the large methane reductions, hazardous air pollutant reductions, and avoidance of the use of limited non–renewable resources such as coal and oil that are more polluting than LFG.

Collecting LFG to produce electricity improves the air quality of the surrounding community by reducing landfill odours. Burning LFG to produce electricity also destroys most of the non–methane organic compounds that are present at low concentrations in uncontrolled LFG, thereby reducing possible health risks from these compounds. Gas collection can also improve safety by reducing explosion hazards from gas accumulation in structures on or near the landfill. Generating electricity from existing MSW landfills is also a relatively cost-effective way to provide new renewable energy generation capacity to supply community power needs.

It benefits the local economy. LFG energy projects generate revenue from the sale of the gas. LFG use can also create jobs associated with the design, construction, and operation of energy recovery systems. LFG energy projects involve engineers, construction firms, equipment vendors, and utilities or end-users of the power produced. Much of this cost is spent locally for drilling, piping, construction, and operational personnel, helping communities to realize economic benefits from increased employment and local sales. Businesses are also realizing the cost savings associated with using LFG as a replacement for more expensive fossil fuels, such as natural gas. Some companies will save millions of dollars over the life of their LFG energy projects. By linking communities with innovative ways to deal with their LFG, LMOP helps communities enjoy increased environmental protection, better waste management, and responsible community planning. For example, the Ecology Club at Pattonville High School in Maryland Heights, Missouri, came up with the idea to use gas from the nearby landfill to heat their school. The school paid $175,000 to run a 3,600-foot pipeline between the landfill and the school's two basement boilers. In turn, the landfill owner donated the methane to the school as a way of "giving back to the community." The school anticipates that it will save $40,000 a year and recapture its investment within five years.

It reduces environmental compliance costs. Current EPA regulations under the Clean Air Act require many larger landfills to collect and combust LFG. There are several compliance options, including flaring the gas or installing an LFG use system. Only LFG energy recovery gives communities and landfill owners the opportunity to reduce the costs associated with regulatory compliance by turning pollution into a valuable community resource.

8.8 INTRODUCTION TO THE RESOURCE CONSERVATION AND RECOVERY ACT (RCRA)

The Resource Conservation and Recovery Act (RCRA), an amendment to the Solid Waste Disposal Act, was enacted in 1976 to address the huge volumes of municipal and industrial solid waste generated nationwide.

The goals set by RCRA are:

- To protect human health and the environment from the potential hazards of waste disposal
- To conserve energy and natural resources
- To reduce the amount of waste generated
- To ensure that wastes are managed in an environmentally sound manner.

RCRA also regulates underground storage tanks (USTs) that store petroleum or certain chemical products.

Requirements exist for the design and operation of these tanks and the development of systems to prevent accidental spills. Examples of facilities using these tanks include petroleum refineries, chemical plants, and commercial gas stations .The Medical Waste Tracking Act of 1988 was a 2-year demonstration program that expired in June1991. It created a Subtitle J programme designed to track medical waste from generation to disposal. At present, no federal EPA tracking regulations are in effect for medical waste, but many states have adopted their own programmes. The Comprehensive Environmental Response, Compensation, and Liability Act (known as Superfund or CERCLA) is a related statute that deals with cleaning up inactive and abandoned hazardous waste sites. RCRA, on the other hand, deals with materials that are currently destined for disposal or recycling.

(i) RCRA: The term RCRA is often used interchangeably to refer to the law, regulations, and EPA policy and guidance. The law describes the waste management programme gave EPA authority to develop the RCRA programme. EPA regulations carry out the Congressional intent by providing explicit, legally enforceable requirements for waste management. These regulations can be found in Title 40 of the Code of Federal Regulations (CFR), Parts 239 through 282. EPA guidance documents and policy directives clarify issues related to the implementation of the regulations. These three elements are the primary parts of the RCRA program.

(ii) Legislative History: In 1965 the **Solid Waste Disposal Act** [Public Law (Pub. L.) 89-72] was enacted to improve solid waste disposal methods. It was amended in 1970 by the **Resource Recovery Act** (Pub. L. 91-512), which provided the Environmental Protection Agency (EPA) with funding for resource recovery programmes. However, that Act had little impact on the management and ultimate disposal of hazardous waste. In 1976 Congress enacted the **Resource Conservation and Recovery Act** (RCRA, Pub. L. 94-580). RCRA established a system for managing non-hazardous and hazardous solid wastes in an environmentally sound manner. Specifically, it provides for the management of hazardous wastes from the point of origin to the point of final disposal (i.e., "cradle to grave"). RCRA also promotes resource recovery and waste minimization.

The **Hazardous and Solid Waste Amendments of 1984** (HSWA, Pub. L. 98-616) both expanded the scope and increased the requirements of RCRA. HSWA addressed congressional concern about the adequacy of existing requirements to prevent uncontrolled releases of hazardous constituents or hazardous wastes from hazardous waste management units. Three of the HSWA initiatives were especially noteworthy in preventing or addressing hazardous waste/constituent releases.

First, Congress directed EPA to develop what is now known as the Land Disposal Restrictions (LDR) Program [Section 3004(d) of RCRA]. Under the LDR Program the land disposal of **untreated** wastes is prohibited. Generally speaking, the LDR Program consists of three elements: the disposal prohibition, the dilution prohibition, and the storage prohibition. The most prominent aspect of the LDR Program is the disposal prohibition, which includes treatment standards, variances, and notification requirements. The EPA must establish treatment standards for all listed and characteristic hazardous wastes destined for land disposal. Treatment standards are either concentration levels for hazardous constituents or treatment technologies that must be applied to the waste in order to substantially diminish the toxicity of wastes and/or reduce the likelihood that wastes will migrate from the disposal site.

Second, facilities were required to satisfy minimum technology requirements (i.e., liners and leachate collection systems) for surface impoundments, waste piles, land treatment units, and landfills (40 *CFR*) 264/265, Subparts K-N) to prevent hazardous wastes and/or constituents from migrating into the ground water and to allow releases to be detected when they occur [Section 3004(o)].

Third, when a facility seeks a RCRA permit, EPA was granted the authority to require corrective action for releases of hazardous waste and hazardous constituents from any solid waste management unit, regardless of when the waste was placed in the unit [Sections 3004(u) and 3004(v)].

The **Medical Waste Tracking Act of 1988** (Pub. L. 100-582) amended RCRA in response to public concern over the degradation of shoreline areas, particularly in Connecticut, New Jersey, and New York, due to the washups of medical debris (e.g., needles, syringes, blood bags, bandages, vials). Among other things, this Act required EPA to establish a two-year demonstration program for tracking medical wastes generated in specified states. On March 24, 1989, EPA issued an interim final rule entitled *Standards for the Tracking and Management of Medical Waste* [54 *Federal Register* (*FR*) 12326].

The **Federal Facility Compliance Act** (FFCAct, Pub. L. 102-386) was enacted on October 6, 1992, and, among other things, specifically waived sovereign immunity with respect to RCRA for federal facilities. Therefore, EPA may impose fines and penalties on federal agencies.

On March 26, 1996, RCRA was amended by the **Land Disposal Program Flexibility Act of 1996** (Pub. L. 104-119).

(iii) Purpose and Organization: RCRA is designed to protect human health and the environment reduce or eliminate the generation of hazardous wastes, and conserve energy and natural resources.

RCRA, as amended, contains ten subtitles. Subtitle C, *Hazardous Waste Management;* Subtitle D, *State or Regional Solid Waste Plans;* Subtitle I, *Regulation of Underground Storage Tanks;* and Subtitle J, *Demonstration Medical Waste Tracking Program;* constitute the regulatory portion of the law. The other subtitles provide the legal and administrative structure for achieving the objectives of the law.

EPA, the Department of Commerce (DOC), DOE, and the Department of the Interior (DOI) each have specific responsibilities under RCRA. EPA issues guidelines and regulations for proper management of solid and hazardous wastes, oversees and approves the development of state waste management plans, and provides financial aid to agencies and firms for research on solid waste. DOC encourages greater commercialization of proven resource recovery technologies. DOE oversees activities involving research and development of new techniques for producing energy from wastes. DOI oversees mineral waste problems, including recovery of metals and minerals and methods for stabilizing mining wastes.

(iv) Materials Regulated under RCRA: Under RCRA no material can be a hazardous waste unless it is a solid waste. RCRA defines a solid waste as any garbage, refuse, sludge from a waste treatment plant, water supply treatment plant, or air pollution control facility and other discarded material, including solid, liquid, semisolid, or contained gaseous material resulting from industrial, commercial or mining and agricultural operations, and from community activities, excluding solid or dissolved materials in domestic sewage, or solid or dissolved materials in irrigation return flows, or industrial discharges. Which are point sources subject to permits under Section 402 of the Federal Water Pollution Control Act, special nuclear, or by product material as defined by the Atomic Energy Act [AEA] of 1954 [Section 1004(27)].

Several categories of waste important to DOE are excluded from the RCRA definition of solid wastes (40 *CFR* 261.2). They include source, special nuclear, or by product material as defined by the AEA [Section 11(e), (z), (aa)]; waste from extraction, beneficiation, and processing of ores and minerals, including overburden from mining uranium ores; utility wastes; oil and gas drilling muds and brines; and some wastes that are reused or recycled.

The statutory definition of a hazardous waste is provided in RCRA as follows: a solid waste, or combination of solid wastes, which because of its quantity, concentration, or physical, chemical, or infectious characteristics may: (A) cause, or significantly contribute to an increase in mortality or an increase in serious irreversible, or incapacitating reversible, illness; or (B) pose a substantial present or potential hazard to human health or the environment when improperly treated, stored, transported, or disposed of, or otherwise managed. [Section 1004(5)]

Furthermore, a solid waste is a hazardous waste if it is not excluded by regulation (40 *CFR* 261.4) and if it is listed (261.30) as a hazardous waste, is a waste mixture containing one or more listed hazardous wastes, or exhibits one or more characteristics of hazardous waste (i.e., ignitability, corrosivity, reactivity, or toxicity) (40 *CFR* 261.21 to 261.24). Listed wastes meet the definition of hazardous waste regardless of the concentration level of hazardous constituents in them. With few exceptions [e.g., spent solvents listed solely because they are ignitable (40 *CFR* 261.31)], the only way to have a listed waste relieved from hazardous waste management requirements is to petition EPA or a state to delist the waste (40 *CFR* 260.22). When listed wastes are mixed with non-hazardous wastes or materials, the mixture must be managed as hazardous waste. Two exceptions to this approach are hazardous debris meeting Land Disposal Restriction (LDR) standards [40 *CFR* 261.3(f)] and residues from processing certain wastes using high temperature metals recovery processing [40 *CFR* 261.3(c)(2)(ii)(C)].

In contrast to listed waste, a characteristic waste remains hazardous only as long as it exhibits a hazardous characteristic. Therefore, a mixture of a waste exhibiting a hazardous waste characteristic and a non-hazardous solid waste is not considered hazardous waste unless the mixture exhibits a hazardous waste characteristic.

(v) Hazardous Waste Generators: Proper identification and management of hazardous wastes are critical to the success of the "cradle-to-grave" programme. Waste generators must determine (40 *CFR* 262.11) if their solid wastes are also hazardous wastes. If the wastes are hazardous, the generator must notify EPA, or an authorized state if appropriate, of their hazardous waste management activities and obtain an EPA identification (ID) number (40 *CFR* 262.12). Generators must manage hazardous wastes in accordance with RCRA generator requirements (40 *CFR* 262) and, if wastes are accumulated for more than 90 days, obtain a RCRA permit [40 *CFR* 262.34(a)] for waste management activities. Generators must also verify that transportation, treatment, storage, and disposal of their wastes are conducted by others with EPA identification numbers who are authorized to manage the wastes.

However, certain waste management units do not require a RCRA permit because their management activities are exempt from permitting. Examples of exempted activities are on-site accumulation of hazardous waste for less than 90 days, totally enclosed treatment, wastewater treatment that only involves tanks that discharge through surface water discharges regulated under the Clean Water Act, and elementary neutralization [40 *CFR* 270.1(c)(2)].

Before a hazardous waste may be shipped off-site, the generator must determine: whether the waste is a hazardous waste (40 *CFR* 262.11); the proper packaging for the hazardous waste (40 *CFR* 262.30); the necessary RCRA and Department of Transportation (DOT) labelling , marking (40 *CFR*262.30 to 262.33); the information necessary to complete and sign the hazardous waste manifest (which includes both DOT and RCRA shipping paper information) for the waste shipment (40 *CFR* 262.20-262.23).

Regulations specifying the documentation required when hazardous waste is to be shipped may be found in 40 *CFR* 262, Subpart B, *The Manifest*. The manifest must include the following information (Appendix to Part 262): name and ID number of generator and facility to which the waste is being sent; DOT description of waste being transported; quantities of waste being transported; and address of the off-site treatment, storage, and disposal facility (TSDF) designated to receive the waste.

The purpose of the manifesting system is to establish accountability for and tracking of hazardous waste shipments. The manifest is an important feature of RCRA's "cradle to grave" system. A generator must keep a copy of each manifest for three years. The generator must also maintain records and report hazardous waste management activity, including the amount of hazardous waste produced, the transporters of the wastes, and the TSDFs in possession of the hazardous waste (40 *CFR* 262.40 to 262.44).

(vi) **Hazardous Waste Transporters:** Regulations governing the transportation of hazardous waste may be found in 40 *CFR* 263. In developing its regulations EPA adopted by reference most of DOT's regulations implementing the Hazardous Materials Transportation Act (HMTA) for the safe transportation of hazardous wastes (49 *CFR* 171 through 179). Because EPA adopted by reference rather than incorporating the specific language of DOT's HMTA regulations, a transporter must comply with both EPA's RCRA and DOT's HMTA regulations to ensure compliance when transporting hazardous waste.

Anyone who transports a hazardous waste off-site via air, rail, highway, or water is subject to the RCRA transporter requirements. However, 40 *CFR* 262.20(f) exempts transport of hazardous wastes within or along the border of contiguous properties under the control of the same person, even if the contiguous properties are divided by a public or private right-of-way. In addition, conditionally exempt small quantity generators are excluded in specified situations [40 *CFR*262.20(e)].

(vii) **Treatment, Storage, and Disposal Permits:** RCRA requires an owner or operator of a TSDF to obtain a permit to operate. Congress, recognizing that a permitting program would take time to implement, established interim status as a mechanism by which existing hazardous waste management facilities could continue operation before obtaining a RCRA permit. Owners and operators of such hazardous waste management facilities had to notify EPA that the facility existed and to file a preliminary (Part A) permit application. Interim status is also available to facilities that first become subject to RCRA permitting requirements because of regulatory changes, such as expansion of the list of solid wastes regulated as hazardous waste. [Note: A facility owner or operator who has fully complied with the requirements for interim status—as defined in section 3005(e) of RCRA and regulated under 40 *CFR* 270.70—must comply with the regulations specified in 40 *CFR* 265 in lieu of the regulations in 40*CFR* 264, until final administrative disposition of his permit application is made.]

Section 3005(c) of HSWA established deadlines for permitting interim-status facilities. Interim-status facilities were required to either file the Part B portion of the permit application by the applicable statutory deadline or file a closure plan and close by November 8, 1992. Interim status ends when a final determination is made to grant or deny the RCRA permit or when the facility closes (for facilities that closed rather than seek a RCRA permit). Owners/operators of facilities denied a permit may appeal the decision under 40 *CFR* 124.19 or an equivalent state appeal processes. New facilities are ineligible for interim status and must receive a permit before construction can begin. Thus, new facilities seeking a permit submit the Part A and Part B portions of the permit application at the same time.

Facilities operating under interim status or permits must comply with the general requirements applicable to all types of facilities and also with waste management-unit specific requirements. General requirements include providing security (40*CFR* 264.14 or 265.14), planning for emergencies (40 *CFR* 264 or 265 Subpart D), training personnel (40 *CFR* 264.16 or 265.16), properly characterizing waste to be managed in the facility (40 *CFR* 264.13 or 265.13), ground water monitoring (as appropriate) (40 *CFR* 264 or 265 Subpart F), planning for eventual closure of the facility (40 *CFR* 264 or 265 Subpart G), and proper recordkeeping and reporting [40 *CFR* 264.11 or 265.11, 264.15(d) or 265.15(d), 264.16(e) or 265.16(e), 264.56(j) or 265.56(j), 264.71 through 264.77 or 265.71 through 265.77, and 265.94].

Criteria for the design, maintenance, closure, and record-keeping requirements for waste management units are specific for each type of unit. For example, tanks must meet special tank design and operation requirements (40 *CFR* 264 or 265 Subpart J). Containers used for storing hazardous wastes must satisfy requirements of 40 *CFR* 264 or 265 Subpart I. Unit-specific requirements that may apply to DOE activities are specified for, among other things, containment buildings (40*CFR* 264 or 265 Subpart DD), waste piles (40 *CFR* 264 or 265 Subpart L), incinerators (40 *CFR* 264 or 265 Subpart O), surface impoundments (40 *CFR* 264 or 265 Subpart K), land treatment units (40 *CFR* 264 or 265 Subpart M), landfills (40*CFR* 264 or 265 Subpart N), and miscellaneous units (40 *CFR* 264 Subpart X) not otherwise specified in the regulations.

RCRA Subtitle D

Subtitle D of RCRA is titled *State or Regional Solid Waste Plans* and is implemented at 40 *CFR* 257 and 258:

40 *CFR* 257 focuses on state and local governments as the primary planning, regulating, and implementing entities for the management of non-hazardous solid waste (e.g., household and non-hazardous industrial wastes).

40 *CFR* 258 establishes minimum national criteria for all municipal solid waste landfill (MSWLF) units. It also addresses location restrictions (Subpart B), Operating Criteria (Subpart C), Design Criteria (Subpart D), Ground-Water Monitoring and Corrective Action (Subpart E), and Closure and Post-Closure Care (Subpart F) for municipal solid waste landfills.

Underground Storage Tanks

Subtitle I (implemented at 40 *CFR* 280) was added by HSWA. It established a programme to regulate the three to five million underground storage tanks (USTs) in the United States to prevent their leaking. Under this subtitle RCRA regulates the storage of a product (e.g., petroleum products, hazardous substances) rather than hazardous waste. Hazardous substances regulated under Subtitle I include all the hazardous substances (except those regulated as a hazardous waste under Subtitle C of RCRA) defined under the Comprehensive Environmental Response, Compensation, and Liability Act (CERCLA). Hazardous substances under CERCLA Section 101(14) encompass a wide variety of pollutants regulated under other federal statutes including the Clean Water Act, Clean Air Act, and Toxic Substances Control Act. Radio nuclides, which are specifically excluded under RCRA's definition of solid waste, are regulated under CERCLA because they are defined as hazardous air pollutants under the Clean Air Act. Thus, radioactive materials and waste stored in USTs are within the scope of RCRA Subtitle I authority. However, within the scope of Subtitle I, tanks containing radioactive materials and wastes (i.e., source, special, and by-product materials) subject to Atomic Energy Act requirements are deferred from having to comply with UST regulations [40 *CFR* 280.10(c)(2)]. On the other hand, tanks containing mixtures of radioactive materials and hazardous wastes (i.e., "mixed" waste) are subject to RCRA Subtitle C requirements due to the presence of the RCRA hazardous waste component.

Federal agencies and departments, including DOE, that own or operate USTs are subject to and must comply with all applicable federal, state, interstate, and local requirements, except when the President determines that exemption of specific tanks from these requirements is in the "paramount" interest of the United States (RCRA, Section 9007).

(viii) States with RCRA Authority: Section 3006 of RCRA authorizes states to develop and enforce their own hazardous waste programmes in place of the federal programme administered by EPA. Before administrating any of the provisions of the statute, authorized states must complete a state program approval process. State authorization is a rule making process through which EPA delegates the primary responsibility for implementing the RCRA hazardous waste program to individual states in lieu of EPA. This process ensures national consistency and minimum standards while providing flexibility to states in implementing rules. State RCRA programmes must always be at least as stringent as the federal requirements, but states can adopt more stringent requirements as well. Once a state has obtained final authorization, it must continually revise its program to maintain its authorized status. An authorized state must revise its program to reflect new federal rule makings as well as any state statutory or regulatory changes that affect the state's hazardous waste program. For more information see EPA's *RCRA Orientation Manual* (described below) or see EPA's Web site on RCRA state authorization.

8.9 GUIDANCE ON THE USE OF SECTION 7003 OF RCRA

In October 1997 EPA's Office of Enforcement and Compliance Assurance issued a guidance document entitled *Guidance on the Use of Section 7003 of RCRA*. This guidance updates, expands, and supersedes previous guidance on this subject, dated September 26, 1984. The guidance document states that Section 7003 of RCRA provides the Agency:

"... with broad and effective enforcement tools that can be used to abate conditions that may present an imminent and substantial endangerment to health or the environment. Section 7003 allows EPA to address situations where the handling, storage, treatment, transportation, or disposal of any solid or hazardous waste may present such an endangerment. In these situations, EPA can initiate judicial action or issue an administrative order to any person who has contributed or is contributing to such handling, storage, treatment, transportation, or disposal to require the person to refrain from those activities or to take any necessary action. Among its many benefits, Section 7003 provides EPA with a strong and effective means of furthering risk-based enforcement and implementing its strategy for addressing the worst RCRA sites first."

8.10 COMPOSTING

Composting organic materials such as yard trimmings, food scraps, wood waste and paper are helps to manage the largest component of our trash the garbage waste we dispose of, making up more than two-thirds of the solid waste disposal stream. The amount of yard waste and its municipal solid waste disposal market share have declined dramatically in the last four decades while the composting rate has soared. Backyard compost piles and "grass cycling" programmes have helped to reduce yard waste generation. Composting serves as an environment friendly waste disposal solution by cutting down on the amount of garbage waste headed to landfills. State and local composting requirements have increased the number of commercial composting operations.

(i) What is composting?

Composting involves the controlled decomposition of plant remains and other organic materials to make an earthy, dark, crumbly substance that is excellent for enriching soil and preventing soil erosion—is a great way to recycle yard and kitchen wastes and reduce the volume of garbage waste sent to landfills or incinerators for disposal.

Natural composting, or biological decomposition, began with the first plants on the earth and has been going on ever since. As vegetation falls to the ground, it slowly decays, providing minerals and nutrients needed for plants, animals, and microorganisms. The production of "mature" (or usable) compost from organic wastes also requires high temperatures to destroy pathogens and seeds that natural decomposition does not destroy.

All garbage waste composting requires three basic ingredients: browns (including materials such as dead leaves, branches, twigs, wood chips or cardboard); greens (materials such as grass clippings, vegetable waste, fruit scraps and coffee grounds) and water.

Mature compost contains "humus," a dark brown or black substance with a soil-like, earthy smell. It is created by combining "greens" with "browns" in the proper ratio with water to accelerate the breakdown of organic materials, and then allowing the finished compost material to fully stabilize and mature.

(ii) Environmental benefits of waste composting

There are a number of environmental benefits to composting garbage waste:

Composting organic matter rather than disposing in landfills reduces the emission of greenhouse gases like methane.

Using compost as fertilizer reduces the need for chemical fertilizers, and promotes higher crop yields.

Compost can regenerate poor soils, and can be used to cleanup (or remediate) contaminated soil by absorbing odours and treating chemicals. Compost has also been shown to bind to heavy metals and prevent them from migrating into groundwater or plants.

Municipal waste composting and home waste composting

Some municipalities offer curbside collection for yard waste in the same way recyclables are collected. The waste is taken to a centralized location where it is turned into compost that may be sold, given away or used in public works projects.

You can create your own home compost from garbage waste as well, either in your backyard or indoors using a compost bin, depending on your available space. Backyard and indoor composting work best for households when focusing on converting small quantities of organic waste materials, such as yard trimmings and food scraps, into compost that can be spread in garden beds and under shrubs, or used as potting soil for outdoor plants. Waste composting can be done with a variety of different systems or containers, including kitchen sized compost bins.

The U.S. EPA's web site features an excellent how-to-guide on home waste composting. You can also check the EPA web site to find out more about municipal waste composting programmes in your state.

8.11 HAZARDOUS WASTE TREATMENT

In the past, hazardous waste management meant removing pollutants and disposing of residuals in a landfill. As industry and business have become more environmentally conscientious, and in order to comply with federal, state and local regulations, the need for alternatives to these historic methods has increased, and so has our ability to offer alternate solutions.

Hazardous wastes are generated as the result of treatment steps performed on industrial wastewaters prior to discharge, or contaminant-containing ground waters. In these cases, the treatment process (for example, by precipitation, evaporation, filtration, or ion exchange technology), produces a residual waste, including sludge and spent media, which must be analyzed, using the toxic characteristics leaching procedure (TCLP) test. The results of the TCLP are used to determine whether a waste is hazardous or non-hazardous.

RCRA is the Resource Conservation and Recovery Act, a set of regulations adopted by Congress and administered by US Environmental Protection Agency (EPA), state and local regulatory agencies. RCRA defines what hazardous wastes are and how they must be legally stored, treated and disposed of. Many US states have been given authority by the EPA to enforce RCRA rules and also establish additional local and regional rules if necessary, however, state laws can never reduce the standards enacted by the EPA. There are significant legal liabilities to companies for not properly managing waste residuals according to RCRA regulations.

Waste may be classified as either listed or characteristic hazardous waste. RCRA regulations provide guidance to assist the waste generator in determining whether the waste falls into either classification.

With few exceptions, the EPA has classified the residuals from the following six manufacturing processes, also called electroplating processes, as listed hazardous wastes which must be dealt with as such. The residuals from these processes are assigned the F006 hazardous waste code.

Electroplating

Coating (chromating, phosphating, coloring)

Electroless Plating

Chemical Etching and Milling

Anodizing

Printed circuit board manufacturing

Other metal finishing processes performed in conjunction with the above 6 listed electroplating processes are also considered F006 Listed Hazardous Wastes.

If any part of the resulting wastewater or waste residual which contains the following metals, then the waste is a characteristic hazardous waste:

Arsenic – 5 ppm

Chromium – 5 ppm

Mercury – 0.2 ppm

Barium – 100 ppm

Cyanides – 250 ppm

Selenium – 1 ppm

Cadmium – 1 ppm

Lead – 5 ppm

Silver – 5 ppm

Hazardous waste treatment involves almost anything that can be done to a hazardous waste prior to disposal. Any method, technique or, process designed to change the physical, chemical, or biological character or composition of the hazardous waste, so as to neutralize the waste or to make the waste less hazardous and thereby safer for transport, increase potential for recovery, reuse or storage, or to reduce waste volume. Waste dewatering is a efficient and effective treatment method used for waste volume reduction. A variety of treatment systems are employed to dewater waste including belt presses, centrifuges, and other devices. Siemens offers a complete range of dewatering systems in a range of sizes and suitable for all forms of reuse application including land application, composting, and incineration. Siemens Water Technologies owns and operates two, fully-permitted facilities with federal RCRA Part B and CERCLA authority and specializing in the treatment of inorganic wastes and wastewater. Wastes are sent to the facilities in drums (of all sizes), portable totes, cubic yard bags or boxes and tanker trucks. Virtually, everything that we take into our facilities is in some way recycled. Recycling reduces liability, which is a significant incentive to many of our customers. The heart of our hazardous waste treatment operations are in facilities located in Roseville, Minnesota and Los Angeles, California. These facilities serve as regional hubs for the treatment of hazardous waste and wastewater and the recovery of reusable by-products collected from the company's hazardous waste customers. Siemens Water Technologies is the only company whose facilities handle and regenerate ion exchange resins and other media. Our goal is to produce zero hazardous waste by extracting all reusable material, to eliminate pollutants from the environment and avoid disposal costs and liability. Most of the treatment residual wastes (F006, etc.) shipped to our facilities are inorganic or trace organic materials. Our advanced processing techniques have allowed us to increase beneficial reuse rate of these wastes to over 90% as reusable chemicals and metallic commodities.

(i) Treatment Technologies for Hazardous Waste:

(a) Sludge

(b) Dewatering

(c) Waste reduction & recycle

(d) Metal treatment technology and hazardous waste solutions

Heavy metal contamination is one of the most common forms of pollution found in and around industrial sites, landfills, abandoned mining operations, and municipal and agricultural outfalls. In addition, heavy metals are used in a variety of industrial processes such as steel production, metal plating and anodizing, automobile and aircraft manufacturing, and general metal fabrication, just to name a few. Ultimately, all heavy metal wastes exceeding strict EPA environmental standards require adequate toxicity reduction and proper disposal. Heavy metal contamination is one the most common forms of pollution found in and around industrial sites, landfills, abandoned mining operations, and municipal and agricultural outfalls. In addition, heavy metals are used in a variety of industrial processes such as steel production, metal plating and anodizing, automobile and aircraft manufacturing, and general metal fabrication, just to name a few. Ultimately, all heavy metal wastes exceeding strict EPA environmental standards require adequate toxicity reduction and proper disposal.

Treatable Waste Stream

Lead

Firing ranges

Battery breaking facilities

Soils – tills, clays, fine sands, coarse sands, loam, silts, etc.

Wastewater sludge – plumbing fixture mfg., TV tube mfg.

Sediments – river and lake bottoms, settling basins

Mineral processing waste – tailings, media separator bottoms

Battery recycling debris

Primary smelter slag

Secondary smelter slag

Steel electromelt furnace dust (K061)

Petroleum refinery sludge – K048, K051, K052, tetraethyl lead

Mirror mfg. waste

Target range backstop soils

Automobile recycling "shredder fluff"

Wire chop "fluff"

Foundry sands

Paint sandblasting debris

Scrap yard soils and storm water run-off basin sediments

Municipal waste incinerator ash

Abrasive mfg. debris

Cadmium

Wastewater sludge – battery mfg.

Wastewater sludge – printed circuit board mfg.

Lead and zinc ore processing waste

Automobile "shredder" fluff

Battery breaking facilities

Arsenic

Soils and sediments at CCA wood treating plants

Veterinary medicine formulation waste.
Ash from thermal destruction of chemical weapons.
Zinc
Zinc ore processing waste
Galvanizing process rinseate sediments
Chromium
Soil at chromium ore staging site
Soil around chrome plating line.
Wastewater treatment sediment at bearing mfg.
Cooling tower basin sediments
Selenium
Agricultural run-off sediment
Ceramic "frit" residue
Antimony
Weapons mfg wastewater sediment
Barium
Waste water treatment lagoon sediments

Hazardous Waste Land Disposal Units (LDUs)

The land disposal restrictions (LDR) require that a hazardous waste meet or be treated to meet specified standards for hazardous constituents before being disposed of on the land. Instead of requiring barriers to separate hazardous contaminants from groundwater like other hazardous waste requirements, the land disposal restrictions require that hazardous waste undergo fundamental physical or chemical changes so that the waste poses less of a threat to groundwater. The obvious advantage of hazardous waste treatment is that it provides a more permanent and lasting form of groundwater protection than does simple hazardous waste containment. In addition to prohibiting the disposal of wastes that do not meet treatment standards, the LDRs also forbid the storage of hazardous waste as substitute for meeting the required treatment standards and forbid the dilution of wastes as a substitute for legitimate treatment. Generators may treat their hazardous wastes to meet one or more applicable LDR treatment standards without obtaining a hazardous waste permit. The permit exemption for generator treatment only extends to treatment activities that share the same standards as storage (i.e., the treatment must occur in tanks or containers). The standards for thermal treatment are different from generator storage requirements because of the inherent dangers of fire, explosion, or evolution of toxic gases; therefore, thermal treatment may not be performed without a hazardous waste permit. If treating their hazardous waste to meet LDR treatment standards, the generator must modify their contingency plan, training plan, security plan, etc., to reflect the change in processes required for treatment to meet the land disposal restrictions. There are also increased requirements for record keeping and a detailed waste analysis plan must be prepared before treatment can begin. Treatment must occur in an accumulation tank or container or in a containment building. If any of these elements are missing, then the generator is conducting illegal treatment of hazardous waste. Generators treating their wastes that are subject to LDR have certain notification and certification requirements under CHWR Section 268.7. Whether or not they treat their waste onsite, the generator is always responsible for notifying subsequent facilities that treat, store, or dispose of the waste of the waste's LDR status. Much like a hazardous waste manifest, the LDR notification and certification paperwork helps hazardous waste handlers and regulatory agencies ensure that wastes are properly managed. Generators may partially treat a waste to meet an LDR standard without treating the waste for all applicable LDR standards. Partially treated wastes must be sent to a permitted facility for additional treatment to meet the rest of the standards before disposal can occur. If a waste is subject to the land disposal restrictions at the point of generation, then all of the LDR requirements continue to apply to the waste even if it is later decharacterized. Even after treatment to meet the appropriate treatment standards, most hazardous wastes must still be disposed of in a hazardous waste facility.

(ii) Types of LDUs for hazardous waste disposal

Landfill

Surface impoundment

Waste pile

Land treatment unit

Injection well

Salt dome formation

Salt bed formation

Underground mine

Underground cave

Landfills are excavated or engineered sites where non-liquid hazardous waste is deposited for final disposal and covered. These units are selected and designed to minimize the chance of release of hazardous waste into the environment. Design standards for hazardous waste landfills require a double liner; double leachate collection and removal systems (LCRS); leak detection system; run on, run-off, and wind dispersal controls; construction quality assurance (CQA) program. Liquid wastes may not be placed in a hazardous waste landfill. Operators must also comply with inspection, monitoring, and release response requirements. Since landfills are permanent disposal sites and are closed with waste in place, closure and post-closure care requirements include installing and maintaining a final cover, continuing operation of the LCRS until leachate is no longer detected, maintaining and monitoring the leak detection system, maintaining ground water monitoring, preventing storm water run on and run-off, and installing and protecting surveyed benchmarks.

Surface impoundments are natural topographic depressions, man-made excavations, or diked areas formed primarily of earthen materials used for temporary storage or treatment of liquid hazardous waste. Examples include holding, storage, settling, aeration pits, ponds, and lagoons. Hazardous waste surface impoundments are required to be constructed with a double liner system, a leachate collection and removal systems (LCRS), and a leak detection system. To ensure proper installation and construction, regulations require the unit to have and follow a construction quality assurance (CQA) programme. The regulations also outline monitoring, inspection, response action, and closure requirements.

Waste piles are non-containerized piles of solid, non-liquid hazardous waste that are used for temporary storage or treatment. In addition to the standard double liner and leachate collection and removal systems (LCRS), waste piles are required to have a second LCRS above the top liner. Waste piles must also have run on and run-off controls, be managed to prevent wind dispersal of waste, and are subject to inspection, monitoring, and release response requirements. When closing a waste pile, all waste residue and contaminated soils and equipment must be removed or decontaminated.

Land Treatment Units use naturally occurring soil microbes and sunlight to treat hazardous waste. This is accomplished by applying the hazardous waste directly on the soil surface or incorporating it into the upper layers of the soil in order to degrade, transform, or immobilize the hazardous constituents. Land treatment units rely upon the physical, chemical, and biological processes occurring in the topsoil layers to contain the waste. Because of this, the units are not required to have liner systems or an leachate collection and removal systems (LCRS). Before hazardous waste can be placed in a land treatment unit, operators must complete a treatment demonstration to demonstrate the unit's effectiveness and ability to treat the hazardous waste. Once operational, operators must monitor the unit (unsaturated zone monitoring) to ensure that all hazardous constituents are being treated adequately. Unit closure consists primarily of placing a vegetative cover over the unit and certifying that hazardous constituent levels in the treatment zone do not exceed background levels. The remaining types of land disposal units are categorized as miscellaneous units.

Injection wells are regulated primarily under the Safe Drinking Water Act and the Underground Injection Control (UIC) Programme.

Salt dome formations, salt bed formations, underground mines, and underground caves are geologic repositories. Because these units vary greatly, they are subject to environmental performance standards, not prescribed technology-based leachate collection. The standards require that these miscellaneous units must be located, designed, constructed, operated, maintained, and closed in a manner that ensures the protection of human health and the environment

(iii) Groundwater Contaminantion and Remediation

Groundwater contamination occurs when man-made products such as gasoline, oil, road salts and chemicals get into the groundwater and cause it to become unsafe and unfit for human use. Some of the major sources of these products, called contaminants, are storage tanks, septic systems, hazardous waste sites, landfills, and the widespread use of road salts, fertilizers, pesticides and other chemicals.

Storage tanks may contain gasoline, oil, chemicals, or other types of liquids and they can either be above or below ground. There are estimated to be over 10 million storage tanks buried in the United States and over time the tanks can corrode, crack and develop leaks. If the contaminants leak out and get into the groundwater, serious contamination can occur.

Septic systems can be another serious contamination source. Septic systems are used by homes, offices or other buildings that are not connected to a city sewer system. Septic systems are designed to slowly drain away human waste underground at a slow, harmless rate. An improperly designed, located, constructed, or maintained septic system can leak bacteria, viruses, household chemicals, and other contaminants into the groundwater causing serious problems.

In the United States today, there are thought to be over 20,000 known abandoned and uncontrolled hazardous waste sites and the numbers grow every year. Hazardous waste sites can lead to groundwater contamination if there are barrels or other containers laying around that are full of hazardous materials. If there is a leak, these contaminants can eventually make their way down through the soil and into the groundwater.

Landfills are another major source of contamination. Landfills are the places that our garbage is taken to be buried. Landfills are supposed to have a protective bottom layer to prevent contaminants from getting into the water. However, if there is no layer or it is cracked, contaminants from the landfill (car battery acid, paint, household cleaners, etc.) can make their way down into the groundwater.

The widespread use of road salts and chemicals is another source of potential groundwater contamination. Road salts are used in the winter-time to put melt ice on roads to keep cars from sliding around. When the ice melts, the salt gets washed off the roads and eventually ends up in the water. Chemicals include products used on lawns and farm fields to kill weeds and insects and to fertilize the plants. When the rain comes, these chemicals get washed into the ground and eventually into the water.

We have to remember that since groundwater is part of the hydrologic cycle, contaminants in other parts of the cycle, such as the atmosphere or bodies of surface water, can eventually be transferred into our groundwater supplies.

Groundwater remediation is the process that is used to remove pollution from groundwater. Groundwater is water present below the ground surface that saturates the pore space in the subsurface. At least one half of the population of the United States depends upon groundwater as a source of drinking water. Groundwater is also used by farmers to irrigate crops and by industries to produce everyday goods. Most groundwater is clean, but groundwater can become polluted, or contaminated as a result of human activities or as a result of natural conditions. The many and diverse activities of man produce innumerable waste materials and by-products; before the 1980s, the regulation of these wastes was less stringent and waste materials were often disposed of or stored on land surfaces where they percolated into the underlying soil and eventually were carried downward, contaminating the underlying groundwater and therefore jeopardizing the natural quality of it. As a result, contaminated groundwater became unsuitable for use. Current practices can still impact groundwater, such as the over-application of fertilizer or pesticides, spills from industrial operations, infiltration from urban run-off, and leaking from landfills. Using contaminated groundwater causes hazards to public health through poisoning or the spread of disease, and the practice of groundwater remediation has been developed to address these issues. Contaminants found in groundwater cover a broad range of physical, inorganic chemical, organic chemical, bacteriological, and radioactive parameters. Pollutants and contaminants can be removed from groundwater by applying various techniques thereby making it safe for use.

(i) **Groundwater remediation techniques:** Groundwater remediation techniques span biological, chemical, and physical treatment technologies. Most groundwater treatment techniques utilize a combination of technologies. Some of the biological treatment techniques include bioaugmentation, bioventing, biosparging, bioslurping, and phytoremediation. Some chemical treatment techniques include ozone and oxygen gas injection, chemical precipitation, membrane separation, ion exchange, carbon absorption, aqueous chemical oxidation, and surfactant enhanced recovery. Physical treatment techniques include, but not limited to, pump and treat, air sparging, and dual phase extraction.

(ii) Biological treatment technologies:

(a) **Bioaugmentation:** If a treatability study shows no degradation (or an extended lab period before significant degradation is achieved) in contamination contained in the groundwater, then inoculation with strains known to be capable of degrading the contaminants may be helpful. This process increases the reactive enzyme concentration within the bioremediation system and subsequently may increase contaminant degradation rates over the non-augmented rates, at least initially after inoculation.

(b) **Bioventing:** Bioventing is an in situ remediation technology that uses microorganisms to biodegrade organic constituents adsorbed in the groundwater. Bioventing enhances the activity of indigenous bacteria and simulates the natural in situ biodegradation of hydrocarbons by inducing air or oxygen flow into the unsaturated zone and, if necessary, by adding nutrients. During bioventing, oxygen may be supplied through direct air injection into residual contamination in soil. Bioventing primarily assists in the degradation of adsorbed fuel residuals, but also assists in the degradation of volatile organic compounds (VOCs) as vapors move slowly through biologically active soil.

(c) **Biosparging:** Biosparging is an in situ remediation technology that uses indigenous microorganisms to biodegrade organic constituents in the saturated zone. In biosparging, air (or oxygen) and nutrients (if needed) are injected into the saturated zone to increase the biological activity of the indigenous microorganisms. Biosparging can be used to reduce concentrations of petroleum constituents that are dissolved in groundwater, adsorbed to soil below the water table, and within the capillary fringe.

(d) **Bioslurping:** Bioslurping combines elements of bioventing and vacuum-enhanced pumping of free-product that is lighter than water (light non-aqueous phase liquid or LNAPL) to recover free-product from the groundwater and soil, and to bioremediate soils. The bioslurper system uses a "slurp" tube that extends into the free-product layer. Much like a straw in a glass draws liquid, the pump draws liquid (including free-product) and soil gas up the tube in the same process stream. Pumping lifts LNAPLs, such as oil, off the top of the water table and from the capillary fringe (i.e., an area just above the saturated zone, where water is held in place by capillary forces). The LNAPL is brought to the surface, where it is separated from water and air. The biological processes in the term "bioslurping" refer to aerobic biological degradation of the hydrocarbons when air is introduced into the unsaturated zone.

(e) **Phytoremediation:** In the phytoremediation process certain plants and trees are planted, whose roots absorb contaminants from groundwater over time, and are harvested and destroyed. This process can be carried out in areas where the roots can tap the groundwater. Few examples of plants that are used in this process are Chinese Ladder fern Pteris vittata, also known as the brake fern, is a highly efficient accumulator of arsenic. Genetically, altered cottonwood trees are good absorbers of mercury and transgenic Indian mustard plants soak up selenium well.

(f) **Permeable Reactive Barriers:** Certain types of permeable reactive barriers utilize biological organisms in order to remediate groundwater.

One particular type of permeable reactive barrier utilizes a swellable, organically-modified silica embedded in iron, which is injected in situ in order to create a permanent soft barrier in the ground. Water filters through the barrier, and the silica material absorbs toxins, such as TCE. The iron dechlorinates the solvents in the groundwater, often reducing toxicity levels below detectable limits with no toxic daughter products, no solid waste removal, and no air pollution. This type of permeable reactive barrier is also more dispersed than others.

(iii) Chemical treatment technologies: There are various chemical treatment technologies presently being used for waste treatment. Some of the major treatment technologies are discussed below.

(a) Chemical precipitation: Chemical precipitation is commonly used in wastewater treatment to remove hardness and heavy metals. In general, the process involves addition of agent to an aqueous waste stream in a stirred reaction vessel, either batch wise or with steady flow. Most metals can be converted to insoluble compounds by chemical reactions between the agent and the dissolved metal ions. The insoluble compounds (precipitates) are removed by settling and/or filtering.

(b) Ion exchange: For groundwater remediation is virtually always carried out by passing the water downward under pressure through a fixed bed of granular medium (either cation exchange media and anion exchange media) or spherical beads. Cations are displaced by certain cations from the solutions and ions are displaced by certain anions from the solution. Ion exchange media most often used for remediation are zeolites (both natural and synthetic) and synthetic resins.

(c) Carbon absorption: The most common activated carbon used for remediation is derived from bituminous coal. Activated carbon absorbs volatile organic compounds from groundwater by chemically binding them to the carbon atoms.

(d) Chemical oxidation: In this process, called In Situ Chemical Oxidation or ISCO, chemical oxidants are delivered in the subsurface to destroy (converted to water and carbon dioxide or to nontoxic substances) the organics molecules. The oxidants are introduced as either liquids or gasses. Oxidants include air or oxygen, ozone, and certain liquid chemicals such as hydrogen peroxide, permanganate and persulfate. Ozone and oxygen gas can be generated on site from air and electricity and directly injected into soil and groundwater contamination. The process has the potential to oxidize and/or enhance naturally occurring aerobic degradation. Chemical oxidation has proven to be an effective techique for dense non-aqueous phase liquid or DNAPL when it is present.

(e) Surfactant enhanced recovery: Surfactant enhanced recovery increases the mobility and solubility of the contaminants absorbed to the saturated soil matrix or present as dense non-aqueous phase liquid. Surfactant-enhanced recovery injects surfactants (surface-active agents that are primary ingredient in soap and detergent) into contaminated groundwater. A typical system uses an extraction pump to remove groundwater downstream from the injection point. The extracted groundwater is treated aboveground to separate the injected surfactants from the contaminants and groundwater. Once the surfactants have separated from the groundwater they are reused. The surfactants used are non-toxic, food-grade, and biodegradable. Surfactant enhanced recovery is used most often when the groundwater is contaminated by dense non-aqueous phase liquids (DNAPLs). These dense compounds, such as trichloroethylene (TCE), sink in groundwater because they have a higher density than water. They then act as a continuous source for contaminant plumes that can stretch for miles within an aquifer. These compounds may biodegrade very slowly. They are commonly found in the vicinity of the original spill or leak where capillary forces have trapped them.

(iv) Physical treatment technologies: Physical treatment technologies mainly involved physical means e.g. force, air flow, etc. for treatment technologies. Some of the widely used techniques are discussed below:

(a) Pump and treat: Pump and treat is one of the most widely used groundwater remediation technologies. In this process groundwater is pumped to the surface and is coupled with either biological or chemical treatments to remove the impurities.

(b) Air sparging: Air sparging is the process of blowing air directly into the ground water. As the bubbles rise, the contaminants are removed from the groundwater by physical contact with the air (i.e., stripping) and are carried up into the unsaturated zone (i.e., soil). As the contaminants move into the soil, a soil vapour extraction system is usually used to remove vapors.

(c) Dual phase vacuum extraction: Dual-phase vacuum extraction (DPVE), also known as multi-phase extraction, is a technology that uses a high-vacuum system to remove both contaminated groundwater and soil vapour. In DPVE systems a high-vacuum extraction well is installed with its screened section in the zone of contaminated soils and groundwater. Fluid/vapour extraction systems depress the water table and water flows faster to the extraction well. DPVE removes contaminants from above and below the water table. As the water table around the well is lowered from pumping, unsaturated soil is exposed. This area, called the capillary fringe, is often highly contaminated, as it holds undissolved chemicals, chemicals that are lighter than water, and vapours that have escaped from the dissolved groundwater below. Contaminants in the newly exposed zone can be removed by vapour extraction. Once above ground, the extracted vapours and liquid-phase organics and groundwater are separated and treated. Use of dual-phase vacuum extraction with these technologies can shorten the cleanup time at a site, because the capillary fringe is often the most contaminated area.

(d) Monitoring-well oil skimming: Monitoring-wells are often drilled for the purpose of collecting groundwater samples for analysis. These wells, which are usually six inches or fewer in diameter, can also be used to remove hydrocarbons from the contaminant plume within a groundwater aquifer by using a belt style oil skimmer. Belt oil skimmers, which are simple in design, are commonly used to remove oil and other floating hydrocarbon contaminants from industrial water systems.

A monitoring-well oil skimmer remediates various oils, ranging from light fuel oils such as petrol, light diesel or kerosene to heavy products such as No. 6 oil, creosote and coal tar. It consists of a continuously moving belt that runs on a pulley system driven by an electric motor. The belt material has a strong affinity for hydrocarbon liquids and for shedding water. The belt, which can have a vertical drop of 100+ feet, is lowered into the monitoring well past the LNAPL/water interface. As the belt moves through this interface it picks up liquid hydrocarbon contaminant, which is removed and collected at ground level as the belt passes through a wiper mechanism. To the extent that DNAPL hydrocarbons settle at the bottom of a monitoring well, and the lower pulley of the belt skimmer reaches them, these contaminants can also be removed by a monitoring-well oil skimmer.

Typically, belt skimmers remove very little water with the contaminant, so simple weir type separators can be used to collect any remaining hydrocarbon liquid, which often makes the water suitable for its return to the aquifer. Because the small electric motor uses little electricity, it can be powered from solar panels or a wind turbine, making the system self-sufficient and eliminating the cost of running electricity to a remote location.

8.12 CONCLUSIONS AND ROLE OF INDIVIDUALS IN PREVENTION OF POLLUTION

The role of an individual in maintaining a pollution free, pure and congenial environment and in preserving its resources is actually the need of the hour. Individuals can, however, play an important role in abatement of air, water, soil or noise pollution in the following simple manners:

(1) Use low-phosphate, phosphate-free or biodegradable dishwashing liquid, laundry detergent, and shampoo.

(2) Don't use water fresheners in toilets.

(3) Use manure or compost instead of commercial inorganic fertilizers to fertilize gardens and yard plant.

(4) Use biological methods or integrated pest management to control garden, yard, and household pests.

(5) Don't pour pesticides, paints, solvents, oils, or other products containing harmful chemicals down drain or on the ground. Contact the authorities responsible for their disposal.

(6) Recycle old motor oil and antifreeze at an auto service centre that has an oil recycling program.

(7) If you get water from a private well or suspect that municipal water is contaminated, have tested by an EPA certified laboratory for lead, nitrates, trihalomethanes, radon, volatile, organic compounds and pesticides.

(8) Run water from taps for several minutes every morning before using the water for drinking or cooking. Save it and use it to water plants.

If you have a septic tank, monitor it yearly and have it cleaned out every three to five years so that it won't contribute to groundwater pollution. Do not use a septic tank cleaner, which contain toxic chemicals that can kill bacteria important to sewage decomposition and that can contaminate groundwater if systems malfunction.

(9) Support ecological land-use planning in your community.

(10) Get to know your local water bodies and form watchdog groups to help monitor, protect, and restore them.

Learning Outcomes

Some of the imperative learning contribution from this chapter were about causes, effects and control measures of urban and industrial wastes; Perspectives & properties, Collection, Transfer & transports, Life cycle assessment, Routes, Disposal in a landfill, Waste to energy, Composting, Resource conservation & recovery for sustainability. Also present chapter provided information on hazardous waste management and treatment technologies

Exercise

1. Write 250 word essay about source of waste in your community.
2. Create a consultancy report for your local municipality council for waste recycle and management.
3. Create a poster showing methods of waste management around the world in different industry.

Chapter **9**

Environment Laws

9.1 INTRODUCTION

In India, in 1980 to ensure a healthy environment for the country, the Department of Environment was established, which later became the Ministry of Environment and Forests in 1985. It is clearly stated in the Constitution of India "it is the duty of state to protect and improve and to safeguard the forests and wildlife of the country. It imposes a duty on every citizen to protect and improve the natural environment including forests, lakes, rivers and wildlife". References to the environment have also been made in the Directive Principles of State Policy as well as the Fundamental Rights. The constitutional provisions are backed by a number of laws—acts, rules and notifications. The EPA (Environment Protection Act), 1986 came into force soon after the Bhopal Gas Tragedy and is considered an umbrella legislation as it fills many gaps in the existing laws. Environmental law is the area of law that seeks to manage human impact on the environment. The "environment" refers to all aspects of the natural environment, including land, air, water, flora and fauna, and the human environment (both indigenous and non-indigenous cultural and built heritage).

Environmental law exists at a local, state, national and international level. It covers topics such as:

- environmental planning and assessment;
- natural resource management;
- pollution management;
- biodiversity conservation; and
- natural and cultural heritage.

Environmental order, understood as the whole gamut of regulations and institutions serving the protection, care and development of the natural foundations of human life, is geared towards the sustainability and standardization of care of the environment. It is more than peripheral short-term environmental protection effected on a case-by-case basis to ward off risks; rather it means blanket, comprehensive and long-term protective care in line with the enduring responsibility we bear for the world around us, for the world we share with others and for the world of the generations to come.

9.1.1 Functions of Environmental Law

The key functions of environmental law are to:

- Establish regulatory structures for environmental management, including regulatory agencies and specialist courts and tribunals.

- Enable regulators to manage environmental impacts using plans, policies, standards, licences and incentives.
- Require those proposing environmentally significant activities to obtain approval from regulators.
- Enable members of the public to take part in environmental planning and environmental assessment.
- Require activities of environmental significance to be assessed before permission can be granted.
- Provide administrative, civil, and criminal penalties for breaches of the law.
- Allow the legality of the decisions of regulators to be challenged by members of the public.
- Allow the merits of certain decisions of regulators to be challenged by members of the public.

9.1.2 Law and Care of the Environment

9.1.2.1 Environmental law: the "ecological nuancing" of law

Law was given an "economic nuance" and thus became economic law during and after the First World War. The newly created environment policy is subsumed in giving an "ecological nuance" to legal norms in public law, private law and international law, designed to protect and look after the natural foundations of human life. Traditional legal areas such as water law and emission control law and more recent fields such as nature conservation law are, under environmental law, linked to new regulatory areas such as protection against hazardous substances and the environmental impacts of genetic engineering, the avoidance, recovery and disposal of waste, soil protection, energy saving, radiological protection and nuclear safety. The environmental law gears legal provisions with an "ecological nuance" to the new-found objective of protecting and caring for the environment, in other words, the environmental media soil, water and air and flora and fauna including climate, landscape and the eco-balance as well as the relationship between these sectors of the environment both among themselves and with man.

9.1.3 Principles and Functions of Environmental Law

9.1.3.1 Principle of prevention, polluter-pays principle, principle of cooperation

Environmental law gears human actions on the basis of the principle of prevention, the polluter-pays principle and the principle of cooperation. Firstly, let us look at the principle of prevention, the material leitmotif for environment policy. The early application of measures founded on this principle is designed not only to ward off danger and remove damage, but also to prevent potential negative environmental impacts from arising at all — if possible at source and by means of risk-reduction — so as to achieve sustainable use of the environment, providing care for the assets at stake. Use of the environment rather extends very much to all benefits brought to mankind both now and in the future from the careful and sparing use of an environment, the quality and quantity of which is secured, protected and cared for. As former Federal President Richard Von Weizsaker noted in 1986, this understanding does not exclude the protection. The capacity to enjoy rights and obligations should also in future be confined to men and women as natural and legal persons since care of the environment is not an end in itself. The second concept, the polluter-pays principle, must be understood as a fundamental allocation of costs but not automatically also as a principle of allocation of responsibility. This also serves to avoid distortions in competition. The third principle that of cooperation means that care of the environment is a common task of both the State and citizens and must, as far as possible, be effected in cooperation between the two. Early involvement of those concerned can also lead to voluntary change in behaviour so that the State need no longer adopt prohibitions or orders in the particular field at hand. This helps avoid proliferation of red-tape. Involvement in environmental decision-making means taking on responsibility for care of the environment.

9.1.3.2 *The principle of immissions and the principle of emissions*

The standards to be met either now or in the future by the environment or a part of a certain environment both as regards quality and quantity. Under the principle of immissions, these standards can be expressly termed environmental quality standards. The standards may also emerge along the lines of the principle of emissions, as a result of the application of geared legal measures to avoid or reduce human impacts on the environment. Standards of immissions and emissions are regularly linked.

(i) ***Threefold function*** Law has a threefold function, the most significant being its existential function in care of the environment. Law is designed to secure mankind's life, survival and health and give men and women an existence worthy of human dignity. On the other hand, the existential function is complemented by the social function whereby environment law helps provide facilities for people to relax in and enjoy the natural environment and the landscape. Thirdly, law also has an aesthetic function when it allows us to experience the diversity, uniqueness and beauty of the environment.

9.2 DEVELOPMENT PHASES AND TRENDS IN ENVIRONMENTAL LAW

Each crisis situation deemed to be able to be resolved by means of environmental law reopens the discussion on the effectiveness and further expansion of environmental law regulations and leads to a renewed evaluation of the application of the "correct" legal instruments and the role of law in care of the environment. It also applies to environmental damage where the time and source cannot be determined, such as the new types of forest damage, large-scale detrimental impacts on the soil, damage to buildings and monuments and global climate change.

9.2.1 Development Phases

(i) ***First legislative phase:*** Warding-off of risks to deal with individual situations, followed by centre stage being given to preventing damage to the environment occurring in the first place, were to become the first two objectives of the threefold goals of environment policy within the framework of the Federal Government's 1971 environment programme, which were in future to be achieved by effective administrative law in this field. To safeguard the environment, man needs to remain healthy and to live a life in human dignity and to protect the soil, water and air, and flora and fauna, from negative human impacts. The removal of damage and detriment caused to the environment by human intervention was the third objective.

The first legislative phase saw the Basic Law (constitution) of the Federal Republic of Germany amended where a number of spheres of "concurring" competence (i.e., between the Federal and Lander governments) were altered in certain environmental areas in favour of the Federation, with the result that environmental legislation has now principally been switched to the Federal Government while the Lander remain fundamentally responsible for enforcement. Moreover, a whole range of new laws were made in individual environmental areas and existing laws were expanded or re-worked and which were almost in their entirety further improved from a legislative point of view in the next legislative phase. Thus was created an admittedly legislatively splintered framework of environment law which to date has no proper plan of construction under the terms of law.

(ii) ***Administration phase:*** The strong proliferation of legislative environmental law engenders a need for sub judicial specification: the rough legislative objectives must be implemented in specific administrative goals. In the ongoing administrative phase of environmental law, it is in particular the executive in the Federation and in the Lander who must meet this need by promulgating administrative environmental law in the form of legal ordinances and also in the internal administrative provisions required to enforce laws and ordinances. In addition, there is the drawing-up of environmental plans such as landscape plans, waste management plans or air pollution abatement plans.

(iii) ***Second legislative phase:*** Environmental law is a dynamic law. New scientific findings, experience from existing environmental law and provisions under European and international law lead, in the administrative phase, to subsequent legislative improvements. The trend towards compaction and specification shows, together with four other development tendencies, that the second legislative phase of environmental law has begun. This phase is characterized by the trend towards the continued ecological development of environmental law, combined with the trend toward standardizing environmental law. There is also the trend towards promoting or emulating supranational and international developments. These trends are the moulding factor of the state of environmental law in the Federal Republic of Germany, and are also similar for the development of environmental law in the European Union too.

9.2.2 Development Trends

(i) ***Trend towards compaction and specification:*** The trend towards compaction and specification strengthens the legislator's environmental law and puts it into more specific terms. Environmental law is continually compacted and specified by sub judicial legislation by the government and by the application of law by the administration, with legal pronouncements on decisions affecting individual points of dispute also playing a role. The increased compaction and specification of environmental law at sub judicial level leads to greater security of law. The growing density of provisions, however, can also affect environmental matters, when, for example, the lengthening of licensing provisions may delay the application of environmentally sound provisions. This risk can be countered by the trend towards standardization and the use of indirect strategies.

(ii) ***Trends towards continued ecological development:*** More stress is laid on the principle of prevention which aims at caring, sustainable environmental use and the corresponding

emissions approach that must be complemented by the development of environmental quality objectives. Thus, ever growing importance is given to measures geared towards preventive, structured and caring planning and measures to manage, allocate or distribute environmental assets as opposed to those which concentrate on the warding-off of risks occurring in individual cases. Environmental measures are also increasingly substance-related. The polluter-pays principle is being strengthened by enhancing the responsibility manufacturers bear for the duration of the entire life cycle of their substances and products as intended by the Closed Substance Cycle and Waste Management Act.

Of particular importance for the continued ecological development of environmental law is the Chapter VII entitled "Environment" added to the Treaty establishing of the European Communities on the basis of the 1986 Single European Act; this chapter provides the legal foundation for comprehensive, supra-medial environment policy and for its integration into other policy areas — now given special focus in Article 6 of the EC Treaty as the environmental integration clause. Further impetus is given by Article 34 of the 1990 Unification Treaty, which elevates the protection of the natural basis for life to an objective of State of the newly unified Germany. Article 20a of the Basic Law of the Federal Republic of Germany, in force since 15 November 1994, made continued ecological development a constitutionally defined principle:

In any case, this is a task the state is justified to carry out even without express mention in the Basic Law and to which, to a certain degree, it feels obliged to undertake in order to protect basic protection obligations, to secure the prerequisites for the basic right of the protection of human dignity, life and physical integrity and to secure property.

The trend towards the continued ecological development of environmental law is also found in the Federal Government 1985 Soil Protection Plan and in the 1998 Soil Protection Act, which make the soil an object of legal protection in the same way as water and air and are also designed to help solve the problem of contaminated sites. The continued economic development of environmental law, however, is most clearly seen in the 1980 Chemicals Act. Continued ecological development applies not only to administrative environmental law, but also to the accompanying private environmental law and criminal environmental law.

A particularly good example of this is the 1990 Environmental Liability Act which introduced multi-medial risk liability for the party responsible for damage caused by the environmental impacts of certain high-risk installations. Worthy of mention here is also the 1980 Act to Combat Environmental Crime as amended in 1994, which removed the main environmental offences from individual pieces of legislation, harmonized them and included them in a new section of the Criminal Code entitled "Offences against the Environment." The trend towards continued ecological development, in general, has brought with it an increased demand for environmental information and environmental data, which in turn requires the development of environmental indicators. This hunger for data is set to intensify the procurement and processing of information, and in particular to strengthen the documentation of environmentally relevant technical procedures and promote continuous, automatic measurement of emissions.

(iii) Trends towards standardization: While the trend towards the continued ecological development of environmental law is linked to its material expansion, the trend towards standardization of environmental law puts this into more formal terms. The aim of this trend is, at all levels — Federation, Lander and European Union — to combat the fragmentation of environmental law by means of an inner harmonization of objectives, principles, measures, organization, procedures and sanctions without losing sight of the intrinsic differences within the three environmental media and the multifarious functions of the environment. Centre stage here is given in particular to aspects aimed at cutting red-tape and standardizing and speeding up licensing procedures. In the longer term, this trend leads to an external harmonization of the central elements of environmental law by collating around thirty pieces of Federal environmental legislation into an Environmental Code with an overall Environment Act

(iv) ***Trends towards applying "indirect strategies":*** Environmental law has been dominated by administrative law. The steering effect is engendered in particular by the application of orders. Both repressive and preventive prohibitions; this effect applies directly to the conduct geared towards care of the environment, not least in order to fulfil the constitutional requirements to protect the natural foundations of life and to protect life itself, physical inviolability and property.

In future, too, centre stage in environmental law will be in constitutional and administrative law fundamentally, these principles cannot be left up to criminal law alone; neither, however, can they be left up solely to private law. "Indirect strategies" can, moreover, also reduce state participation in monitoring environmentally-damaging activities by strengthening self-monitoring within companies. Instruments of indirect strategies also include sanctions under private and criminal law with liability for damage caused by environmental impacts and the possibility of punishment if obligations under environmental law are violated. Mention should also be made of the efforts at organizing self-monitoring by several firms.

(v) ***Trends to promote or adopt supranational and international developments:*** The administrative framework of environmental law is not only moulded by the legal and administrative provisions of the Federal Government and the Länder. The trans-boundary structure of most environmental problems and the increasing globalization of economic activities mean that even greater importance will be attached in future. The development of more far-reaching regional and global international law in the field of the environment also should be promoted, to reduce distortions in competition as a result of harmonized environmental norms and to reinforce the integration of environmental concerns into the international economic order. The environmental law of the European Community has developed, forcing domestic environmental law to harmonize or ousting it altogether. At the end of 1992, legal harmonization in the completion of the European Single Market in particular accelerated the harmonization of environmental law. The danger here is that if legal pronouncements are transferred to European level, standards of environmental care will be reduced to the lowest common denominator.

The Montevideo environmental law programme has also brought to light a worldwide consensus on the core areas of environmental law and facilitated its recognition as a legal area throughout the world. The primary topics mentioned include land-based marine pollution, the protection of the stratospheric ozone layer and the transport, treatment and disposal of toxic and hazardous waste; secondary topics include international cooperation in the case of environmental emergencies, the management of coastal zones, soil protection, trans-boundary air pollution, international trade in potentially hazardous chemicals, the protection against pollution of rivers and other water bodies, the prevention and restoration of environmental damage and the overall development of environmental law. Almost all of these fields are now the subject of either international conventions on the one hand or directives on conduct on the other, developed along the lines of "soft law."

Further issues are those of protecting the earth's atmosphere, in particular the stemming of climate change, safeguarding biodiversity and conserving forests. The continued development of national environmental law is made necessary, which drew attention to the importance of globally effective international environmental law.

List of the environmental legislations that have come into effect:

- General
- Forest and wildlife
- Water
- Air

Table 9.1 Chronologically arranged table for environmental acts.

General		
Year	*Act*	*Aim*
1986	**The Environment (Protection) Act**	It authorizes the central government to protect and improve environmental quality, control and reduce pollution from all sources, and prohibit or restrict the setting and/or operation of any industrial facility on environmental grounds.
1986	**The Environment (Protection) Rules**	These rules lay down procedures for setting standards of emission or discharge of environmental pollutants.
1989	**The objective of Hazardous Waste (Management and Handling) Rules**	These rules control the generation, collection, treatment, import, storage, and handling of hazardous waste.
1989	**The Manufacture, Storage, and Import of Hazardous Rules**	These rules define the terms used in this context, and sets up an authority to inspect, once a year, the industrial activity connected with hazardous chemicals and isolated storage facilities.
1989	**The Manufacture, Use, Import, Export, and Storage of hazardous Microorganisms/ Genetically Engineered Organisms or Cells Rules**	It was introduced with a view to protect the environment, nature, and health, in connection with the application of gene technology and microorganisms.
1991	**The Public Liability Insurance Act and Rules and Amendment, 1992**	It was drawn up to provide for public liability insurance for the purpose of providing immediate relief to the persons affected by accident while handling any hazardous substance.
1995	**The National Environmental Tribunal Act**	It has been created to award compensation for damages to persons, property, and the environment arising from any activity involving hazardous substances.
1997	**The National Environment Appellate Authority Act**	It has been created to hear appeals with respect to restrictions of areas in which classes of industries, etc. are carried out or prescribed subject to certain safeguards under the EPA.

Contd...

Contd...

1998	**The Biomedical Waste (Management and Handling) Rules**	These rules are a legal binding on the health care institutions to streamline the process of proper handling of hospital waste such as segregation, disposal, collection, and treatment.
1999	**The Environment (Siting for Industrial Projects) Rules**	These rules lay down detailed provisions relating to areas to be avoided for siting of industries, precautionary measures to be taken for site selecting as also the aspects of environmental protection which should have been incorporated during the implementation of the industrial development projects.
2000	**The Municipal Solid Wastes (Management and Handling) Rules**	These rules apply to every municipal authority responsible for the collection, segregation, storage, transportation, processing, and disposal of municipal solid wastes.
2000	**The Ozone Depleting Substances (Regulation and Control) Rules**	These rules have been laid down for the regulation of production and consumption of ozone depleting substances.
2001	**The Batteries (Management and Handling) Rules, 2001**	These rules shall apply to every manufacturer, importer, re-conditioner, assembler, dealer, auctioneer, consumer, and bulk consumer involved in the manufacture, processing, sale, purchase, and use of batteries or components so as to regulate and ensure the environmentally safe disposal of used batteries.
2002	**The Noise Pollution (Regulation and Control) (Amendment) Rules**	These rules lay down such terms and conditions as are necessary to reduce noise pollution, permit use of loudspeakers or public address systems during night hours (between 10:00 p.m. and 12:00 midnight) on or during any cultural or religious festive occasion.
2002	**The Biological Diversity Act**	It is an act to provide for the conservation of biological diversity, sustainable use of its components, and fair and equitable sharing of the benefits arising out of the use of biological resources and knowledge associated with it.
Forest and wildlife		
1927	**The Indian Forest Act and Amendment, 1984**	It is one of the many surviving colonial statutes. It was enacted to 'consolidate the law related to forest, the transit of forest produce, and the duty leviable on timber and other forest produce'.
1972	**The Wildlife Protection Act, Rules 1973 and Amendment 1991**	It provides for the protection of birds and animals and for all matters that are connected to it whether it be their habitat or the waterhole or the forests that sustain them.
1980	**The Forest (Conservation) Act and Rules, 1981**	It provides for the protection and conservation of the forests.
Water		

Contd...

Contd...

1882	**The Easement Act**	It allows private rights to use a resource that is, groundwater, by viewing it as an attachment to the land. It also states that all surface water belongs to the state and is a state property.
1897	**The Indian Fisheries Act**	It establishes two sets of penal offences whereby the government can sue any person who uses dynamite or other explosive substance in any way (whether coastal or inland) with intent to catch or destroy any fish or poisonous fish in order to kill.
1956	**The River Boards Act**	It enables the states to enroll the central government in setting up an Advisory River Board to resolve issues in inter-state cooperation
1970	**The Merchant Shipping Act**	It aims to deal with waste arising from ships along the coastal areas within a specified radius.
1974	**The Water (Prevention and Control of Pollution) Act**	It establishes an institutional structure for preventing and abating water pollution. It establishes standards for water quality and effluent. Polluting industries must seek permission to discharge waste into effluent bodies. The CPCB (Central Pollution Control Board) was constituted under this act.
1977	**The Water (Prevention and Control of Pollution) Cess Act**	It provides for the levy and collection of cess or fees on water consuming industries and local authorities.
1978	**The Water (Prevention and Control of Pollution) Cess Rules**	It contains the standard definitions and indicates the kind of and location of meters that every consumer of water is required to affix.
1991	**The Coastal Regulation Zone Notification**	It puts regulations on various activities, including construction, are regulated. It gives some protection to the backwaters and estuaries.
Air		
1948	**The Factories Act and Amendment in 1987**	It was the first to express concern for the working environment of the workers. The amendment of 1987 has sharpened its environmental focus and expanded its application to hazardous processes.
1981	**The Air (Prevention and Control of Pollution) Act**	It provides for the control and abatement of air pollution. It entrusts the power of enforcing this act to the CPCB.
1982	**The Air (Prevention and Control of Pollution) Rules**	It defines the procedures of the meetings of the Boards and the powers entrusted to them.
1982	**The Atomic Energy Act**	It deals with the radioactive waste.
1987	**The Air (Prevention and Control of Pollution) Amendment Act**	It empowers the central and state pollution control boards to meet with grave emergencies of air pollution.
1988	**The Motor Vehicles Act**	It states that all hazardous waste is to be properly packaged, labelled, and transported.

9.3 ENVIRONMENT (PROTECTION) ACT, 1986

No. 29 of 1986 [23rd May, 1986.]

An Act to provide for the protection and improvement of environment and for matters connected therewith: whereas the decisions were taken at the United Nations Conference on the Human Environment held at Stockholm in June, 1972, in which India participated, to take appropriate steps for the protection and improvement of human environment; and whereas it is considered necessary further to implement the decisions aforesaid in so far as they relate to the protection and improvement of environment and the prevention of hazards to human beings, other living creatures, plants and property; be it enacted by Parliament in the Thirty-seventh Year of the Republic of India as follows:

1. Short Title, Extend and Commencement

(1) This Act may be called the Environment (Protection) Act, 1986.

(2) It extends to the whole of India.

(3) It shall come into force on such date as the Central Government may, by notification in the Official Gazette, appoint and different dates may be appointed for different provisions of this Act and for different areas.

2. Definitions

In this Act, unless the context otherwise requires–

(a) "environment" includes water, air and land and the interrelationship which exists among and between water, air and land, and human beings, other living creatures, plants, microorganism and property;

(b) "environmental pollutant" means any solid, liquid or gaseous substance present in such concentration as may be, or tend to be, injurious to environment;

(c) "environmental pollution" means the presence in the environment of any environmental pollutant;

(d) "handling", in relation to any substance, means the manufacture, processing, treatment, package, storage, transportation, use, collection, destruction, conversion, offering for sale, transfer or the like of such substance;

(e) "hazardous substance" means any substance or preparation which, by reason of its chemical or physico-chemical properties or handling, is liable to cause harm to human beings, other living creatures, plant, microorganism, property or the environment;

(f) "occupier", in relation to any factory or premises, means a person who has, control over the affairs of the factory or the premises and includes in relation to any substance, the person in possession of the substance;

(g) "prescribed" means prescribed by rules made under this Act.

3. Power of Central Government to take Measures to Protect and Improve Environment

(1) Subject to the provisions of this Act, the Central Government shall have the power to take all such measures as it deems necessary or expedient for the purpose of protecting and improving the quality of the environment and preventing controlling and abating environmental pollution.

(2) In particular, and without prejudice to the generality of the provisions of sub-section (1), such measures may include measures with respect to all or any of the following matters, namely:

(i) co-ordination of actions by the State Governments, officers and other authorities–
   (a) under this Act, or the rules made there under, or
   (b) under any other law for the time being in force which is relatable to the objects of this Act;

(ii) planning and execution of a nation-wide programme for the prevention, control and abatement of environmental pollution;

(iii) laying down standards for the quality of environment in its various aspects;

(iv) laying down standards for emission or discharge of environmental pollutants from various sources whatsoever:

Provided that different standards for emission or discharge may be laid down under this clause from different sources having regard to the quality or composition of the emission or discharge of environmental pollutants from such sources;

(v) restriction of areas in which any industries, operations or processes or class of industries, operations or processes shall not be carried out or carried out subject to certain safeguards;

(vi) laying down procedures and safeguards for the prevention of accidents which may cause environmental pollution and remedial measures for such accidents;

(vii) laying down procedures and safeguards for the handling of hazardous substances;

(viii) examination of such manufacturing processes, materials and substances as are likely to cause environmental pollution;

(ix) carrying out and sponsoring investigations and research relating to problems of environmental pollution;

(x) inspection of any premises, plant, equipment, machinery, manufacturing or other processes, materials or substances and giving, by order, of such directions to such authorities, officers or persons as it may consider necessary to take steps for the prevention, control and abatement of environmental pollution;

(xi) establishment or recognition of environmental laboratories and institutes to carry out the functions entrusted to such environmental laboratories and institutes under this Act;

(xii) collection and dissemination of information in respect of matters relating to environmental pollution;

(xiii) preparation of manuals, codes or guides relating to the prevention, control and abatement of environmental pollution;

(xiv) such other matters as the Central Government deems necessary or expedient for the purpose of securing the effective implementation of the provisions of this Act.

(3) The Central Government may, if it considers it necessary or expedient so to do for the purpose of this Act, by order, published in the Official Gazette, constitute an authority or authorities by such name or names as may be specified in the order for the purpose of exercising and performing such of the powers and functions (including the power to issue directions under section 5) of the Central Government under this Act and for taking measures with respect to such matters referred to in sub-section (2) as may be mentioned in the order and subject to the supervision and control of the Central Government and

the provisions of such order, such authority or authorities may exercise any powers or perform the functions or take the measures so mentioned in the order as if such authority or authorities had been empowered by this Act to exercise those powers or perform those functions or take such measures.

4. Appointment of Officers and their Powers and Functions

(1) Without prejudice to the provisions of sub-section (3) of section 3, the Central Government may appoint officers with such designation as it thinks fit for the purposes of this Act and may entrust to them such of the powers and functions under this Act as it may deem fit.

(2) The officers appointed under sub-section (1) shall be subject to the general control and direction of the Central Government or, if so directed by that Government, also of the authority or authorities, if any, constituted under sub-section (3) of section 3 or of any other authority or officer.

5. Power to give Directions

Notwithstanding anything contained in any other law but subject to the provisions of this Act, the Central Government may, in the exercise of its powers and performance of its functions under this Act, issue directions in writing to any person, officer or any authority and such person, officer or authority shall be bound to comply with such directions.

Explanation — For the avoidance of doubts, it is hereby declared that the power to issue directions under this section includes the power to direct–

(a) the closure, prohibition or regulation of any industry, operation or process; or

(b) stoppage or regulation of the supply of electricity or water or any other service.

6. Rules to Regulate Environmental Pollution

(1) The Central Government may, by notification in the Official Gazette, make rules in respect of all or any of the matters referred to in section 3.

(2) In particular, and without prejudice to the generality of the foregoing power, such rules may provide for all or any of the following matters, namely:

- (a) the standards of quality of air, water or soil for various areas and purposes;
- (b) the maximum allowable limits of concentration of various environmental pollutants (including noise) for different areas;
- (c) the procedures and safeguards for the handling of hazardous substances;
- (d) the prohibition and restrictions on the handling of hazardous substances in different areas;
- (e) the prohibition and restriction on the location of industries and the carrying on process and operations in different areas;
- (f) the procedures and safeguards for the prevention of accidents which may cause environmental pollution and for providing for remedial measures for such accidents.

7. Persons Carrying on Industry Operation, etc., not to Allow Emission or Discharge of Environmental Pollutants in Excess of the Standards

No person carrying on any industry, operation or process shall discharge or emit or permit to be discharged or emitted any environmental pollutants in excess of such standards as may be prescribed.

8. Persons Handling Hazardous Substances to Comply with Procedural Safeguards

No person shall handle or cause to be handled any hazardous substance except in accordance with such procedure and after complying with such safeguards as may be prescribed.

9. Furnishing of Information to Authorities and Agencies in Certain Cases

(1) Where the discharge of any environmental pollutant in excess of the prescribed standards occurs or is apprehended to occur due to any accident or other unforeseen act or event, the person responsible for such discharge and the person in charge of the place at which such discharge occurs or is apprehended to occur shall be bound to prevent or mitigate the environmental pollution caused as a result of such discharge and shall also forthwith–

(a) intimate the fact of such occurrence or apprehension of such occurrence; and

(b) be bound, if called upon, to render all assistance, to such authorities or agencies as may be prescribed.

(2) On receipt of information with respect to the fact or apprehension on any occurrence of the nature referred to in sub-section (1), whether through intimation under that sub-section or otherwise, the authorities or agencies referred to in sub-section (1) shall, as early as practicable, cause such remedial measures to be taken as necessary to prevent or mitigate the environmental pollution.

(3) The expenses, if any, incurred by any authority or agency with respect to the remedial measures referred to in sub-section (2), together with interest (at such reasonable rate as the Government may, by order, fix) from the date when a demand for the expenses is made until it is paid, may be recovered by such authority or agency from the person concerned as arrears of land revenue or of public demand.

10. Powers of Entry and Inspection

(1) Subject to the provisions of this section, any person empowered by the Central Government in this behalf shall have a right to enter, at all reasonable times with such assistance as he considers necessary, any place–

(a) for the purpose of performing any of the functions of the Central Government entrusted to him;

(b) for the purpose of determining whether and if so in what manner, any such functions are to be performed or whether any provisions of this Act or the rules made there under or any notice, order, direction or authorization served, made, given or granted under this Act is being or has been complied with;

(c) for the purpose of examining and testing any equipment, industrial plant, record, register, document or any other material object or for conducting a search of any building in which he has reason to believe that an offence under this Act or the rules made there under has been or is being or is about to be committed and for seizing any such equipment, industrial plant, record, register, document or other material object if he has reason to believe that it may furnish evidence of the commission of an offence punishable under this Act or the rules made there under or that such seizure is necessary to prevent or mitigate environmental pollution.

(2) Every person carrying on any industry, operation or process of handling any hazardous substance shall be bound to render all assistance to the person empowered by the Central Government under sub-section (1) for carrying out the functions under that

sub-section and if he fails to do so without any reasonable cause or excuse, he shall be guilty of an offence under this Act.

(3) If any person willfully delays or obstructs any persons empowered by the Central Government under sub-section (1) in the performance of his functions, he shall be guilty of an offence under this Act.

(4) The provisions of the Code of Criminal Procedure, 1973, or, in relation to the State of Jammu and Kashmir, or an area in which that Code is not in force, the provisions of any corresponding law in force in that State or area shall, so far as may be, apply to any search or seizures under this section as they apply to any search or seizure made under the authority of a warrant issued under section 94 of the said Code or as the case may be, under the corresponding provision of the said law.

11. Power to Take Sample and Procedure to be Followed in Connection Therewith

(1) The Central Government or any officer empowered by it in this behalf, shall have power to take, for the purpose of analysis, samples of air, water, soil or other substance from any factory, premises or other place in such manner as may be prescribed.

(2) The result of any analysis of a sample taken under sub-section (1) shall not be admissible in evidence in any legal proceeding unless the provisions of sub-sections (3) and (4) are complied with.

(3) Subject to the provisions of sub-section (4), the person taking the sample under sub-section (1) shall–

(a) serve on the occupier or his agent or person in charge of the place, a notice, then and there, in such form as may be prescribed, of his intention to have it so analyzed;

(b) in the presence of the occupier of his agent or person, collect a sample for analysis;

(c) cause the sample to be placed in a container or containers which shall be marked and sealed and shall also be signed both by the person taking the sample and the occupier or his agent or person;

(d) send without delay, the container or the containers to the laboratory established or recognized by the Central Government under section 12.

(4) When a sample is taken for analysis under sub-section (1) and the person taking the sample serves on the occupier or his agent or person, a notice under clause (a) of sub-section (3), then,–

(a) in a case where the occupier, his agent or person will fully absents himself, the person taking the sample shall collect the sample for analysis to be placed in a container or containers which shall be marked and sealed and shall also be signed by the person taking the sample, and

(b) in a case where the occupier or his agent or person present at the time of taking the sample refuses to sign the marked and sealed container or containers of the sample as required under clause (c) of sub-section (3), the marked and sealed container or containers shall be signed by the person taking the samples, and the container or containers shall be sent without delay by the person taking the sample for analysis to the laboratory established or recognized under section 12 and such person shall inform the Government Analyst appointed or recognized under section 12 in writing, about the willful absence of the occupier or his agent or person, or, as the case may be, his refusal to sign the container or containers

12. Environmental Laboratories

(1) The Central Government may, by notification in the Official Gazette:

(a) establish one or more environmental laboratories;

(b) recognize one or more laboratories or institutes as environmental laboratories to carry out the functions entrusted to an environmental laboratory under this Act.

(2) The Central Government may, by notification in the Official Gazette, make rules specifying–

(a) the functions of the environmental laboratory;

(b) the procedure for the submission to the said laboratory of samples of air, water, soil or other substance for analysis or tests, the form of the laboratory report thereon and the fees payable for such report;

(c) such other matters as may be necessary or expedient to enable that laboratory to carry out its functions.

13. Government Analysts

The Central Government may by notification in the Official Gazette, appoint or recognize such persons as it thinks fit and having the prescribed qualifications to be Government Analysts for the purpose of analysis of samples of air, water, soil or other substance sent for analysis to any environmental laboratory established or recognized under sub-section (1) of section 12.

14. Reports of Government Analysts

Any document purporting to be a report signed by a Government analyst may be used as evidence of the facts stated therein in any proceeding under this Act.

15. Penalty for Contravention of the Provisiòns of the Act and the Rules, Orders and Directions

(1) Whoever fails to comply with or contravenes any of the provisions of this Act, or the rules made or orders or directions issued there under, shall, in respect of each such failure or contravention, be punishable with imprisonment for a term which may extend to five years with fine which may extend to one lakh rupees, or with both, and in case the failure or contravention continues, with additional fine which may extend to five thousand rupees for every day during which such failure or contravention continues after the conviction for the first such failure or contravention.

(2) If the failure or contravention referred to in sub-section (1) continues beyond a period of one year after the date of conviction, the offender shall be punishable with imprisonment for a term which may extend to seven years.

16. Offences by Companies

(1) Where any offence under this Act has been committed by a company, every person who, at the time the offence was committed, was directly in charge of, and was responsible to, the company for the conduct of the business of the company, as well as the company, shall be deemed to be guilty of the offence and shall be liable to be proceeded against and punished accordingly:

Provided that nothing contained in this sub-section shall render any such person liable to any punishment provided in this Act, if he proves that the offence was committed without his knowledge or that he exercised all due diligence to prevent the commission of such offence.

(2) Notwithstanding anything contained in sub-section (1), where an offence under this Act has been committed by a company and it is proved that the offence has been committed with the consent or connivance of, or is attributable to any neglect on the part of, any director, manager, secretary or other officer of the company, such director, manager, secretary or other officer shall also deemed to be guilty of that offence and shall be liable to be proceeded against and punished accordingly.

Explanation–For the purpose of this section,–

(a) "company" means any body corporate and includes a firm or other association of individuals;

(b) "director", in relation to a firm, means a partner in the firm.

17. Offences by Government Departments

(1) Where an offence under this Act has been committed by any Department of Government, the Head of the Department shall be deemed to be guilty of the offence and shall be liable to be proceeded against and punished accordingly.

Provided that nothing contained in this section shall render such Head of the Department liable to any punishment if he proves that the offence was committed without his knowledge or that he exercise all due diligence to prevent the commission of such offence.

(2) Notwithstanding anything contained in sub-section (1), where an offence under this Act has been committed by a Department of Government and it is proved that the offence has been committed with the consent or connivance of, or is attributable to any neglect on the part of, any officer, other than the Head of the Department, such officer shall also be deemed to be guilty of that offence and shall be liable to be proceeded against and punished accordingly.

18. Protection of Action Taken in Good Faith

No suit, prosecution or other legal proceeding shall lie against the Government or any officer or other employee of the Government or any authority constituted under this Act or any member, officer or other employee of such authority in respect of anything which is done or intended to be done in good faith in pursuance of this Act or the rules made or orders or directions issued there under.

19. Cognizance of Offences

No court shall take cognizance of any offence under this Act except on a complaint made by–

(a) the Central Government or any authority or officer authorized in this behalf by that Government, or

(b) any person who has given notice of not less than sixty days, in the manner prescribed, of the alleged offence and of his intention to make a complaint, to the Central Government or the authority or officer authorized as aforesaid.

20. Information, Reports or Returns

The Central Government may, in relation to its function under this Act, from time to time, require any person, officer, State Government or other authority to furnish to it or any prescribed authority or officer any reports, returns, statistics, accounts and other information and such person, officer, State Government or other authority shall be bound to do so.

21. Members, Officers and Employees of the Authority Constituted under Section 3 to be Public Servants

All the members of the authority, constituted, if any, under section 3 and all officers and other employees of such authority when acting or purporting to act in pursuance of any provisions of this Act or the rules made or orders or directions issued there under shall be deemed to be public servants within the meaning of section 21 of the Indian Penal Code (45 of 1860).

22. Bar of Jurisdiction

No civil court shall have jurisdiction to entertain any suit or proceeding in respect of anything done, action taken or order or direction issued by the Central Government or any other authority or officer in pursuance of any power conferred by or in relation to its or his functions under this Act.

23. Powers to Delegate

Without prejudice to the provisions of sub-section (3) of section 3, the Central Government may, by notification in the Official Gazette, delegate, subject to such conditions and limitations as may be specified in the notifications, such of its powers and functions under this Act [except the powers to constitute an authority under sub-section (3) of section 3 and to make rules under section 25 as it may deem necessary or expedient, to any officer, State Government or other authority.

24. Effect of other Laws

(1) Subject to the provisions of sub-section (2), the provisions of this Act and the rules or orders made therein shall have effect notwithstanding anything inconsistent therewith contained in any enactment other than this Act.

(2) Where any act or omission constitutes an offence punishable under this Act and also under any other Act then the offender found guilty of such offence shall be liable to be punished under the other Act and not under this Act.

25. Power to make Rules

(1) The Central Government may, by notification in the Official Gazette, make rules for carrying out the purposes of this Act.

(2) In particular, and without prejudice to the generality of the foregoing power, such rules may provide for all or any of the following matters, namely–

(a) the standards in excess of which environmental pollutants shall not be discharged or emitted under section 7;

(b) the procedure in accordance with and the safeguards in compliance with which hazardous substances shall be handled or caused to be handled under section 8;

(c) the authorities or agencies to which intimation of the fact of occurrence or apprehension of occurrence of the discharge of any environmental pollutant in excess of the prescribed standards shall be given and to whom all assistance shall be bound to be rendered under sub-section (1) of section 9;

(d) the manner in which samples of air, water, soil or other substance for the purpose of analysis shall be taken under sub-section (1) of section 11;

(e) the form in which notice of intention to have a sample analyzed shall be served under clause (a) of sub-section (3) of section 11;

(f) the functions of the environmental laboratories, the procedure for the submission to such laboratories of samples of air, water, soil and other substances for analysis or test; the form of laboratory report; the fees payable for such report and other matters to enable such laboratories to carry out their functions under sub-section (2) of section 12;

(g) the qualifications of Government Analyst appointed or recognized for the purpose of analysis of samples of air, water, soil or other substances under section 13;

(h) the manner in which notice of the offence and of the intention to make a complaint to the Central Government shall be given under clause (b) of section 19;

(i) the authority of officer to whom any reports, returns, statistics, accounts and other information shall be furnished under section 20;

(j) any other matter which is required to be, or may be, prescribed.

9.4 THE WATER (PREVENTION & CONTROL OF POLLUTION) RULES, 1975

[Dated the 27th February. 1975]

[G.S.R. 58(E).-In exercise of the powers conferred by section 63 of the Water (Prevention and Control of Pollution) Act, 1974 (6 of 1974) the Central Government after consultation with Central Board for the Prevention and Control of Water Pollution, hereby makes the following rules, namely:-

1. Short Title and Commencement

(1) These rules may be called the Water (Prevention and Control of Pollution) Rules, 1975.

(2) They shall come into force on the date of their publication in the Official Gazette.

2. Definitions

In these rules unless the context otherwise requires:

(a) "Act" means the Water (Prevention and Control of Pollution) Act,

(b) "Chairman" means the Chairman of the Central Board;

(c) "Central Board Laboratory" means a laboratory established or recognized as such under sub-section (3) of section 16;

(d) "Central Water Laboratory" means a laboratory established or specified as such under sub-section (I) of section 51;

(e) "Form" means a form set out in Schedule I;

(f) "Member" means a member of the Central Board and includes the Chairman thereof;

(g) "Member-Secretary" means the Member-Secretary of the Central Board;

(h) "Section" means a section of the Act;

(i) "Schedule" means a schedule appended to these rules;

(j) "Year" means the financial year commencing on the first day of April.

3. Salaries, Allowances and Other Conditions of Service of the Chairman

(1) The Chairman shall be paid a fixed monthly salary of Rs. 3000.

(2) (a) in addition to the salary specified in sub-rule (1), the chairman shall be entitled to city compensatory allowance and house rent allowance as admissible to a Central government servant in terms of Ministry of Finance (Department of Expenditure) O.M. No. 2(3)-E. II(B)/64, dated the 27th November, 1965

Provided that where the Chairman is allotted an accommodation by the Central Government, he shall not be entitled to house rent allowance and shall be required to pay ten per cent of the emoluments drawn by him as house rent or licence fee.

(b) in respect of journeys undertaken by the Chairman in connection with his duties as Chairman, he shall be entitled to travelling allowance and daily allowance at the rates permissible in the case of a government officer of equivalent rank under the Supplementary Rules of the Central Government:

(c) the Central Board shall provide to the Chairman, medical facilities as admissible to an office of the Central Government drawing salary of Rs. 3000/- per mensem.]

(3) Notwithstanding anything contained in sub-rules (1) and (2), where a Government servant is appointed as Chairman, the terms and conditions of his service shall be such as may be specified by the Central Government from time to time.

4. Salaries, Allowances and Other Conditions of Service of Member-Secretary

(l) The Member-Secretary shall be paid a monthly pay in the scale of Rs. 2250-125-2500.

(2) (a) In addition to the salary specified in sub-rule (1) the Member-Secretary shall be entitled to city compensatory allowance and house rent allowance as admissible to a Central Government servant in terms of Ministry of Finance (Department of Expenditure) O.M. No. 2(3)-E. II(b)/64, dated the 27th November, 1965:

Provided that where the Member-Secretary is allotted an accommodation by the Central Government, he shall not be entitled to house rent allowance and shall be required to pay ten per cent of the emolument drawn by him as house rent or license fee.

(b) in respect of journeys undertaken by the Member-Secretary in connection with his duties as Member-Secretary he shall be entitled to travelling allowance and daily allowance at the rates permissible in the case of a Government Officer of equivalent rank under the Supplementary Rules of the Central Government;

(c) the Central Board shall provide to the Member-Secretary, medical facilities as admissible to an officer of the Central Government drawing salary in the scale of Rs. 2250-2500 per mensem.

(d) the other terms and conditions of service of the Member-Secretary including allowances payable to him, shall be, as far as may be, the same as are applicable to a Grade I Officer of corresponding status of the Central Government.

(3) Notwithstanding anything contained in sub-rules (1) and (2) where a Government servant is appointed as Member-Secretary, the terms and conditions of his service shall be such as may be specified by the Central Government from time to time.

5. Terms and Conditions of Service of Members of the Central Board

(1) Non-official members of the Central Board resident in Delhi shall be paid an allowance of rupees seventy five per day for each day of the actual meetings of the Central Board.

(2) Non-official members of the Central Board, not resident in Delhi shall be paid an allowance of rupees seventy-five per day (inclusive of daily allowance) for each day of the actual meetings of the Central Board and also travelling allowance at such rate as is admissible to a Grade I officer of the Central Government.

Provided that in case of a member of Parliament who is also a member of the Central Board, the said daily and travelling allowance will be admissible when the Parliament is not in session and on production of a certificate by the member that he has not drawn any such allowance for the same journey and halts from any other Government source.

6. Fees and Allowances to be paid to such Members of a Committee of the Central Board as are not Members of the Board Under Sub-section (3) of Section 9

A member of a committee of the Central Board shall be paid in respect of meetings of the committee travelling and daily allowances, if he is a non-official, at the rates specified in rule 5, as if he were a member of the Central Board and, if he is a Government servant, at the rates, admissible under the relevant rules of the respective government under whom he is serving.

7. Powers and Duties of the Chairman

(1) The Chairman shall have overall control over the day-to-day activities of the Central Board.

(2) (i) The Chairman may undertake tours within India for carrying out the functions of the Central Board:

Provided that he shall keep the Central Government (Secretary to the Ministry of Works and Housing) and the Central Board, informed of his tours.

(ii) The Chairman may, with the prior approval of the Central Government, visit any country outside India.

(3) In the matter of acceptance to tenders, the Chairman shall have full powers subject to the concurrence of the Central board:

Provided that no such occurrence is required for acceptance of tenders upto an amount of rupees one thousand in each case.

(4) Subject to overall sanctioned budget provision, the Chairman shall have full powers to administratively approve and sanction all estimates.

8. Creation and Abolition of Posts

The Central Board may create such posts as it considers necessary for the efficient performance of its functions and may abolish any post, so created:

Provided that for the creation of, and appointment to, posts, the maximum of the scale of which is above Rs. 1,600 per month, the Central Board shall obtain prior sanction of the Central Government.

9. Powers and Duties of the Member-Secretary

The Member-Secretary shall be subordinate to the Chairman and shall, subject to the control of the Chairman, exercise the following powers, namely:

(1) The Member-Secretary shall be in charge of all the confidential papers of the Board and shall be responsible for preserving them.

(2) The Member-Secretary shall produce such papers whenever so directed by the Chairman or by the Central Board.

(3) The Member-Secretary shall make available to any member of the Central Board, for his perusal, any record of the Board.

(4) The Member-Secretary shall be entitled to call for the services of any officer or employee of the Central Board, and files, papers and documents for study from any department of the Board, as also to carry out inspection of any department at any time including checking of accounts, vouchers, bills and others records and stores pertaining to the Board or regional offices there under.

(5) The Member-Secretary may withhold any payment:

Provided that as soon as may be [and in any case not later than three months] after such withholding of payment the matter shall be placed before the Central Board for its approval:

(6) The Member-Secretary shall make all arrangements for holding meetings of the Central Board and meetings of the Committees constituted by the Central Board.

(7) All orders or instructions to be issued by the Central Board shall be over the signature of the Member-Secretary or of any other officer authorized in this behalf by the Chairman.

(8) The Member-Secretary shall authorize, sanction or pass all payments against allotments made or estimates sanctioned.

(9) The Member-Secretary shall write and maintain confidential reports of all Class I and Class II officers of the Central Board and shall get them countersigned by the Chairman.

(10) The Member-Secretary shall countersign the confidential reports of all the Class III employees of the Central Board.

(11) (i) The Member-Secretary shall sanction the annual increments of the Class I and Class II officers of the Central Board:

Provided that the increment of Class I and Class II officers shall be withheld only with the approval of the Chairman.

(ii) The annual increment of other employees of the Central Board not referred to in clause (i) shall be sanctioned by officers authorized in this behalf by the Member-Secretary.

(12) The Member-Secretary shall have full powers for according technical sanction to all estimates.

(13) The Member-Secretary shall exercise such other powers and perform such other functions as may be delegated to him from time to time either by the Board or by the Chairman.

10. Manner and Purpose of Association of Persons with Central Board

(1) The Central Board may invite any persons, whose assistance or advice it considers useful to obtain in performing any of its functions, to participate in the deliverations of any of its meetings.

(2) If the person associated with the Board under sub-rule (1) happens to be a non-official, resident in Delhi, he shall be entitled to get an allowance of [rupees seventy-five] per day for each day of actual meeting of the Central Board in which he is so associated.

(3) If such person is non-resident in Delhi, he shall be entitled to get an allowance of rupees seventy-five per day (inclusive of daily allowance) for each day of actual meeting of the Central Board in which he is so associated and also to travelling allowance at such rates as admissible to a Grade I Officer of the Central Government.

(4) Notwithstanding anything in sub-rules (2) and (3), if such person is Government servant or an employee in a Government undertaking he shall be entitled to travelling and daily allowances only at the rates admissible under the relevant rules applicable to him.

11. Appointment of Consulting Engineer

For the purpose of assisting the Central Board in the performance of its functions, the Board may appoint a consulting engineer to the Board for a specified period not exceeding four months:

Provided that the Board may, with the prior approval of the Central Government extend the period of the appointment from time to time:

Provided further that if at the time of the initial appointment the Central Board had reason to believe that the services of the consulting engineer would be required for a period of more than four months, the Central Board shall not make the appointment without the prior approval of the Central Government.

12. Power to Terminate Appointment

Notwithstanding the appointment of a consulting engineer for a specified period under rule 11, the Central Board shall have the right to terminate the services of the consulting engineer before the expiry of the specified period, if, in the opinion of the Board, the consulting engineer is not discharging his duties properly or to the satisfaction of the Board or such a course of action is necessary in the public interest.

[Provided that the services of a consulting engineer shall not be terminated under this rule by the Central Board except after giving him a reasonable opportunity of showing cause against the proposed action.]

13. Emoluments of the Consulting Engineer

The Central Board may pay the consulting engineer suitable emoluments or fees depending on the nature of work, and the qualifications and experience of the consulting engineer:

Provided that the Central Board shall not appoint any person as consulting engineer without the prior approval of the Central Government if the emoluments or fees payable to him exceeds rupees two thousand per month.

14. Tours by Consulting Engineer

The consulting engineer may undertake tours within the country for the performance of the duties entrusted to him by the Central Board and in respect of such tours he shall be entitled to travelling and daily allowances as admissible to a Grade I officer of the Central Government. He shall, however, get the prior approval of the Member-Secretary to his tour programme.

15. Consulting Engineer not to Disclose Information

The consulting engineer shall not disclose any information either given by the Central Board or obtained during the performance of the duties assigned to him either from the Central Board or otherwise, to any person other than the Central Board without the written permission of the Board.

16. Duties and Functions of the Consulting Engineer

The consulting engineer shall discharge such duties and perform such functions as are assigned to him, by the Central Board and it will be his duty to advice the Board on all technical matters referred to him by the Board.

17. Form of Budget Estimates

(1) The budget in respect of the year next ensuing showing the estimates receipts and expenditure of the Central Board shall be prepared in Forms I, II, III and IV and submitted to the Central government.

(2) The estimated receipts and expenditure shall be accompanied by the revised budget estimates for the current year.

(3) The budget shall, as far as may be, based on the account heads specified in Schedule II.

18. Submission of Budget Estimates to the Central Board

(1) The budget estimates as compiled in accordance with rule 17 shall be placed by the Member-Secretary before the Central Board by the 5th October each year for approval.

(2) After approval of the budget estimates by the Central Board, four copies of the final budget proposals incorporating therein such modifications as have been decided upon by the Central Board shall be submitted to the Central Government by the 15th October each year.

19. Estimates of Establishment Expenditure and Fixed Recurring Charges

(1) The estimates of expenditure on fixed establishment as well as fixed monthly recurring charges on account of rent, allowances, etc., shall provide for the gross sanctioned pay without deductions of any kind.

(2) To the estimates referred to in sub-rule (I) shall be added a suitable provision for leave salary based on past experience with due regard to the intention of the members of the staff in regard to leave as far as the same can be ascertained.

(3) If experience indicates that the total estimate for fixed charges referred to in sub-rules (1) and (2) is not likely to be fully utilized, a suitable lump deduction shall be made from the total amount estimated.

20. Re-appropriations and Emergent Expenditure

No expenditure which is not covered by a provision in the sanctioned budget estimates, or which is likely to be in excess over the amount provide under any head, shall be incurred by the Central Board without provision being made by re-appropriation from some other head under which saving are firmly established and available.

21. Power to Incur Expenditure

The Central Board shall incur expenditure out of the funds received by it in accordance with the instructions laid down under the General Financial Rules of the Central Government and other instructions issued by that Government from time to time.

22. Operation of Fund of the Central Board

The fund of the Central Board shall be operated by the Member-Secretary of the Central Board or in his absence by any officer of the Central Board who may, subject to the approval of the Central Government, be so empowered by the Central Board.

23. Saving

Nothing in this Chapter shall apply to a budget already finalised before the commencement of these rules.

24. Form of the Annual Report

The annual report in respect of the year last ended giving a true and full account of the activities of the Central Board during the previous financial year shall contain the particulars specified in the [Schedule III].

25. Form of Annual Statement of Accounts of the Central Board

The annual statement of accounts of the Central Board shall be in Forms V to IX.

26. Form of Report of Central Board Analyst

When a sample or any water, sewage or trade effluent has been sent for analysis to a laboratory established or recognized by the Central Board, the Central Board analyst appointed under sub-section (3) of section 53 shall analyze the sample and submit to the Central Board a report in triplicate in Form X of the result of such analysis.

26A. Persons possessing the following qualifications are eligible for appointment as Government analysts or as Board analysts under sub-section (I) or sub-section (3), as the case may be, of section 53, namely:

Essential:–

(i) M.Sc. Degree in Chemistry or equivalent or Bachelor Degree in Chemical Engineering or Bio-chemical Engineering; and

(ii) 10 years' experience in analysis of water, sewage or industrial wastes in teaching, research or Government Laboratory.

Desirable.–

(i) Doctorate Degree in Chemistry or Bio-chemistry or Master's Degree in Chemical Engineering or Bio-chemical Engineering;

(ii) Evidence of published papers in the field.

27. Functions of the Central Water Laboratory

The Central Water Laboratory shall cause to be analyzed any samples of water, sewage or trade effluent received by it from any officer authorized by the Central Board for the purpose, and the findings shall be recorded in triplicate in Form XI.

28. Fee for Report

The fees payable in respect of each report of the Central Water Laboratory on any analysis or test mentioned in column (2) of Schedule IV shall be at the rates specified in the corresponding entry in column (3) thereof.

29. Central Board to Act as State Board for Union Territories

The Central Board shall act as State as State Board for Union territories under sub-section (4) of section 4.

30. Power to Take Samples

The Central Board or any officer empowered by it in this behalf, shall have power to take for the purpose of analysis samples of water from any stream or well or samples of sewage or trade effluent which is passing from any plant or vessel or from or over any place into any such stream or well in any Union territory.

31. Form of Notice

A notice under clause (a) of sub-section (3) of section 21 shall, in the case of a Union territory, be in Form XII.

32. Application of Consent

An application for obtaining the consent of the Central Board for establishing or taking any steps to establish any industry, operation or process or any treatment and disposal system or any extension or addition thereto, which is likely to discharge sewage or trade effluent into a stream or well or sewer or on land (such discharge being hereinafter in this Rule referred to as discharge of sewage); or for bringing into use any new or altered outlet for the discharge of sewage or beginning to make any new discharge of sewage under section 25 or for continuing an existing discharge of sewage under section 26 shall be made to the Central Board in Form XIII.

33. Procedure for Making Inquiry into Application for Consent

(1) On receipt of an application for consent under section 25 or section 26, the Central Board may depute any of its Officers, accompanied by as many assistants as may be necessary, to visit to the premises of the applicant, to which such application relates, for the purpose of verifying the correctness or otherwise of the particulars furnished in the application or for obtaining such further particulars or information as such officer may consider necessary. Such officer may for that purpose inspect any place where water or sewage or trade effluent is discharged by the applicant, or treatment plants, purification works or disposal systems of the applicant and may require the applicant to furnish to him any plans, specifications and other data relating to such treatment plants, purification works or disposal systems or any part thereof, that he considers necessary.

(2) Such officer shall before visiting any premises of the applicant for the purpose of inspection under sub-rule (1) above, give notice to the applicant of his intention to do so in Form XIV. The applicant shall furnish to such officer all facilities that such officer may legitimately require for the purpose.

(3) An officer of the Central Board may, before or officer carrying out an inspection under sub-rule (1) above, require the applicant to furnish to him, orally or in writing such additional information or clarification, or to produce before him such documents, as he may consider necessary for the purpose of investigation of the application and may, for that purpose, summon the applicant or his authorized agent to the office of the Central Board.

34. Directions

(1) Any direction issued under section 33A shall be in writing.

(2) The direction shall specify the nature of action to be taken and the time within which it shall be complied with by the person, officer or the authority to whom such direction is given.

(3) The person, officer or authority to whom any direction is sought to be issued shall be served with a copy of the proposed direction and shall be given an opportunity of not less than fifteen days from the date of service of a notice to file with an officer designated in this behalf the objections, if any, to the issue of the proposed direction.

(4) Where the proposed direction is for the stoppage or regulation of electricity or water or any other services affecting the carrying on an industry, operation or process and is sought to be issue to an officer or an authority, a copy of the proposed direction shall also be endorsed to the occupier of the industry, operation or process, as the case may be, and objections, if any, filed by the occupier with an officer designed in this behalf shall be dealt with in accordance with the procedures under sub-rules (3) and (5) of this rule:

Provided that no opportunity of being heard shall be given to the occupier, if he had already been heard earlier and the proposed direction referred to in sub-rule (3) above for the stoppage or regulation of electricity or water or any other service was the resultant decision of the Central Board after such earlier hearing.

(5) The Central Board shall within a period of 45 days from the date of receipt of objections, if any, or from the date up to which an opportunity is given to the person, officer or authority to file objections whichever is earlier, after considering the objections, if any, sought to be directed and for reasons to be recorded in writing, confirm, modify or decide not to issue the proposed direction.

(6) In a case where the Central Board is of the opinion that in view of the likelihood of the grave injury to the environment, it is not expedient to provide an opportunity to file objections against the proposed direction, it may, for reasons to be recorded in writing, issue directions without providing such an opportunity.

(7) Every notice or direction required to be issued under this rule shall be deemed to be duly served:-

(a) Where the person to be served is a company, if the document is addressed in the name of the company and its registered office or at its principal office or place of business and is either–

(i) sent by registered post; or

(ii) delivered at its registered office or at the principal office or place or business;

(b) Where the person to be served is an officer serving Government, if the document is addressed to the person and a copy thereof is endorsed to his Head of the Department and also to the Secretary to the Government as the case may be, in charge of the Department in which for the time being the business relating to the Department in which the officer is employed is transacted and is either-:

(i) sent by registered post, or

(ii) is given or tendered to him;

(c) in any other case, if the document is addressed to the person to be served and-

(i) is given or tendered to him, or

(ii) if such persons cannot be found, is affixed on some conspicuous part of his last known place of residence or business or is given or tendered to some adult members of his family or is affixed on some conspicuous part or the land or building, if any, to which it relates, or

(iii) is sent by registered post to that person.

Explanation:- For the purpose of this sub-rule:-

(a) "company" means by body corporate and includes a firm or other association of individuals;

(b) "a servant" is not a member of the family.

35. Manner of Giving Notice

The manner of giving notice under clause (b) of sub-section (1) of section 49 shall be as follows, namely:–

(1) The notice shall be in writing in Form XV;

(2) If the alleged offence has taken place in a Union Territory, the person giving notice may sent notice to:-

(i) the Central Board, and

(ii) the Ministry of Environmental and Forests (represented by the Secretary to the Government of India);

(iii) the Administrator of the Union Territory (represented by the Secretary Head in charge of Environment).

(3) The notice shall be sent by registered post acknowledgement due; and

(4) The period of sixty days mentioned in clause (b) of sub-section (l) of section 49 of the Act, shall be reckoned from the date it is first received by one of the authorities mentioned in sub-rule (2).]

9.5 MINISTRY OF ENVIRONMENT AND FORESTS

OZONE DEPLETING SUBSTANCES(REGULATION) RULES, 2000

Notification New Delhi, dated, the July 17, 2000

S.O. 670 (E) - Whereas the draft Ozone Depleting Substances (Regulation) Rules, 2000 were published, under the notification of the Government of India in the Ministry of Environment and Forests number S.O. 69(E), dated, the 25th January, 2000, in the Gazette of India, Extra-ordinary, Part II, section 3, sub-section (ii) at pages 39-96 on the same date, inviting objections and suggestions from all persons likely to be affected thereby, before the expiry of the period of forty-five days from the date on which the copies of the Gazette containing the said notification are made available to the public;

And whereas copies of the said Gazette were made available to the public on 26.01.2000;

And whereas the objections and suggestions received from the public in respect of the said draft rules have been duly considered by the Central Government;

Now, therefore, in exercise of the powers conferred by sections 6, 8 and 25 of the Environment (Protection) Act, 1986, the Central Government hereby makes the following rules for regulating ozone depleting substances, namely:

1. Short Title and Commencement

(1) These rules may be called the Ozone Depleting Substances (Regulation and Control) Rules, 2000.

(2) They shall come into force on the date of their publication in the Official Gazette.

2. Definitions

In these rules unless the context otherwise requires,–

(a) "Act" means the Environment (Protection) Act, 1986 (29 of 1986);

(b) "authority" means an authority mentioned in columns (4) and (6) of Schedule V;

(c) "base level" means the quantity of ozone depleting substance produced or consumed, as the case may be, in the year or average of the years listed in column (3) of Schedule II and Schedule III;

(d) "consumption" with respect to any ozone depleting substance means the amount of that substance produced in India in addition to the amount imported, less the amount exported;

(e) "calculated level of production, sale, import or export", as the case may be, means level determined by multiplying quantity of the ozone depleting substance by its ozone depleting potential specified in column (5) of Schedule I;

(f) "calculated level of consumption" shall be determined by adding together calculated levels of production and imports and subtracting calculated level of exports;

(g) "Group" means collection of one or more ozone depleting substances as specified in column (4) of Schedule I;

(h) "manufacture" in relation to any ozone depleting substance includes-

(i) any process or part of a process for making, altering, finishing, packing, labelling, blending or otherwise treating or any ozone depleting substance with a view to sell, distribute or use but does not include the repacking or breaking up of any ozone depleting substance in the ordinary course of retail business; and

(ii) any process in which a preparation containing ozone depleting substance is formulated;

(i) "ozone depleting substance" means the ozone depleting substances specified in column (2) of Schedule I, whether existing by itself or in a mixture, excluding any such substance or mixture (blend) which is in a manufactured product other than a container used for the transportation or storage of such substance;

"parties" means, unless the text otherwise indicates, parties to the protocol;

(j) "pre-shipment applications" are those treatments applied directly preceding and in relation to export, to meet the phytosanitary or sanitary requirements of the importing country or existing phytosanitary or sanitary requirements of the exporting country;

"production" in relation to any ozone depleting substance means the manufacture of an ozone depleting substance from any raw material or feedstock chemicals, but does not include-

(i) the manufacture of a substance that is used and entirely consumed (except for trace quantity) in the manufacture of other chemicals; or

(ii) quantities which are produced incidentally in the manufacture of other chemical substances; or

(iii) quantities which are recycled or reused; or

(iv) quantities which are destroyed by technologies to be specified by the Central Government;

(k) "protocol" means the Montreal Protocol on Substances that Deplete the Ozone Layer, adopted on 16th September 1987;

(l) "quarantine applications", with respect to Group VIII of Schedule I ozone depleting substance, are treatments to prevent the introduction, establishment and or spread of quarantine pests (including diseases), or to ensure their control as specified by the Central Government;

(m) "recovery" means collection and storage of ozone depleting substances from machinery, equipment, or containment vessel during servicing or prior to disposal;

(n) "reclamation" means reprocessing and upgrading of a recovered ozone depleting substance through such methods as filtering, drying, distillation and, or chemical treatment in order to restore the substance to a specified standard of performance.

(o) "schedule" means a schedule annexed to these rules.

3. Regulation of Production and Consumption of Ozone Depleting Substances

(1) No person shall produce or cause to produce any ozone depleting substance after the date specified in column (5) of Schedule V unless he is registered with the authority specified in column (4) of that Schedule:

Provided that for the twelve month period commencing on the date specified in column (6) of Schedule II, and in each twelve month period thereafter, no person shall produce or cause to be produced any group of ozone depleting substances in excess of the corresponding percentage of his calculated base level of production specified in column (4) of that Schedule:

Provided further that calculated level of consumption of such substances in India shall, as a percentage of calculated level of consumption in base years does not exceed the number specified in column (5) of Schedule II.

(2) No person shall produce or cause to produce ozone depleting substances specified as Group I and Group III in column (4) of Schedule I during the period from 1, August, 2000 to 1st January, 2010 in excess of the quantity specified in column (4) of Schedule III and the calculated level of consumption of such substances in India shall as a percentage of calculated level of consumption in base year does not exceed the number specified in column (5) of that Schedule.

(3) A person having received financial assistance from the Multilateral Fund in accordance with article 10 and 10 A of the protocol to which the Central Government is a party for gradual reduction of production of ozone depleting substances specified as Group I and Group III in column (4) of Schedule I shall, limit the production of ozone depleting substances as specified in Group I and Group III in column (4) of Schedule I in each year from 1st August, 2000 to January 1, 2010 to quantities specified in column (4) for each year given in column (6) of Schedule III as per the agreement approved, by the Executive Committee of the Multilateral Fund.

(4) In order to implement the agreement, referred to in sub-rule (3), the Central Government shall introduce implementation modalities, such as, quota system for producing chlorofluorocarbons and the non-compliance with such modalities shall result in consequential penalties laid out in the agreement.

4. Prohibition on Export to or Import from Countries not Specified in Schedule VI

No person shall import or cause to import from or export or cause to export to any country not specified in Schedule VI any ozone depleting substance after the commencement of these rules.

5. Ozone Depleting Substances are to be Exported to or Imported from Countries Specified in Schedule VI Under a Licence

(1) No person shall import or cause to import from or export or cause to export to, any country specified in Schedule VI, any ozone depleting substance unless he obtains a license issued by the authority.

(2) No license shall be issued under sub-rule (1) unless the said authority is satisfied that the grant of license shall not cause calculated level of consumption of that group of ozone depleting substances (except Group I and Group III given in column (4) of schedule I in the relevant twelve month period, as a percentage of corresponding calculated consumption in base years, to exceed the number specified in column (5) of Schedule II).

(3) No license shall be issued under sub-rule (1) unless the said authority is satisfied that the grant of license shall not cause calculated level of consumption of ozone depleting substance given in Group 1 and III in column (4) of Schedule I in the relevant twelve months period as specified in column (6) of Schedule III, as a percentage of

calculated consumption in base years to exceed the number specified in column (5) of that Schedule.

(4) The calculated base level of consumption and the calculated base level of production for India as a whole for each group of ozone depleting substances shall be notified by the Central Government.

6. Regulation of the Sale of Ozone Depleting Substances

(1) No person shall either himself or by any other person on his behalf or enterprise sell, stock or exhibit for sale or distribute any ozone depleting substance after the date specified in column (5) of Schedule V unless he is registered with the authority specified in column (4) of that Schedule

Provided that no person or enterprise shall sell ozone depleting substances specified in column (3) of Schedule IV for activities specified in column (2) of that Schedule unless the person engaged in that activity has got himself registered with the authority and has given a declaration in accordance with these rules and the person selling ozone depleting substances has verified particulars of the registration given in the declaration with the certificate of registration as per procedure specified in Part II of Schedule XII :

Provided further that after the date specified in column (4) of Schedule IV, no person or enterprise shall sell, stock, distribute or exhibit or cause to be sold, stocked, distributed or exhibited ozone depleting substances specified in column (3) for activities specified in column (2) of that Schedule.

(2) No person shall either himself or by any person on his behalf, or enterprise sell, stock or exhibit for sale or distribute any ozone depleting substance to any person or enterprise who has informed the Central Government that he or that enterprise shall not use the specified ozone depleting substances in manufacturing or other activities after the date specified by such person or as the case may be, the enterprise.

(3) The Central Government shall notify the list of persons, ozone depleting substances and dates informed to it under sub-rule (2).

7. Regulation on the Purchase of Ozone Depleting Substances

No person shall either himself or by any person on his behalf or enterprise, purchase ozone depleting substances specified in column (3) of Schedule IV from any person for making stock or for using such ozone depleting substances for activities specified in column (2) of that Schedule unless he has given the declaration specified in Part I of Schedule XII to the seller of such substances within the time period specified in Serial number 4 of column (5) of Schedule V.

8. Regulation on the use of Ozone Depleting Substance

(1) No person or enterprise shall engage in any activity specified in column (2) of Schedule IV that uses ozone depleting substances specified in column (3) of that Schedule after the date specified in column (5) of Schedule V unless he is registered with the authority specified in column (4) of that Schedule.

(2) No person shall engage in any activity specified in column (2) of Schedule IV using ozone depleting substances specified in column (3) of that Schedule after the date specified in column (5) of Schedule V unless the products are labeled to indicate the ozone depleting substance they contain.

(3) No person shall engage in any activity specified in column (2) of Schedule IV using ozone depleting substances specified in column (3) after the date specified in column (4) of that Schedule

(4) No person shall engage in any activity specified in column (2) of Schedule IV without using label indicating absence of use of ozone depleting substance mentioned in column (3) after the date specified in column (4) of that Schedule.

(5) A person, having received financial and technical assistance from the Multilateral Fund in accordance with the Article 10 and 10 A of the Montreal Protocol on Substances that Deplete the Ozone Layer, to which the Central Government is a Party for phasing out of ozone depleting substances specified in column (2) of Schedule II used in activities specified in column (2) of Schedule IV, either himself or by any person on his behalf or through any enterprise, shall not engage in such activity as specified in column (2) of Schedule IV using ozone depleting substances specified in column (3) of the Schedule, after the date of completion of the conversion work or signing of the Handing Over Protocol, or the submission of the completion report to change from ozone depleting substance technology to non ozone depleting substance technology and the said date be registered with the authority specified in column (4) of the Schedule V.

(6) Any person or enterprise having received, financial assistance from the Multilateral Fund in accordance with the Article 10 and 10 A of the Montreal Protocol on Substances that Deplete the Ozone Layer shall submit an affidavit or declaration with the authority specified in column (4) of Schedule V stating that replaced equipment, resulted from completion of conversion process from ozone depleting substance technology to non ozone depleting substance technology, have been destroyed, dismantled, rendered unusable and that no ozone depleting substance should be used after the date of the completion of project and the said date be registered with the authority specified in the column (4) of the Schedule V.

9. Prohibition on new Investments with Ozone Depleting Substances

(1) No person shall establish or expand or cause to establish or expand any manufacturing facility for production of any ozone depleting substance after the date specified in column (7) of Schedule II and III.

(2) No person shall establish or expand or cause to establish or expand any manufacturing facility, with a view to manufacturing products which contain, or are made with, any ozone depleting substance after the date specified in column (8) of Schedule II & III.

(3) A person having received financial and technical assistance from the Multilateral Fund in accordance with the Article 10 and 10A of the Montreal Protocol on Substances that Deplete the Ozone Layer for phasing out of ozone depleting substances specified in column (2) of Schedule II used in activities specified in column (2) of Schedule IV to which the Central Government is a Party, shall not establish or expand or cause to establish or expand the manufacturing facility for production of any ozone depleting substance or with a view to manufacturing products which contain or are made with any ozone depleting substances after the approval of the project for conversion and date of completion of the conversion work from the ozone depleting substance technology to non ozone depleting substance technology.

10. Regulation of Import, Export and Sale of Products made with or Containing Ozone Depleting Substances

(1) No person shall import or cause to import any product specified in column (2) of Schedule VII which are made with or contain ozone depleting substances specified in column (3) after the date specified in column (4) of that Schedule unless he obtains a license issued by the authority:

Provided that such products which do not contain such ozone depleting substances shall carry a label to that effect before its import is allowed after the date specified in Column 4 of Schedule VII.

(2) No person or enterprise shall export or cause to export any product specified in column (2) of Schedule VII unless such product carries a label specifying whether or not the product has been made with or contains, as the case may be, ozone depleting substances specified in column (3) of that Schedule, after the date specified in column (5) of that Schedule.

(3) No person shall either himself or by any other person or enterprise on his behalf sell, stock or exhibit for sale or distribute any product resulting out of activities, or provide services, specified in column (2) of Schedule IV using ozone depleting substances specified in column (3) after the date specified in column (4) of that Schedule.

11. Regulation on Reclamation and Destruction of Ozone Depleting Substances

(1) No person shall reclaim or cause to reclaim any ozone depleting substance after the date specified in column (5) of Schedule V unless he has registered with the authority specified in column (4) of that Schedule.

(2) No person shall destroy or cause to destroy any ozone depleting substance after the date specified in column (5) of Schedule V unless he has registered with the authority specified in column (4) of that Schedule.

12. Regulation on Manufacture, Import and Export of Compressors

(1) No person shall manufacture, import or export compressors after the date specified in column (5) of Schedule V unless he is registered with the authority specified in column (4) of that Schedule.

13. Procedure for Registration, Cancellation of Registration and Appeal Against such Orders

(1) The procedure for registration and conditions of registration under various provisions of these rules shall be as specified in Schedule IX.

(2) The registering authority shall not register if he is not satisfied that the procedure for registration or conditions of registration is fulfilled.

(3) The registering authority shall cancel the registration if he is satisfied that condition(s) of registration have been violated.

(4) The registering authority shall give the concerned person an opportunity of being heard before passing orders under sub-rules (2) and (3) and the orders shall be made in writing.

(5) An appeal against an order of the registering authority shall lie with the authority specified in column (6) of Schedule V within thirty days of communication of such order.

14. Monitoring and Reporting Requirements

(1) Every person who produces, imports, exports or sells any ozone depleting substance shall maintain records and file reports in the manner specified in Part I of Schedule X.

(2) Every person stocking or purchasing any ozone depleting substance for use in activities specified in column (2) of Schedule IV shall maintain records and file reports in the manner specified in Part II of Schedule X.

(3) Every person who has received technical or financial assistance from any international organization or any financial assistance, which includes concession or exemption from payment of duties, from the Central Government, shall maintain records and file reports in the manner specified in Part III of Schedule X and the list of such persons shall be notified by the Central Government.

(4) Every person who has facility to reclaim an ozone depleting substance shall maintain records and file reports in the manner specified in Part IV of Schedule X.

(5) Every person who has facility to destroy any ozone depleting substance shall maintain records and file reports in the manner specified in Part V of Schedule X.

(6) Every person who manufactures, imports, exports or sells compressors shall maintain records and file reports in the manner specified in Part VI of Schedule X.

(7) The records maintained in accordance with the above sub-rules shall be made available for inspection as specified in Part VII of Schedule X.

15. Exemption

(1) Nothing contained in these rules shall apply to applications or circumstances specified in Schedule VIII.

Schedule – I: List of Ozone Depleting Substances

Schedule – II: Regulation on production and consumption of group of ozone depleting substances

Schedule – III: Regulation on production and consumption of Group I & Group III ozone depleting substances Specified in column (4) of Schedule I.

Schedule – IV: Regulation on consumption of ozone depleting substances on end use basis

Schedule – V: List of authorities , their functions and last date for registration

Schedule – VI: List of countries which are party to the 1987 Montreal Protocol

Schedule – VII: Regulation on import and export products containing ozone depleting substances

Schedule – VIII: Exemption

Schedule – IX: Procedure for Registration

Schedule – X: Records to be maintained and Reports to be submitted

Schedule – XI: Report on production of ozone depleting substances

Schedule – XII: End-use declaration

9.6 MINISTRY OF ENVIRONMENT AND FORESTS THE MUNICIPAL SOLID WASTE (MANAGEMENT AND HANDLING) RULES, 2000

Notification New Delhi, the 25th September, 2000

S.O. 908(E).- Whereas the draft of the Municipal Solid Wastes (Management and Handling) Rules, 1999 were published under the notification of the Government of India in the Ministry of Environment and Forests number S.O. 783(E), dated, the 27th September, 1999 in the Gazette of India, Part II, Section 3, Sub-section (ii) of the same date inviting objections and suggestions from the persons likely to be affected thereby, before the expiry of the period of sixty days from the date on which the copies of the Gazette containing the said notification are made available to the public;

And whereas copies of the said Gazette were made available to the public on the 5th October, 1999;

And whereas the objections and suggestions received from the public in respect of the said draft rules have been duly considered by the Central Government;

Now, therefore, in exercise of the powers conferred by section 3, 6 and 25 of the Environment (Protection) Act, 1986 (29 of 1986), the Central Government hereby makes the following rules to regulate the management and handling of the municipal solid wastes, namely:

1. Short Title and Commencement

These rules may be called the Municipal Solid Wastes (Management and Handling) Rules, 2000.

Save as otherwise provided in these rules, they shall come into force on the date of their publication in the Official Gazette.

2. Application

These rules shall apply to every municipal authority responsible for collection, segregation, storage, transportation, processing and disposal of municipal solid wastes.

3. Definitions

In these rules, unless the context otherwise requires,

"anaerobic digestion" means a controlled process involving microbial decomposition of organic matter in the absence of oxygen;

"authorization" means the consent given by the Board or Committee to the "operator of a facility";

"biodegradable substance" means a substance that can be degraded by microorganisms;

"biomethanation" means a process which entails enzymatic decomposition of the organic matter by microbial action to produce methane rich biogas;

"collection" means lifting and removal of solid wastes from collection points or any other location;

"composting" means a controlled process involving microbial decomposition of organic matter;

"demolition and construction waste" means wastes from building materials debris and rubble resulting from construction, re-modeling, repair and demolition operation;

"disposal" means final disposal of municipal solid wastes in terms of the specified measures to prevent contamination of groundwater, surface water and ambient air quality;

"form" means a Form appended to these rules;

"generator of wastes" means persons or establishments generating municipal solid wastes;

"land filling" means disposal of residual solid wastes on land in a facility designed with protective measures against pollution of ground water, surface water and air fugitive dust, wind-blown litter, bad odour, fire hazard, bird menace, pests or rodents, greenhouse gas emissions, slope instability and erosion;

"leachate" means liquid that seeps through solid wastes or other medium and has extracts of dissolved or suspended material from it;

"lysimeter" is a device used to measure rate of movement of water through or from a soil layer or is used to collect percolated water for quality analysis;

"municipal authority" means Municipal Corporation, Municipality, Nagar Palika, Nagar Nigam, Nagar Panchayat, Municipal Council including notified area committee (NAC) or any other local body constituted under the relevant statutes and, where the management and handling of municipal solid waste is entrusted to such agency;

"municipal solid waste" includes commercial and residential wastes generated in a municipal or notified areas in either solid or semi-solid form excluding industrial hazardous wastes but including treated bio-medical wastes;

"operator of a facility" means a person who owns or operates a facility for collection, segregation, storage, transportation, processing and disposal of municipal solid wastes and also includes any other agency appointed as such by the municipal authority for the management and handling of municipal solid wastes in the respective areas;

"pelletisation" means a process whereby pellets are prepared which are small cubes or cylindrical pieces made out of solid wastes and includes fuel pellets which are also referred as refuse derived fuel;

"processing" means the process by which solid wastes are transformed into new or recycled products;

"recycling" means the process of transforming segregated solid wastes into raw materials for producing new products, which may or may not be similar to the original products;

"Schedule" means a Schedule appended to these rules;

"segregation" means to separate the municipal solid wastes into the groups of organic, inorganic, recyclables and hazardous wastes;

"State Board or the Committee" means the State Pollution Control Board of a State, or as the case may be, the Pollution Control Committee of a Union territory;

"storage" means the temporary containment of municipal solid wastes in a manner so as to prevent littering, attraction to vectors, stray animals and excessive foul odour;

"transportation" means conveyance of municipal solid wastes from place to place hygienically through specially designed transport system so as to prevent foul odour, littering, unsightly conditions and accessibility to vectors;

"vadose water" water which occurs between the ground, surface and the water table that is the unsaturated zone;

"vermi-composting" is a process of using earthworms for conversion of bio-degradable wastes into compost.

4. Responsibility of Municipal Authority

Every municipal authority shall, within the territorial area of the municipality, be responsible for the implementation of the provisions of these rules, and for any infrastructure development for collection, storage, segregation, transportation, processing and disposal of municipal solid wastes.

The municipal authority or an operator of a facility shall make an application in Form-I, for grant of authorization for setting up waste processing and disposal facility including landfills from the State Board or the Committee in order to comply with the implementation programme laid down in Schedule I.

The municipal authority shall comply with these rules as per the implementation schedule laid down in Schedule I.

(4) The municipal authority shall furnish its annual report in Form-II,- to the Secretary-incharge of the Department of Urban Development of the concerned State or as the case may be of the Union territory, in case of a metropolitan city; or to the District Magistrate or the Deputy Commissioner concerned in case of all other towns and cities, with a copy to the State Board or the Committee on or before the 30th day of June every year.

5. Responsibility of the State Government and the Union Territory Administrations

(1) The Secretary-incharge of the Department of Urban Development of the concerned State or the Union territory, as the case may be, shall have the overall responsibility for the enforcement of the provisions of these rules in the metropolitan cities.

(2) The District Magistrate or the Deputy Commissioner of the concerned district shall have the overall responsibility for the enforcement of the provisions of these rules within the territorial limits of their jurisdiction.

6. Responsibility of the Central Pollution Control Board and the State Board or the Committees

The State Board or the Committee shall monitor the compliance of the standards regarding ground water, ambient air, leachate quality and the compost quality including incineration standards as specified under Schedules II, III and IV.

The State Board or the Committee, after the receipt of application from the municipal authority or the operator of a facility in Form I, for grant of authorization for setting up waste processing and disposal facility including landfills, shall examine the proposal taking into consideration the views of other agencies like the State Urban Development Department, the Town and Country Planning Department, Airport or Air Base Authority, the Ground Water Board or any such other agency prior to issuing the authorization.

The State Board or the Committee shall issue the authorization in Form-III to the municipal authority or an operator of a facility within forty-five days stipulating compliance criteria and standards as specified in Schedules II, III and IV including such other conditions, as may be necessary.

The authorization shall be valid for a given period and after the validity is over, a fresh authorization shall be required.

(5) The Central Pollution Control Board shall co-ordinate with the State Boards and the Committees with particular reference to implementation and review of standards and guidelines and compilation of monitoring data.

7. Management of Municipal Solid Wastes

(1) Any municipal solid waste generated in a city or a town, shall be managed and handled in accordance with the compliance criteria and the procedure laid down in Schedule-II.

(2) The waste processing and disposal facilities to be set up by the municipal authority on their own or through an operator of a facility shall meet the specifications and standards as specified in Schedules III and IV.

8. Annual Reports

The State Boards and the Committees shall prepare and submit to the Central Pollution Control Board an annual report with regard to the implementation of these rules by the 15th of September every year in Form-IV.

The Central Pollution Control Board shall prepare the consolidated annual review report on management of municipal solid wastes and forward it to the Central Government along with its recommendations before the 15th of December every year.

9. Accident Reporting

When an accident occurs at any municipal solid wastes collection, segregation, storage, processing, treatment and disposal facility or landfill site or during the transportation of such wastes, the municipal authority shall forthwith report the accident in Form-V to the Secretary in-charge of the Urban Development Department in metropolitan cities, and to District Collector or Deputy Commissioner in all other cases.

Management of Municipal Solid Wastes

S. No.	Parameters	Compliance criteria
1.	**Collection of municipal solid wastes**	1. Littering of municipal solid waste shall be prohibited in cities, towns and in urban areas notified by the State Governments. To prohibit littering and facilitate compliance, the following steps shall be taken by the municipal authority, namely: Organizing house-to-house collection of municipal solid wastes through any of the methods, like community bin collection (central bin), house-to-house collection, collection on regular pre-informed timings and scheduling by using bell ringing of musical vehicle (without exceeding permissible noise levels); Devising collection of waste from slums and squatter areas or localities including hotels, restaurants, office complexes and commercial areas; Wastes from slaughter houses, meat and fish markets, fruits and vegetable markets, which are biodegradable in nature, shall be managed to make use of such wastes; Bio-medical wastes and industrial wastes shall not be mixed with municipal solid wastes and such wastes shall follow the rules separately specified for the purpose; Collected waste from residential and other areas shall be transferred to community bin by hand-driven containerized carts or other small vehicles;

Horticultural and construction or demolition wastes or debris shall be separately collected and disposed off following proper norms. Similarly, wastes generated at dairies shall be regulated in accordance with the State laws;

Waste (garbage, dry leaves) shall not be burnt;

Stray animals shall not be allowed to move around waste storage facilities or at any other place in the city or town and shall be managed in accordance with the State laws.

2. The municipal authority shall notify waste collection schedule and the likely method to be adopted for public benefit in a city or town.
3. It shall be the responsibility of generator of wastes to avoid littering and ensure delivery of wastes in accordance with the collection and segregation system to be notified by the municipal authority as per para 1(2) of this Schedule.

2. Segregation of municipal solid wastes

In order to encourage the citizens, municipal authority shall organize awareness programmes for segregation of wastes and shall promote recycling or reuse of segregated materials.

The municipal authority shall undertake phased programme to ensure community participation in waste segregation. For this purpose, regular meetings at quarterly intervals shall be arranged by the municipal authorities with representatives of local resident welfare associations and non-governmental organizations.

3. Storage of municipal solid wastes

Municipal authorities shall establish and maintain storage facilities in such a manner as they do not create unhygienic and insanitary conditions around it. Following criteria shall be taken into account while establishing and maintaining storage facilities, namely :-

Storage facilities shall be created and established by taking into account quantities of waste generation in a given area and the population densities. A storage facility shall be so placed that it is accessible to users;

Storage facilities to be set up by municipal authorities or any other agency shall be so designed that wastes stored are not exposed to open atmosphere and shall be aesthetically acceptable and user-friendly;

Storage facilities or bins shall have easy to operate design for handling, transfer and transportation of waste. Bins for storage of biodegradable wastes shall be painted green, those for storage of recyclable wastes shall be painted white and those for storage of other wastes shall be painted black;

		Manual handling of waste shall be prohibited. If unavoidable due to constraints, manual handling shall be carried out under proper precaution with due care for safety of workers.
4.	**Transportation of municipal solid wastes**	Vehicles used for transportation of wastes shall be covered. Waste should not be visible to public, nor exposed to open environment preventing their scattering. The following criteria shall be met, namely:- The storage facilities set up by municipal authorities shall be daily attended for clearing of wastes. The bins or containers wherever placed shall be cleaned before they start overflowing; Transportation vehicles shall be so designed that multiple handling of wastes, prior to final disposal, is avoided.
5.	**Processing of municipal solid wastes**	Municipal authorities shall adopt suitable technology or combination of such technologies to make use of wastes so as to minimize burden on landfill. Following criteria shall be adopted, namely:- (i) The biodegradable wastes shall be processed by composting, vermicomposting, anaerobic digestion or any other appropriate biological processing for stabilization of wastes. It shall be ensured that compost or any other end product shall comply with standards as specified in Schedule-IV; Mixed waste containing recoverable resources shall follow the route of recycling. Incineration with or without energy recovery including pelletisation can also be used for processing wastes in specific cases. Municipal authority or the operator of a facility wishing to use other state-of-the-art technologies shall approach the Central Pollution Control Board to get the standards laid down before applying for grant of authorization.
6.	**Disposal of municipal solid wastes**	Land filling shall be restricted to non-biodegradable, inert waste and other waste that are not suitable either for recycling or for biological processing. Land filling shall also be carried out for residues of waste processing facilities as well as pre-processing rejects from waste processing facilities. Land filling of mixed waste shall be avoided unless the same is found unsuitable for waste processing. Under unavoidable circumstances or till installation of alternate facilities, land filling shall be done following proper norms. Landfill sites shall meet the specifications as given in Schedule III.

10. Specifications for Landfill Sites

1. Site selection

In areas falling under the jurisdiction of Development Authorities it shall be the responsibility of such Development Authorities to identify the landfill sites and hand over the sites to the concerned municipal authority for development, operation and maintenance. Elsewhere, this responsibility shall lie with the concerned municipal authority.

Selection of landfill sites shall be based on examination of environmental issues. The Department of Urban Development of the State or the Union territory shall co-ordinate with the concerned organizations for obtaining the necessary approvals and clearances.

The landfill site shall be planned and designed with proper documentation of a phased construction plan as well as a closure plan.

The landfill sites shall be selected to make use of nearby wastes processing facility. Otherwise, wastes processing facility shall be planned as an integral part of the landfill site.

The existing landfill sites which continue to be used for more than five years, shall be improved in accordance of the specifications given in this Schedule.

Biomedical wastes shall be disposed off in accordance with the Bio-medical Wastes (Management and Handling) Rules, 1998 and hazardous wastes shall be managed in accordance with the Hazardous Wastes (Management and Handling) Rules, 1989, as amended from time to time.

The landfill site shall be large enough to last for 20-25 years.

The landfill site shall be away from habitation clusters, forest areas, water bodies monuments, National Parks, Wetlands and places of important cultural, historical or religious interest.

A buffer zone of no-development shall be maintained around landfill site and shall be incorporated in the Town Planning Departments land-use plans.

Landfill site shall be away from airport including airbase. Necessary approval of airport or airbase authorities prior to the setting up of the landfill site shall be obtained in cases where the site is to be located within 20 km of an airport or airbase.

2. Facilities at the site

Landfill site shall be fenced or hedged and provided with proper gate to monitor incoming vehicles or other modes of transportation.

The landfill site shall be well protected to prevent entry of unauthorized persons and stray animals.

Approach and other internal roads for free movement of vehicles and other machinery shall exist at the landfill site.

The landfill site shall have wastes inspection facility to monitor wastes brought in for landfill, office facility for record keeping and shelter for keeping equipment and machinery including pollution monitoring equipments.

Provisions like weigh bridge to measure quantity of waste brought at landfill site, fire protection equipments and other facilities as may be required shall be provided.

Utilities such as drinking water (preferably bathing facilities for workers) and lighting arrangements for easy landfill operations when carried out in night hours shall be provided.

Safety provisions including health inspections of workers at landfill site shall be periodically made.

3. Specifications for land filling

Wastes subjected to land filling shall be compacted in thin layers using landfill compactors to achieve high density of the wastes. In high rainfall areas where heavy compactors cannot be used alternative measures shall be adopted.

Wastes shall be covered immediately or at the end of each working day with minimum 10 cm of soil, inert debris or construction material till such time waste processing facilities for composting or recycling or energy recovery are set up as per Schedule I.

Prior to the commencement of monsoon season, an intermediate cover of 40-65 cm thickness of soil shall be placed on the landfill with proper compaction and grading to prevent infiltration during monsoon. Proper drainage berms shall be constructed to divert run-off away from the active cell of the landfill.

After completion of landfill, a final cover shall be designed to minimize infiltration and erosion. The final cover shall meet the following specifications, namely:

The final cover shall have a barrier soil layer comprising of 60 cm of clay or amended soil with permeability coefficient less that 1×10^{-7} cm/sec.

On top of the barrier soil layer there shall be a drainage layer of 15 cm.

On top of the drainage layer there shall be a vegetative layer of 45 cm to support natural plant growth and to minimize erosion.

11. Pollution Prevention

In order to prevent pollution problems from landfill operations, the following provisions shall be made, namely:

Diversion of storm water drains to minimize leachate generation and prevent pollution of surface water and also for avoiding flooding and creation of marshy conditions;

Construction of a non-permeable lining system at the base and walls of waste disposal area. For landfill receiving residues of waste processing facilities or mixed waste or waste having contamination of hazardous materials (such as aerosols, bleaches, polishes, batteries, waste oils, paint products and pesticides) minimum liner specifications shall be a composite barrier having 1.5 mm high density polyethylene (HDPE) geomembrane, or equivalent, overlying 90 cm of soil (clay or amended soil) having permeability coefficient not greater than 1×10^{-7} cm/sec. The highest level of water table shall be at least two meters below the base of clay or amended soil barrier layer;

Provisions for management of leachates collection and treatment shall be made. The treated leachates shall meet the standards specified in Schedule- IV;

Prevention of run-off from landfill area entering any stream, river, lake or pond.

12. Water Quality Monitoring

Before establishing any landfill site, baseline data of ground water quality in the area shall be collected and kept in record for future reference. The ground water quality within 50 metres of the periphery of landfill site shall be periodically monitored to ensure that the ground water is not contaminated beyond acceptable limit as decided by the Ground Water Board or the State Board or the Committee. Such monitoring shall be carried out to cover different seasons in a year that is, summer, monsoon and post-monsoon period.

Usage of groundwater in and around landfill sites for any purpose (including drinking and irrigation) is to be considered after ensuring its quality. The following specifications for drinking water quality shall apply for monitoring purpose, namely:–

S. No.	Parameters	IS 10500: 1991 Desirable limit (mg/l except for pH)
1.	Arsenic	0.05
2.	Cadmium	0.01
3	Chromium	0.05
4.	Copper	0.05
5.	Cyanide	0.05
6.	Lead	0.05
7.	Mercury	0.001
8.	Nickel	–
9.	Nitrate as NO_3	45.0
10	pH	6.5-8.5
11.	Iron	0.3
12.	Total hardness (as $CaCO_3$)	300.0
13.	Chlorides	250
14.	Dissolved solids	500
15.	Phenolic compounds (as C_6H_5OH)	0.001
16.	Zinc	5.0
17.	Sulphate (as SO_4)	200

13. Ambient Air Quality Monitoring

Installation of landfill gas control system including gas collection system shall be made at landfill site to minimize odour generation, prevent offsite migration of gases and to protect vegetation planted on the rehabilitated landfill surface.

The concentration of methane gas generated at landfill site shall not exceed 25 per cent of the lower explosive limit (LEL).

The landfill gas from the collection facility at a landfill site shall be utilized for either direct thermal applications or power generation, as per viability. Otherwise, landfill gas shall be burnt (flared) and shall not be allowed to directly escape to the atmosphere or for illegal tapping. Passive venting shall be allowed if its utilization or flaring is not possible.

Ambient air quality at the landfill site and at the vicinity shall be monitored to meet the following specified standards, namely:

S. No.	Parameters	Acceptable levels
(i)	Sulphur dioxide	120 mg/m^3 (24 hours)
(ii)	Suspended Particulate Matter	500 mg/m^3 (24 hours)
(iii)	Methane	Not to exceed 25 per cent of the lower explosive limit (equivalent to 650 mg/m^3)

(iv)	Ammonia	daily average (Sample duration 24 hrs)	0.4 mg/m^3 (400 mg/m^3)
(v)	Carbon monoxide		1 hour average : 2 mg/m^3 8 hour average : 1 mg/m^3

The ambient air quality monitoring shall be carried out by the concerned authority as per the following schedule, namely:

(a) Six times in a year for cities having population of more than fifty lakhs;

(b) Four times in a year for cities having population between ten and fifty lakhs;

(c) Two times in a year for town or cities having population between one and ten lakhs.

Plantation at Landfill Site

A vegetative cover shall be provided over the completed site in accordance with the following specifications, namely:

(a) Selection of locally adopted non-edible perennial plants that are resistant to drought and extreme temperatures shall be allowed to grow;

(b) The plants grown be such that their roots do not penetrate more than 30 cms. This condition shall apply till the landfill is stabilized;

(c) Selected plants shall have ability to thrive on low-nutrient soil with minimum nutrient addition;

(d) Plantation to be made in sufficient density to minimize soil erosion.

14. Closure of Landfill Site and Post-Care

The post-closure care of landfill site shall be conducted for at least fifteen years and long term monitoring or care plan shall consist of the following, namely:–

(a) Maintaining the integrity and effectiveness of final cover, making repairs and preventing run-on and run-off from eroding or otherwise damaging the final cover;

(b) Monitoring leachate collection system in accordance with the requirement;

(c) Monitoring of ground water in accordance with requirements and maintaining ground-water quality;

(d) Maintaining and operating the landfill gas collection system to meet the standards.

Use of closed landfill sites after fifteen years of post-closure monitoring can be considered for human settlement or otherwise only after ensuring that gaseous and leachate analysis comply with the specified standards.

15. Special Provisions for Hilly Areas

Cities and towns located on hills shall have location-specific methods evolved for final disposal of solid wastes by the municipal authority with the approval of the concerned State Board or the Committee. The municipal authority shall set up processing facilities for utilization of biodegradable organic wastes. The inert and non-biodegradable waste shall be used for building roads or filling-up of appropriate areas on hills. Because of constraints in finding adequate land in hilly areas, wastes not suitable for road-laying or filling up shall be disposed of in specially designed landfills.

16. Standards for Composting, Treated Leachates and Incineration

The waste processing or disposal facilities shall include composting, incineration, pelletisation, energy recovery or any other facility based on state-of-the-art technology duly approved by the Central Pollution Control Board.

In case of engagement of private agency by the municipal authority, a specific agreement between the municipal authority and the private agency shall be made particularly, for supply of solid waste and other relevant terms and conditions.

In order to prevent pollution problems from compost plant and other processing facilities, the following shall be complied with, namely:

The incoming wastes at site shall be maintained prior to further processing. To the extent possible, the waste storage area should be covered. If such storage is done in an open area, it shall be provided with impermeable base with facility for collection of leachate and surface water run-off into lined drains leading to a leachate treatment and disposal facility;

Necessary precautions shall be taken to minimize nuisance of odour, flies, rodents, bird menace and fire hazard;

In case of breakdown or maintenance of plant, waste intake shall be stopped and arrangements be worked out for diversion of wastes to the landfill site;

Pre-process and post-process rejects shall be removed from the processing facility on regular basis and shall not be allowed to pile at the site. Recyclables shall be routed through appropriate vendors. The non-recyclables shall be sent for well designed landfill site(s).

In case of compost plant, the windrow area shall be provided with impermeable base. Such a base shall be made of concrete or compacted clay, 50 cm thick, having permeability coefficient less than 107 cm/sec. The base shall be provided with 1 to 2 per cent slope and circled by lined drains for collection of leachate or surface run-off;

Ambient air quality monitoring shall be regularly carried out particularly for checking odour nuisance at down-wind direction on the boundary of processing plant.

In order to ensure safe application of compost, the following specifications for compost quality shall be met, namely:–

Parameters	Concentration not to exceed * (mg/kg dry basis, except pH value and C/N ratio)
Arsenic	10.00
Cadmium	5.00
Chromium	50.00
Copper	300.00
Lead	100.00
Mercury	0.15
Nickel	50.00
Zinc	1000.00
C/N ratio	20-40
pH	5.5–8.5

* Compost (final product) exceeding the above stated concentration limits shall not be used for food crops. However, it may be utilized for purposes other than growing food crops.

The disposal of treated leachates shall follow the following standards, namely:

S. No.	Parameter	Standards (Mode of Disposal)		
		Inland surface water	**Public sewers**	**Land disposal**
1.	Suspended solids, mg/l, max	100	600	200
2.	Dissolved solids (inorganic) mg/l, max.	2100	2100	2100
3.	pH value	5.5 to 9.0	5.5 to 9.0	5.5 to 9.0
4.	Ammonical nitrogen (as N), mg/l, max.	50	50	–
5.	Total Kjeldahl nitrogen (as N), mg/l, max.	100	–	–
6.	Biochemical oxygen demand (3 days at 27°C) max.(mg/l)	30	350	100
7.	Chemical oxygen demand, mg/l, max.	250	–	–
8.	Arsenic (as As), mg/l, max	0.2	0.2	0.2
9.	Mercury (as Hg), mg/l, max	0.01	0.01	–
10.	Lead (as Pb), mg/l, max	0.1	1.0	–
11.	Cadmium (as Cd), mg/l, max	2.0	1.0	–
12.	Total Chromium (as Cr), mg/l, max.	2.0	2.0	–
13.	Copper (as Cu), mg/l, max.	3.0	3.0	–
14.	Zinc (as Zn), mg/l, max.	5.0	15	–
15.	Nickel (as Ni), mg/l, max	3.0	3.0	–
16.	Cyanide (as CN), mg/l, max.	0.2	2.0	0.2
17.	Chloride (as Cl), mg/l, max.	1000	1000	600
18.	Fluoride (as F), mg/l, max	2.0	1.5	--
19.	Phenolic compounds (as C_6H_5OH) mg/l, max.	1.0	5.0	–

Note: While discharging treated leachates into inland surface waters, quantity of leachates being discharged and the quantity of dilution water available in the receiving water body shall be given due consideration.

The incinerators shall meet the following operating and emission standards, namely:

Operating Standards

(1) The combustion efficiency (CE) shall be at least 99.00%.

(2) The combustion efficiency is computed as follows:

$$\text{C.E.} = \frac{\%CO_2}{\%CO_2 + \%CO} \times 100$$

Emission Standards

Parameters	Concentration mg/Nm^3 at (12% CO_2 correction)
(1) Particulate matter	150

(2) Nitrogen Oxides	450
(3) HCl	50
(4) Minimum stack height shall be 30 metres above ground.	
(5) Volatile organic compounds in ash shall not be more than	0.01%

Note: Suitably designed pollution control devices shall be installed or retrofitted with the incinerator to achieve the above emission limits, if necessary.

Learning Outcomes

The main learning outcomes from this chapter should be fundamental understanding of: environmental laws, Environmental law related to land, water and air, law related to landfill and waste.

Readers are also advised to carry out following exercise:

Exercise

1. Find an example where environmental issues had been put aside while planning the project.
2. Describe how citizens and local government are responsible by law for not polluting the environment.
3. State how environmental laws help in protecting the cause and where they had been ignored in past. Consider taking case studies around the world.

Chapter 10

Unsustainable to Sustainable Developments: Social Issues

10.1 INTRODUCTION

Until two decades ago the world looked at economic status alone as a measure of human development. Thus, countries that were economically well developed and where people were relatively richer were called advanced nations while the rest where poverty was widespread and were economically backward were called developing countries. Most countries of North America and Europe which had become industrialized at an earlier stage became economically more advanced. They not only exploited their own natural resources rapidly but also used the natural resources of developing countries to grow even larger economies. Thus, the way development progressed, the rich countries got richer while the poor nations got poorer. However, even the developed world has begun to realize that their lives were being seriously affected by the environmental consequences of development based on economic growth alone. This form of development did not add to the quality of life as the environmental conditions had begun to deteriorate.

By the 1970s most development specialists began to appreciate the fact that economic growth alone could not bring about a better way of life for people unless environmental conditions were improved. Development strategies in which only economic considerations were used, had begun to suffer from serious environmental problems due to air and water pollution, waste management, deforestation and a variety of other ill effects that seriously affected peoples' well-being and health. There were also serious equity issues between the "have's and the have not's" in society, at the global and national levels. The disparity in the lifestyles between the rich and the poor was made worse by these unsustainable development strategies.

A growing realization of the development strategy that Mahatma Gandhi had suggested many decades earlier is now accepted by experts on development across the world. This is based on his concept that the world could support people's needs but not their greed. It has become obvious that the quality of human life has worsened as economies grew. The world now appears to be at crossroads. It has taken the path of short-term economic growth and now suffers the consequences of environmental degradation at the cost of loss of 'quality of human life'. The earth cannot supply the amount of resources used and wasted by the economically well-off sectors of society as well as that required for day-to-day sustenance of the ever growing population in less developed countries. Society must thus change its unsustainable development strategy to a new form where development will not destroy the environment. This form of sustainable development can only be brought about if each individual practices a sustainable lifestyle based on caring for the earth. It was also realized that these were not simple issues. Indira Gandhi said in the Stockholm Conference in 1972 that poverty was the greatest polluter. This meant that while the super rich nations had serious environmental problems, the underdeveloped in Asia, Africa and South America had a different set of environmental problems linked to poverty. Developing countries were suffering the consequences of a rapidly expanding human population with all its effects on the over-utilization of natural resources. Thus, increasingly the world began to see the need for a more equitable use of earth resources. The control over natural resources and the wealth that it produces also begins to create tensions between people that can eventually lead to both strife within a country and wars between nations. This is also a major cause for the loss of quality of life. How then could a new form of development be brought about that could solve the growing discontent in the world? It has become obvious that development must begin to change from aiming at short-term economic gains to a long-term sustainable growth that would not only support the well-being and quality of life of all people living in the world today but that of future generations as well. The current strategies of economic development are using up resources of the world so rapidly that our future generations, the young people of the world, would have serious environmental problems, much worse than those that wear facing at present. Thus, current development strategies have come to be considered unsustainable for the world's long-term development. The newer concept of development has come to be known as "sustainable development". The nations of the world came to clearly understand these issues at the Rio Conference in 1992. Several documents were created for the United Nations Conference on Environment and Development (UNSET), which brought out the fact that environment and development were closely connected and that there was a need to 'care for the earth'.

Sustainable development is defined as development that meets the needs of the present without compromising the ability of future generations to meet their own needs.

To ensure sustainable development, any activity that is expected to bring about economic growth must also consider its environmental impacts so that it is more consistent with long-term growth and development. Many 'development projects', such as dams, mines, roads, industries and tourism development, have severe environmental consequences that must be studied before they are even begun. Thus for every project, in a strategy that looks at sustainable development. There must be a scientifically and honestly done ELIAN, without which the project must not be cleared. The economic benefits of a project must be weighed against the possible environmental costs before a project is permitted. We as citizens of our nation, and increasingly as citizens of one common future at the global level, must constantly monitor the pattern of development. If we see that a development project or an industry is leading to serious environmental problems, it is our duty to bring this to the attention of authorities such as the local administration, the Forest Department or the Pollution Control Board, to look into the issue. Further, if new development projects are being planned in and around the place where we live is our duty to see that this is brought about in accordance with environmental safeguards. While we all need to think globally, we need to act locally. We have to see to it that we change development from its present mandate of rapid economic growth without a thought for future ecological integrity, to a more sustainable ecologically appropriate strategy. We cannot support the economic growth of one sector of society while we permit environmental degradation to destroy the lives of the less fortunate.

10.2 URBAN PROBLEMS RELATED TO ENERGY

Urban centres use enormous quantities of energy. In the past, urban housing required relatively smaller amounts of energy than we use at present. Traditional housing in India required very little temperature adjustments as the materials used, such as wood and bricks handled temperature changes better than the current concrete, glass and steel of ultra modern buildings.

10.2.1 Embodied Energy

Materials like iron, glass, aluminium, steel, cement, marble and burnt bricks, which are used in urban housing, are very energy intensive. The process of extraction, refinement, fabrication and delivery are all energy consuming and add to pollution of earth, air and water. This energy consumed in the processes is called embodied energy. Until the 1950s many urban kitchens were based on fuelwood or charcoal. This was possible and practical when homes had chimneys and kitchens were isolated from the rest of the house. Smoke became a problem once this changed to apartment blocks. Kerosene thus became urban fuel. This changed to electrical energy and increasingly to natural gas by the1970s in most parts of urban India. Urban centres in hot climates need energy for cooling. The early systems of fans changed into air-conditioning, which consumes enormous quantities of energy. New buildings in our country have taken to using large areas covered by glass. While in cold climates this uses the greenhouse effect to trap the warmth of the sun inside, in our hot climate this adds several degrees to the temperature inside. Thus, it requires ever more energy to run large central air-condition units. High-rise buildings in urban areas depend on energy to operate lifts and an enormous number of lights. Urban transport depends on energy mainly from fossil fuels. Most urban people use their own individual transport rather than public transport systems for a variety of reasons. This means more and more vehicles on the road which leads to traffic congestion, waste of time for all the commuters, and a great load of particulate matter and carbon monoxide from the exhaust of vehicles. This causes a rise in the number of people having serious respiratory diseases. Thus, there is a need to develop a more efficient public transport system and discourage the use of individual vehicles in all our urban areas. Each of us as an environmentally conscious individual must reduce our use of energy.

10.3 WATER CONSERVATION, RAINWATER HARVESTING AND WATERSHED MANAGEMENT

10.3.1 Water Conservation

Conserving water has become a prime environmental concern. Clean water is becoming increasingly scarce globally. With deforestation surface run-off increases and the subsoil water table drops as water has no time to seep slowly into the ground once vegetation is cleared. As many areas depend on wells, it has become necessary to go on making deeper and deeper wells. This adds to the cost and further deplete underground stores of water. This could take years to recharge even if the present rate of extraction is reduced which seems hardly possible in most situations. As deforestation and desertification spreads due to extensive changes in land use the once perennial rivers are becoming increasingly seasonal. In many areas the small streams run dry soon after the monsoon as the water table drops further and further below the surface. It's over use and misuse due to various activities that waste water or cause pollution that led to a serious shortage of potable water. Thus, water conservation is linked closely with overall human well-being.

CASE STUDY: Water Conservation

Villages all over the country had one or more common 'talabs' or tanks from which people collected or used water carefully. As women had to carry water to their homes over long distances, this was a time consuming and laborious activity, thus water could not be wasted. Many homes had a kitchen garden that was watered by the waste water. Conservation of water was done in traditional homes through a conscious effort. During the British period many dams were built across the country to supply water especially to growing urban areas. Post-independence, India's policy on water changed towards building large dams for expanding agriculture to support the green revolution. While this reduced the need to import food material and removed starvation in the country, the country began to see the effects of serious water shortages and problems related to its distribution. The newer forms of irrigated agriculture such as sugar cane and other water hungry cash crops required enormous quantities of water. Finally, however, such irrigated areas become waterlogged and unproductive. As excess water evaporates rapidly from the surface of heavily irrigated crop lands, it pulls up subsoil water along with salts to the surface of the soil. This leads to stabilization by which the land becomes unproductive. Reducing the high salinity levels in soil is extremely expensive and frequently impossible. With all these ill effects of the poorly conceived management of water at the national and local level there is a need to consider a new water policy for the country. Saving water in agriculture: Drip irrigation supplies water to plants near its roots through system of tubes, thus saving water. Small percolation tanks and rainwater harvesting can provide water for agriculture and domestic use. Rainwater collected from rooftops can be stored or used to effectively recharge subsoil aquifers. Saving water in urban settings: Urban people waste large amounts of water. Leaking taps and pipes are a major source of loss of water. Canals and pipes carrying water from dams to the consumer lead to nearly 50% loss during transfer. Reducing the demand for water by saving its more appropriate than trying to meet growing demands.

CASE STUDY: Pain Patency – Pine District, Matriarchal

Mah village in Pun District of Maharanee situated in a drought-prone area. People were not able to grow a good crop in most years. Clean drinking water was also scarce. Vilasrao Salunkhe initiated a movement known as Pani Panchayat, to conserve water in this drought-prone area. Watershed development was initiated on a barren and uncultivated piece of land belonging to a temple. Conservation of soil and water harvesting through a comprehensive micro watershed management programme gradually led to a surplus of water. Out of the 16 hectares of land in the village, 9.6 hectares were brought under irrigation, 2.4 hectares was afforested and 4 hectares was converted into percolation tanks. Wells and field bunds were built. While 200 quintals of grains were produced on 24 acres of Salunkhe's land, 40 acres in the same area yielded only 10 quintals. This made other villagers follow suit. The area rapidly turned green and productive.

10.3.2 Rainwater Harvesting

As our world faces serious water shortages, every drop of water we can use efficiently becomes of great value. One method is to manage rainwater in such a way that it is used at the source. If as much water as possible is collected and stored this can be used after the rainy season is over. However the stored water has to be kept pollution free and clean so that it can be used as drinking water. Stored water can grow algae and zooplankton (microscopic animals). This can be pathogenic and cause infections. Thus, keeping the water uncontaminated is of great importance. Current technologies of rainwater harvesting require that all roof and terrace water passes down into a covered tank where it can be stored for use after the monsoon. This is most advantageous in arid areas where clean water is scarce. However, there are practical difficulties such as constructing large storage tanks which are expensive.

CASE STUDY: Mewar, Rajasthan

The Mewar region of Rajasthan has a rich legacy of traditional water harvesting systems to share the available water for cultivation. There are various types of rainwater harvesting systems.

Medhbandi: This is a stone embankment built on a hill slope to help create a level field for cultivation. It controls erosion and conserves moisture.

Naada/bandha: These are stone check dams across streams or gullies that are constructed to capture run-off on a stretch of fertile land that is submerged in water during the monsoon. The land not only becomes more fertile after trapping silt, but also retains substantial quantities of water in the soil. These dams are constructed in phases over several years. The height is slowly increased up to the right height of the check dam which determines the size of the naada.

Hembar: These are small temporary dams constructed with stones, twigs and mud over seasonal stream when water flows in it is reduced to a point that it cannot be taken directly to the fields for irrigation.

Chak: Chak is a big plot of land, usually acharnot or village pasture land, enclosed by as tone boundary wall called kot. Tree plantations, seeding of grass for fodder, contourbunds with trenches and loose stone check dams are developed in the chak. The chak isused for fodder and fuelwood. It reduces soil erosion and enhances recharge of groundwater.

Talab: The Mewar region is well-known for its built reservoirs (talabs). Udaipur city is famous for its large number of talabs, and is called the lake city. A small reservoir of less than five bighas is called talai, a medium sized lake is called bands or talab and a bigger lake is called sagar or samand. Sazakuva: This is an open dug well which has several owners. In Mewari language, saza means partner. This is an important method for irrigation in the Aravalli Hills. About 70,000 wells in the Udaipur District provide water for 80% of the area under irrigation and provide water for their owners. These are considered common property resources. Another way of using rooftop rainwater harvesting is to collect it so that it percolates into the ground to recharge wells instead of flowing over the ground into rivers. Thus, by recharging groundwater harvested from rooftops, the water table rises and the surrounding wells retain water throughout the year.

10.3.3 Watershed Management

Rivers originate in streams that flow down mountains and hill slopes. A group of small streams flow down hillsides to meet larger streams in the valley which forms the tributaries of major rivers. The management of a single unit of land with its water drainage system is called watershed management. It is a technique that has several components. This includes soil and water management and developing vegetative cover. The natural drainage pattern of a watershed unit if managed appropriately can bring about local prosperity by a year round abundance of water that improves the quality of human life in the area. As it provides water throughout the year, this improves health in the community, as clean water becomes available. Watershed management enhances the growth of agricultural crops and even makes it possible to grow more than one crop in a year in dry areas. Watershed management begins by taking control over a degraded site through local participation. People must appreciate the need to improve the availability of water both in quantity and quality for their own area. Once this is adequately demonstrated, the community begins to understand the project, people begin to work together in the activities that lead to good watershed management. The first technical step is to take appropriate soil conservation measures. This is done by constructing a series of long trenches and mounds along contours of the hill to hold the rainwater and allow it to percolate into the ground. This ensures that underground stores of water care fully recharged. This is enhanced by allowing grasses and shrubs to grow and by planting trees (mainly local species) which hold the soil and prevent it from being washed away in the monsoon. Local grass cover can however only increase if free grazing of domestic animals is prevented by stall feeding. The next measure is to make 'nala' plugs in the streams so that the water is held in the stream and does not rush down the hillside. In selected sites, several small check dams are built which together hold back larger amounts of water. All these measures constitute sound watershed management. It improves the water table and keeps the streams and nalas flowing throughout the year.

Deforestation is a major cause of poor water supply. Afforesting such degraded areas is an important aspect of watershed management.

10.4 RESETTLEMENT AND REHABILITATION OF PEOPLE: ITS PROBLEMS AND CONCERNS

Major projects such as dams, mines, expressways, or the notification of a national park disrupts the lives of the people who live there and may also require moving them to an alternative site. Uprooting people is a serious issue. It reduces their ability to subsist on the transitional natural resource base and also creates great psychological pressures. Especially tribal people, whose lives are woven closely around their own natural resources, cannot adapt to a new way of life in a new place. Thus, no major project that is likely to displace people can be carried out without the consent of the local people. In India, lakhs of people have been unfairly displaced by dams created since independence to drive the green revolution. The dams have been built virtually at the cost of these poor local people who have been powerless to resist the government's will. The government is expected to find 'good' arable land and to resettle displaced persons and provide them with an adequate rehabilitation package to recover from the disruption. This has rarely occurred to the satisfaction of the project affected individuals. In many cases across the country, this has not been implemented satisfactorily for decades. However, in our overpopulated country, there is no arable high quality land available. Thus, most project affected persons are given unusable wasteland. Rehabilitation involves more than just giving land. In most cases this is also not adequately done. The greatest battle to save their own precious land has been carried out by the tribal people of the Narmada River. They have fought to save their lands for decades. Then Narmada Bachao Andolan has shown how bitter people can get over this issue. Resettlement not only puts pressure on the project affected people but also on the people who have been living in the area that has been selected for resettlement. Thus, both the communities suffer and conflict over resources is a distinct possibility in future.

CASE STUDY: The Tehri Project

The Tehri Dam in the outer Himalayas in Uttar Pradesh, when finished will submerge Tehri town and nearly 100 villages. Since the dam was sanctioned in 1972, local people have been opposing the dam and resisting its construction. Scientists, environmentalists and other groups have also opposed this dam.

Little is done to ensure proper rehabilitation and compensation for nearly a lakh of people who will be uprooted from their homes as result of this dam, with little hope of rehabilitation, as no alternative land is available. There is also emotional and psychological trauma caused by forcibly removing people from their homeland where their families have lived for centuries.

CASE STUDY: Indigenous Tribes

It is not flora and fauna alone that is under the threat of extinction. Among the many tribes across the globe, the Jarawa of the Andamans in the Indian Ocean are dwindling. Dispossession of their customary rights over land has put their survival at risk. They have been compelled to give up their traditional lifestyles resulting in rapidly diminishing indigenous population. There are however situations where communities request for shifting to a new site. This is often observed where people live inside or on the periphery of a national park or wildlife sanctuary. In these situations, such as the Gir in Gujarat, the local people have asked to be given alternate land where they could live peacefully away from lions that kill their cattle, but the government has been unable to find suitable areas where they can be shifted for decades.

10.5 ENVIRONMENTAL ETHICS: ISSUES AND POSSIBLE SOLUTIONS

Environmental ethics deals with issues related to the rights of individuals that are fundamental to life and well-being. This concerns not only the needs of each person today, but also those who will come after us. It also deals with the rights of other living creatures that inhabit our earth.

10.5.1 Resource Consumption Patterns and the Need for Their Equitable Utilisation

Environmental ethics deals with issues that are related to how we utilise and distribute resources. Can individuals justifiably use resources so differently that one individual uses resources many times more lavishly than other individuals who have barely enough to survive? In a just world, there has to be a more equitable sharing of resources than we encounter at present. The just distribution of resources has global, national and local concerns that we need to address. There are rich and poor nations. There are rich and poor communities in every country. And there are rich and poor families. In this era of modern economic development, the disparity between the have's and have-not's is widening. Our human environments in the urban, rural and wilderness sectors use natural resources that shift from the wilderness (forests, grasslands, wetlands, etc.) to the rural sector, and from there to the urban sector. Wealth also shifts in the same direction. This unequal distribution of wealth and access to land and its resources is serious environmental concern. An equitable sharing of resources forms the basis of sustainable development for urban, rural and wilderness welling communities. As the political power base is in the urban centres, this itself leads to in equalities and a subsequent loss of sustainability in resource management in the rural and even more so for forest dwelling people.

In 1985, Anil Agarwal published the first report on the Status of India's Environment. Anil Agarwal brought forth a set of 8 propositions which are of great relevance to the ethical issues that are related to environmental concerns. These include:

1. Environmental destruction is largely caused by the consumption of the rich.
2. The worst sufferers of environmental destruction are the poor.
3. Even where nature is being 'recreated', as in aforestation, it is being transformed away from the needs of the poor and towards those of the rich.
4. Even among the poor, the worst sufferers are the marginalized cultures and occupations, and most of all, women.
5. There cannot be proper economic and social development without a holistic understanding of society and nature.
6. If we care for the poor, we cannot allowt he gross nature product to be destroyed any further. Conserving and recreating nature has become our highest priority.
7. Gross nature product will be enhanced only if we can arrest and reverse the growing alienation between the people and the common property resources. In this we will have to learn a lot from our traditional cultures.
8. It is totally inadequate to talk only of sustainable rural development, as the world conservation strategy does. We cannot save the rural environment or rural people dependent on it, unless we can bring about sustainable urban development.

10.5.2 Equitable Use of Forest Resources

We think of forests as being degraded due to fuelwood collection by poor rural communities, but forget that the rich use much greater quantities of timber. Biomass based industries include cotton textiles, paper, plywood, rubber, soap, sugar, tobacco, jute, chocolate, food processing and packaging. These need land, energy, irrigation and forest resources. Do each of us realize this when we utilize, use excessively or waste these resources that we get indirectly from the forests?

Who pays for the cost of environmental degradation?

Most sections of society do not feel the direct effects of degradation of the environment till it is too late. Those who suffer most are the poor, especially rural women, and tribal people who are dependent on forests. Traditional fishermen who are dependent on streams and rivers, and coastal people who fish and catch crustacean, are seriously affected by the degradation of aquatic ecosystems. Fuelwood gatherers from different types of forests, and pastoralists who are dependent on common grazing lands suffer when their resources are depleted. Several marginalized sectors of society are most affected by deforestation, or the loss of grassland tracts, or the deterioration of perennial water sources. All these effects can be linked to unsustainable increasing pressures on land and natural resources."I am often amazed and extremely angry, when people talk about Environment Education for the villagers. It is the so-called, educated people who need Environment Education more than anyone else".– Anil Agarwal, 'Human-Nature Interactions in a Third World Country'. The well to do educated urban dweller consumes much larger quantities of resources and energy, than the traditional rural individual. Urban dwellers who are far removed from the source of natural resources that sustain their lives thus require exposure to a well-designed environment education programme to appreciate these issues. While the rural people have a deep insight on the need for sustainable use of space natural resorces and know about methods of conservation, there are however several newer environmental concerns that are frequently outside their sphere of life experiences. Their traditional knowledge of environmental concerns cannot be expected to bring about an understanding of issues such as global warming, or problems created by pollution, pesticides, etc. These people thus require a different pattern of environment education that is related to their gaps in information. With the rapidly changing rural scenario the development that is thrust on unsuspecting rural communities needs to be addressed through local specific environment awareness programmes designed specifically for rural school children and adults. This must also use their local traditional knowledge systems as a base on which modern concepts can be built, rather than by fostering concepts that are completely alien to their own knowledge systems. In the past, in traditional villages that were managed by local panchayats, there were well defined rules about managing grazing lands, collecting forest resources, protecting sacred groves, etc., that supported conservation. There was a more or less equitable distribution that was controlled by traditional mechanisms to prevent misuse of common property resources. Any infringement was quickly dealt with by the panchayat and the offender was punished. Common property resources were thus locally protected by communities. As land use patterns changed, these mechanisms were lost and unsustainable practices evolved, frequently as a result of an inadequately planned development strategy.

10.5.3 Equity – Disparity in the Northern and Southern Countries

Environmental ethics are concerned with, who owns resources and how they are distributed. This can be looked upon at different levels. At the global level it deals with the great North-South divide between the rich industrialized nations of North America and Europe, as against the needs of developing countries of the South such as in South and Southeast Asia and South America. People living in the economically advanced nations use greater amounts of resources and energy per individual and also waste more resources. This is at the cost of poor people who are resource dependent and live in developing nations. The economically advanced West has exploited its own natural resources to such an extent that they have exhausted them nearly everywhere. They now buy these resources from resource-rich but economically deprived nations at a low cost. This depletes the developing nations of natural resources on which their poor depend for their livelihood. Changing this unfair economic practice to a more just and fair way in managing trade would require a new thinking on the part of people who live in the super rich countries.

(i) **Urban-rural equity issues:** The common property of rural communities has increasingly been used to supply the needs of the urban sector. Land itself that was once held as a common property resource of villages is being taken over by the urban and industrial sectors as it expands. The rural sector not only supplies food, but also a part of the energy needs (mainly fuelwood) to most towns and cities in India, at a pittance. As a result, the commons of the rural sector are being depleted of their resources. Thus, while the cities get richer, the rural sector, especially the landless, get poorer. The urban rich must appreciate where their resources are derived from and be willing to pay a fair price for using them.

(ii) **The need for gender equity:** All over India, especially in the rural sector, women work on the whole longer hours than men. The life of a woman is enmeshed in an inextricable cycle of poverty. In attempting to eke out a living from their environment, they must constantly collect fuelwood for their homes and for sale to nearby urban areas. They laboriously collect fodder for their cattle. They have to trudge several kilometres to reach a reasonably clean water source. And finally must cook meals in a smoky unhealthy at mosphere on crop waste or other inefficient sources of energy. All this can take 10 to 12 hours a day of very hardwork, every day of the year. There is thus the question of who should control the environmental resources of a rural community. Unfortunately it is the men who play a decisive role in managing the village commons and its resources where as it should be the local women whose lives are deeply linked with the utilization and conservation patterns of natural resources, who should be decision-makers at the local level. Unfortunately women have not been given an equal opportunity to develop and better their lot. This begins with the lack of attention given to girls whose education is always given less attention than the boys in the family. Unless society begins to see that development cannot be planned by a male dominated society from the male perspective alone, will we be able to create a better living environment for women and their children?

The great divide between women and men is most apparent in communities that live near forests and have by tradition made the woman play a greater role than men in collection of natural resources. Women fetch water, collect fuelwood, fruit, medicinal products, etc., day in and day out, while the men work only sporadically in the fields. This disparity in the lives of women and men has also led to a lower access to education and health care for girl children. This has deep implications for the rate of utilization of natural resources and its conservation. Rural women who are intimately connected to resources, appreciate the value of conserving natural resources more deeply than men. Thus several environmental movements such as Chipko Movement have been more strongly supported by local women folk rather than men.

10.5.4 Preserving Resources for Future Generations

Can we use up all the resources of the world, leaving nothing for our future generations? This ethical issue must be considered when we use resources unsustainable. If we overuse and misuse resources and energy from fossil fuels, our future generations would find survival much more difficult. A critical concern is to preserve species and natural undisturbed ecosystems that are linked with bioresources, which must be protected for the use of future generations. Our generation does not own the world's resources to do whatever we please with them. Just as our ancestors have left resources for us, it is our duty to leave them behind for our future generations. These unborn people have a right to these resources. We only hold the world as trustees so that future generations can also survive. Our current development strategies have led to environmental resources being overused and misused by our present generation, without a thought for the needs of future generations.We need to appreciate that the next generation and those that will come later also have a right to the earth's natural resources. As they are not here today to exercise their rights, it is our responsibility to appreciate the needs of future generations.

10.5.5 The Rights of Animals

Can man, a single species, use and severely exploit the earth's resources which we share with billions of other plant and animal species? Within our world there are a variety of living beings. The plants and animals that share the earth with us too have a right to live and share our earth's resources and living space. We have no right to push a species that has taken millions of years to evolve towards extinction. Not only do wild and domesticated animals have a right to life, but have the right to a dignified existence. Cruelty to an animal is no different ethically from cruelty to another human being. Mahatma Gandhi's philosophy was based on the assumption that human beings were not masters of the other forms of life. He believed that humans were 'trustees of the lower animal kingdom'. Every plant and animal has a right to life as a part of our earth's community of living things. While nature by itself has natural prey-predator relationships, left to itself, nature maintains a balance in each ecosystem. While evolution has developed a system whereby species become extinct and new ones evolve to fill the world's ecosystems with new plant and animal species, it is man alone that has been responsible for the recent rapid decline in the number of species on the earth. Much more important man is now reducing the abundance levels of so many species that in the near future we will in all probability create a major extinction spasm on the earth that will seriously endanger the existence of mankind. Thus endangering the existence of wild plants and animals and bringing them close to the brink of extinction is not only unfair to a species but also to future generation of people who may find them of great use. There is a growing awareness of animal rights in our country and cruelty to animals is being increasingly regarded as a criminal offence.

CASE STUDY: Chipko Movement

About 300 years ago, a ruler in Rajasthan decided to fell 'khejri' trees in his State to create lime. Local women led by a Bishnoi woman, Amrita Devi, clung to the trees to prevent the felling of the trees that formed the basis of the scarce resources on which they were dependent. The women were ruthlessly massacred. It is said that the ruler later realised his mistake. The story however has been remembered and was revived in the 1970s when severe tree felling for timber in the Himalayas prompted local women, supported by people such as Sunderlal ji Bahuguna and Chandi Prasad Bhat, led a people's movement to prevent deforestation by timber contractors. They called their movement the 'Chipko' movement in memory of the event during which women had clung to their trees and given up their lives. The movement followed the path the 300 Bishnoi women had taken three centuries ago in Rajasthan. Chipko is a movement primarily begun and supported by local women in the hills of Uttarakhand and Garhwal, where the women have had to bear the brunt of deforestation. They have not only realised that their fuelwood and fodder resources have receded away from their 'resource use areas' around their settlements due to commercial timber extraction, but that this has led to serious floods and loss of precious soil. Chipko activists have made long padyatras across the Himalayas protesting against deforestation. The movement has been highly successful and has been primarily supported by empowering local women's groups who are the most seriously affected segment of society by deforestation. The movement has proved to the world that the forests of the hills are the life support systems of local communities of immense value in terms of local produce that is essential for the survival of local people and that the forest has less quantifiable but even more important ecological services such as soil conservation and the maintenance of the natural water regime of the whole region. The ability of local women to band themselves together in the foothills of the Himalayas goes back to the pre-independence days when women such as Miraben, a disciple of Gandhiji, moved to this region and understood that it was the deforestation that led to floods and devastation of villages in the valleys and in the Gangetic plains below. They also appreciated that substitutions of oak and other broad-leaved forests of the Himalayas by planting fast growing pine for timber and resin was an ecological and social disaster which reduced the forest resources used by traditional hill communities.

10.6 THE ETHICAL BASIS OF ENVIRONMENT EDUCATION AND AWARENESS

Perhaps the most important concern is related to creating an ethos that will support a sustainable lifestyle in society. This brings us to the need for environmental education. The Honorary Supreme Court of our country has thus ordered that every young individual at school and college level be exposed to a course on environment. It is not to create only an awareness of environmental issues, but also to bring about pro-environmental action. Among the variety of tools that can bring home the ethical issues of the environment; no solution is as powerful as real life experiences in nature. Creating a love for nature brings about strong pro-environmental action. Our current educational processes at school and college level are being reoriented to bring this about.

10.6.1 The Gandhian Way of Life

Mahatma Gandhi had deep insights into the need to conserve resources. 'Man's needs but not his greed can be supported by our earth'was an important concept that was initiated by him when people had not realized how short the world would be of resources in future. At the time natural resources seemed to be limitless to most people. This was thus a new concept and suggested the need for a uniquely different pattern of living. Gandhiji believed in simplistic living to save our earth's resources. He once said that if India was to become an industrial nation on the lines of England, the world itself would be stripped bare of its resources by India's people alone. There are two aspects that are closely connected with ethical issues that are related to our environment. These are based on valuing nature and appreciating the beauty of nature and treasuring the magnificence of the wilderness.

In today's world where many of us are far removed from nature, we need to remind ourselves that everything we use, if tracked back to its source, has come from nature. We depend on an intact unpolluted world which is based on nature's goods and services. No life is possible without this. If we as citizens begin to again respect nature and its entire varied species forming a complex web of life, and appreciate nature's functions and services, it will continue to support our lives. If we disrespect nature one cannot expect her to continue to support our well being. Nature's resources that we all use and depend on can only be optimized if they are equitably shared by all of us. If the disparity is too great it can only result in anarchy.

10.6.2 Education in Nature – The Shantiniketan Model

Rabindranath Tagore founded Shantiniketan and taught an environment-based philosophy. Tagore's philosophy of education focused attention on the need for a harmonious association between human beings and their environment. To achieve this he relied on exposing young people to nature.

This went back to our roots where in ancient India, learning centres were established in remote forests. Tagore linked these concepts with celebrations of nature through music, dance, drama and poetry. At Shantiniketan, there were celebrations for each season and ceremonial tree planting. He started Vriksharopan way back in 1928. In fact much of what was initiated in Shantiniketan is now accepted as the route to environment education and sustainable living and is essentially based on preserving nature. Bringing back an ethic for nature conservation requires environment education and conservation awareness. The best way to do so is to expose young people not only to our dependence on natural resources from the wilderness, but by bringing about an appreciation of the beauty and wondrous aspects of nature. This forms a sharp contrast to the sad plight of degraded areas and polluted sites in which most of humanity now lives in the developed and developing world. The concept of 'Karma' is based on a thinking that the soul moves from man to animal and in reverse depending on one's actions. This itself brings about a concept of the oneness of all forms of life. Ahimsa or non-violence towards life which includes all plants and animals provides India with its basic philosophy which early Hindu philosophers and later sages such as Buddha, Mahavir and Mahatma Gandhi spoke of. Buddhist and Jain philosophy is intrinsically woven around non-violence and the great value of all forms of 'life'. It brings in the notion that animals are not to be viewed purely for their utility value but are a part of the earth's oneness which is linked with our own lives as well. In Hindu philosophy the earth itself is respected and venerated. In contrast, in Western thought nature is to be subjugated and used. These are basic differences in thinking processes. Several modern philosophers in the West have now begun to see these eastern patterns of thought as a new basis for human development. This shift however, from a purely utilitarian or scientific exploitation of nature, to one of harmony with nature, can only occur if each of us loves and respects nature's great 'oneness'.

10.6.3 The Conservation Ethic and Traditional Value Systems of India

In ancient Indian traditions people have always valued mountains, rivers, forests, trees and several animals. Thus much of nature was venerated and protected. Forests have been associated with the names of forest gods and goddesses both in the Hindu religion as well as in tribal cultures. 'Tree' goddesses have been associated with specific plant species. Ficus religious, the peepal tree, is venerated and is thus not to be cut down. The Banyan tree in some regions such as Maharashtra is venerated once a year by tying a thread around it as asymbol of respect. The Tulsi plant is grown on the doorstep outside every home. Patches of forest have been dedicated to a deity in many Indian cultures especially in tribal areas. These traditionally protected forest patches depict the true nature of undisturbed vegetation and have a large number of indigenous plant species as their exploitation has been controlled through local sentiments. Certain species of trees have been protected as they are valued for their fruit or flowers. The mango tree is protected for its fruit around most farms even when wood becomes scarce. The Mohua tree (Madhu caindica) is protected by tribal people as it provides edible flowers, oil from its seeds and is used to make a potent alcohol. Many plants, shrubs and herbs have been used in Indian medicines which were once available in the wild in plenty. These are now rapidly vanishing. Many species of animals are venerated as being the 'vahan' or vehicle of different gods on which they are said to travel through the cosmos. In Indian mythology, the elephant is associated with Ganesha. The elephant headed Ganesha is also linked to the rat. Vishnu is associated with the eagle. Rama is linked to monkeys. In mythology, Hanuman, the monkey god, rendered in valuable help to Rama during his travels to Lanka. The Sun god, Surya, rides a horse and has a superb chariot on which he moves through the sky. The lion is linked to Durga and the black buck to the moon goddess. The cow is associated with Krishna. Vishnu's incarnation shave been represented as taking various animal forms which serially include, fish, tortoise, a boar and a dwarf, and a half man half lion form. In traditional societies of the past, these examples were all a part of ethical values that protected nature. As modern science based on the exploitation on nature spread into India, many of these traditions began to lose their effectivenessas measures that led to conserving nature. Concepts that support nature's integrity must thus become a part of our modern educational systems. This constitutes a key solution to bring about a new ethic of conserving nature and living sustainable life styles.

10.7 CLIMATE CHANGE, GLOBAL WARMING, ACID RAIN, OZONE LAYER DEPLETION, NUCLEAR ACCIDENTS AND HOLOCAUST

10.7.1 Climate Change

The average temperature in many regions has been increasing in recent decades. The global average surface temperature has increased by 0.6° + 0.2° C over the last century. Globally, 1998 was the warmest year and the 1990s the warmest decade on record. Many countries have experienced increases in rainfall, particularly in the countries situated in the mid to high latitudes. In some regions, such as parts of Asia and Africa, the frequency and intensity of droughts have been observed to increase in recent decades. Episodes of El Niño, which creates great storms, have been more frequent, persistent and intense since mid-1970s compared with the previous 100 years. All these are signs that the earth is sick. Its climate is changing, making it more difficult for mankind to survive. The earth is losing its ability to balance itself due to the imbalance created by human activities. Projections of future climate change are derived from a series of experiments made by computer based global climate models. These are worked out on estimates of aspects such as future population growth and energy use. Climatologists of the intergovernmental panel on climate change (IPCC) have reviewed the results of several experiments in order to estimate changes in climate in the course of this century. These studies have shown that in the near future, the global mean surface temperature will rise by 1.4° to 5.8°C. Warming will be greatest over land areas and at high latitudes. The projected rate of warming is greater than has occurred in the last 10,000 years. The frequency of weather extremes is likely to increase leading to floods or drought. There will be fewer cold spells but more heat waves. The frequency and intensity of El Niño is likely to increase. Global mean sea level is projected to rise by 9 to 88 cm by the year 2100. More than half of the world's population now lives within 60 km of the sea. They are likely to be seriously impacted by an ingress of salt water and by the rising sea. Some of the most vulnerable regions are the Nile delta in Egypt, the Ganges-Brahmaputra delta in Bangladesh, and many small islands including the Marshall Islands and the Maldives, (WHO, 2001). Human societies will be seriously affected by extremes of climate such as droughts and floods. A changing climate would bring about changes in the frequency and/or intensity of these extremes. This is a major concern for human health. To a large extent, public health depends on safe drinking water, sufficient food, secure shelter, and good social conditions. All these factors are affected by climate change. Freshwater supplies may be seriously affected, reducing the availability of clean water for drinking and washing during drought as well as floods. Water can be contaminated and sewage systems may be damaged. The risk of spread of infectious diseases such as diarrheal diseases will increase. Food production will be seriously reduced in vulnerable regions directly and also indirectly through an increase in pests and plant or animal diseases. The local reduction in food production would lead to starvation and malnutrition with long-term health consequences, especially for children. Food and water shortages may lead to conflicts in vulnerable regions, with serious implications for public health. Climate change related impacts on human health could lead to displacement of a large number of people, creating environmental refugees and lead to further health issues. Changes in climate may affect the distribution of vector species (e.g. mosquitoes) which in turn will increase the spread of disease, such as malaria and filariasis, to new areas which lack a strong public health infrastructure. The seasonal transmission and distribution of many diseases that are transmitted by mosquitoes (dengue, yellow fever) and by ticks (Lyme disease, Tick borne encephalitis) may spread due to climate change. Strategies aimed at reducing potential health impacts of anticipated climate changes should include monitoring of infectious diseases and disease vectors to detect early changes in the incidence of diseases and the geographical distribution of vectors; environmental management measures to reduce risk; disaster preparedness for floods or droughts; and their health related consequences. It will be necessary to create early warning systems and education for epidemic preparedness. Improved water and air pollution control will become increasingly essential for human health. Public education will have to be directed at changes in personal behaviour. Training of researchers and health professionals must become an essential part of the world becoming more responsible towards the expected outcome of Global Climate Change (GCC).

10.7.2 Global Warming

About 75% of the solar energy reaching the earth is absorbed on the earth's surface which increases its temperature. The rest of the heat radiates back to the atmosphere. Some of the heat is trapped by greenhouse gases, mostly carbon dioxide. As carbon dioxide is released by various human activities, it is rapidly increasing. This is causing global warming. The average surface temperature is about 15°C. This is about 33°C higher than it would be in the absence of the greenhouse effect. Without such gases most of the earth's surface would be frozen with a mean air temperature of -18°C. Human activities during the last few decades of industrialization and population growth have polluted the atmosphere to the extent that it has begun to seriously affect the climate. Carbon dioxide in the atmosphere has increased by 31% since pre-industrial times, causing more heat to be trapped in the lower atmosphere.

CASE STUDIES

(i) **Damage to coral reefs, pacific:** The severity of periodic warming due to El Nino in 1997 in the Pacific led to the most serious death in coral ever known. It is estimated that about 10% of the earth's coral reefs were dead, another 30% were seriously affected and another 30% were degraded. The Global Coral Reef Monitoring Network Townsville, Australia, has predicted that all the reefs could be dead by 2050.

(ii) **Butterfly populations in the United Kingdom:** Global warming is leading to an early arrival of butterflies in Britain. Scientists say that butterflies can now be spotted much earlier every year in the last two decades. Some, like the red admiral, can now be seen a month earlier than was the case in the mid – 1970s. Others, like the peacock and the orange tipare appearing between 15 and 25 days earlier than in the past. Future rise in temperature is likely to have a detrimental effect on these butterflies. Some butterflies which need cooler temperature might suffer. A Task Group set up by WHO has warned that climate change may have serious impacts on human health. Climate change will increase various current health problems, and may also bring levels are still increasing. Many countries have signed a convention to reduce greenhouse gases under the United Nations Convention on Climate Change. Current international agreements are however not still effective to prevent the significant changes in climate and a rise in sea levels. Global warming is accelerating faster than what climatologists had calculated a few years ago. In 1995, the Intergovernmental Panel on Climate Change predicted that global warming would rise temperatures by 3.5 to 10 degrees centigrade during the 21st century, if the present trends continue. It is now believed that this could be much greater. This would lead to not only temperature changes but in the amount of rainfall. India may see great annual fluctuations in rainfall leading to floods and drought.

10.7.3 Acid Rain

When fossil fuels such as coal, oil and natural gas are burnt, chemicals like sulfur dioxide and nitrogen oxides are produced. These chemicals react with water and other chemicals in the air to form sulfuric acid, nitric acid and other harmful pollutants like sulfates and nitrates. These acid pollutants spread upwards into the atmosphere, and are carried by air currents, to finally return to the ground in the form of acid rain, fog or snow. The corrosive nature of acid rain causes many forms of environmental damage. Acid pollutants also occur as dry particles and gases, which when washed from the ground by rain, add to the acids in the rain to form a more corrosive solution. This is called acid deposition. Damage from acid rain is widespread in North America, Europe, Japan, China and Southeast Asia. In the US coal burning power plants contribute to about 70% of sulfur dioxide. In Canada oil refining, metal smelting and other industrial activities account for 61% of sulfur dioxide pollution. Motor vehicle exhaust fumes are the main source of nitrogen oxides. The acids in acid rain chemically react with any object they come in contact with. Acids react with other chemicals by giving up hydrogen atoms.

1. Acid rain dissolves and washes away nutrients in the soil which are needed by plants. It can also dissolve naturally occurring toxic substances like aluminum and mercury, freeing them to pollute water or poison plants.
2. Acid rain indirectly affects plants by removing nutrients from the soil in which they grow. It affects trees more directly by creating holes in the waxy coating of leaves, causing brown dead spots which affect the plant's photosynthesis. Such trees are also more vulnerable to insect infestations, drought and cold. Spruce and fir forests at higher elevations seem to be most at risk. Farm crops are less affected by acid rain than forests.
3. Acid rain that falls or flows as groundwater to reach rivers, lakes and wetlands, causes the water in them to become acidic. This affects plant and animal life in aquatic ecosystems.
4. Acid rain also has far reaching effects on wildlife. By adversely affecting one species, the entire food chain is disrupted, ultimately endangering the entire ecosystem. Different aquatic species can tolerate different levels of acidity. For instance clams and may flies have a high mortality when water has a pH of 6.0, while frogs can tolerate more acidic water, although with the decline in supply of mayflies, frog populations may also decline. Land animals that are dependent on aquatic organisms are also affected.
5. Acid rain and dry acid deposition damages buildings, automobiles, and other structures made of stone or metal. The acid corrodes the materials causing extensive damage and ruins historic buildings. For instance the Parthenon in Greece and the Taj Mahal in India have been affected by acid rain.
6. Although surface water polluted by acid rain does not directly harm people, the toxic substances leached from soil can pollute water supply. Fish caught in these waters may be harmful for human consumption. Acid, along with other chemicals in the air, produces urban smog, which causes respiratory problems.

The best way to stop the formation of acid rain is to reduce the emissions of sulfur dioxide and nitrogen oxides into the atmosphere. This can be achieved by using less energy from fossil fuels in power plants, vehicles and industry. Switching to cleaner burning fuels is also a way out. For instance using natural gas which is cleaner than coal, using coal with lower sulfur content, and developing more efficient vehicles. If the pollutants have already been formed by burning fossil fuels, they can be prevented from entering the atmosphere by using scrubbers in smoke stacks in industry. These spray mixture of water and limestone into the polluting gases, recapturing the sulfur. In catalytic converters, the gases are passed over metal coated beads that convert harmful chemicals into less harmful ones. These are used in cars to reduce the effects of exhaust fumes on the atmosphere. Once acid rain has affected soil, powdered limestone can be added to the soil by a process known as liming to neutralize the acidity of the soil.

10.7.4 Ozone Layer Depletion

Ozone is formed by the action of sunlight on oxygen. It forms a layer 20 to 50 km above the surface of the earth. This action takes place naturally in the atmosphere, but is very slow. Ozone is a highly poisonous gas with a strong odour. It is a form of oxygen that has three atoms in each molecule. It is considered a pollutant at ground level and constitutes a health hazard by causing respiratory ailments like asthma and bronchitis. It also causes harm to vegetation and leads to a deterioration of certain materials like plastic and rubber. Ozone in the upper atmosphere however, is vital to all life as it protects the earth from the sun's harmful ultraviolet radiation. The ozone layer in the upper atmosphere absorbs the sun's ultraviole tradiation, preventing it from reaching the earth's surface.

This layer in the atmosphere protects life on the earth from the dangerous UV radiation from the sun. In the 1970s, scientists discovered that chemicals called chlorofluorocarbons or CFCs, which were used as refrigerants and aerosol spray propellants, posed a threat to the ozone layer. The CFC molecules are virtually indestructible until they reach the strato sphere, where UV radiation breaks them down to release chlorine atoms. The chlorine atoms react with ozone molecules which break down into oxygen molecules, which do not absorb UV radiations. Since the early 1980s, scientists detected a thinning of the ozone layer in the atmosphere above Antarctica. This phenomenon is now being detected in other places as well including Australia. Although the use of CFCs has been reduced and now banned in most countries, other chemicals and industrial compounds such as bromine, halocarbons and nitrous oxides from fertilizers may also attack the ozone layer. The destruction of the ozone layer is seen to cause increased cases of skin cancer and cataracts. It also causes damage to certain crops and to plankton, thus affecting nature's food chains and food webs. This in turn causes an increase in carbon dioxide due to the decrease in vegetation. With the signing of the Montreal Protocol in 1987, a treaty for the protection of the ozone layer, the use of CFCs was to be banned by the year 2000. After 2000, the ozone layer is expected to recover slowly over a period of about 50 years.

10.7.5 Nuclear Accidents and Nuclear Holocaust

Nuclear energy was researched and discovered by man as a source of alternate energy which would be clean and cheap compared to fossil fuels. And although this did happen, along with the benefits of nuclear energy came its downfalls. In the short history of nuclear energy there have been accidents that have surpassed any natural calamity or other energy source extraction in their impacts. A single nuclear accident can cause loss of life, long-term illness and destruction of property on a large scale for a long period of time. Radio activity and radio active fallout leads to cancer, genetic disorders and death in the affected area for decades after, thus affecting all forms of life for generations to come.

CASE STUDY

(i) **Nuclear disasters and leakages:** In 1986 the Nuclear Power Station at Chernobyl in USSR developed a problem that led to a fire and a number of explosions in its nuclear reactor. The radioactive dust spread over many kilometres and covered not only Europe but North America as well. Three people died in the explosion and 28 shortly after due to radiation exposure. Some 259 sick were hospitalized. As the area had to be evacuated 135,000 people had to be moved immediately and another 1.5 lakh by 1991. As radioactive fallout continued even more people had to be moved. An estimated 6.5 lakh people may have been seriously affected. They may get cancer, thyroid tumonrs, and cataracts, and suffer from a lowered immune mechanism. As radioactivity passes from grass to herbivores, sheep in Scotland and Reindeer in Lapland were affected and were unfit for human consumption. Vegetable, fruit and milk were contaminated in Europe. A French Nuclear Waste Processing Centerin Normandy may have affected the lives of children playing nearby. They may developleukemia (blood cancer) in later life.

10.7.6 Nuclear Holocaust

The use of nuclear energy in war has had devastating effects on man and earth. The Hiroshima and Nagasaki incident during World War II, the only use of nuclear power in war in history, is one of the worst disasters in history. In 1945, the United States dropped atomicbombs in Japan over the towns of Hiroshima and Nagasaki. These two atomic bombs killed thousands of people, left many thousands injured and devastated everything for miles around. The effects of the radiation from these nuclear bombs can still be seen today in the form of cancer and genetic mutations in the affected children and survivors of the incident.

10.8 WASTELAND RECLAMATION

Loss of vegetation cover leads to loss of soil through erosion, which ultimately creates wastelands. This is one of the pressing problems of the country. Loss of soil has already ruined a large amount of cultivable land in our country. If it remains unchecked, it will affect the remaining land. Unless we adequately safeguard our 'good' lands, we may eventually face a serious shortage of food grains, vegetables, fruit, fodder and fuelwood. Hence, conservation of soil, protecting the existing cultivable land and reclaiming the already depleted wastelands figures prominently among the priority tasks of planning for the future. Some of the wasteland and reclamation programmes have been unsuccessful because after sometime the land reverts to its original poor condition due to mismanagement and unscientific ways in which the reclamation has been carried out. In choosing wasteland reclamation methods attention must be paid to the cost factor. This has to be taken into account before deciding on a particular method for reclamation of wastelands. Proper studies of environmental aspects and human impacts which are responsible for the development of wastelands have to be looked into.

Wasteland can be classified into three forms:

(1) Easily reclaimable,
(2) Reclaimable with some difficulty,
(3) Reclaimable with extreme difficulty.

Easily reclaimable wastelands can be used for agricultural purposes. Those which can be reclaimed with some difficulty can be utilized for agro forestry. Wastelands that are reclaimed with extreme difficulty can be used for forestry or to recreate natural ecosystems.

Agriculture: Wasteland can be reclaimed for agriculture by reducing the salt content which can be done by leaching and flushing. Gypsum, urea, potash and compost are added before planting crops in such areas.

Agro forestry: This involves putting land to multiple uses. Its main purpose is to have trees and crops inter- and/or under-planted to form an integrated system of biological production within a certain area. Thus, agro forestry implies integration of trees with agricultural crops or livestock management simultaneously.

Forestry: Attempts to grow trees in highly non-alkaline saline soils have been largely unsuccessful. Field experiments have shown that species like Eucalyptus, Prosopis and Acacia Nilotica could not be grown in highly alkaline soil. Studies have shown that if tree seedlings are planted with a mixture of original soil, gypsum, and manure, better growth can be achieved. It is however important to use indigenous species of trees so that the program recreates the local ecosystem with all its species.

10.8.1 Need for Wasteland Development

Wasteland development provides a source of income for the rural poor. It ensures a constant supply of fuel, fodder and timber for local use. It makes the soil fertile by preventing soil erosion and conserving moisture. The program helps maintain an ecological balance in the area. The increasing forest cover helps in maintain in glocal climatic conditions. Regenerated vegetation cover helps in attracting birds which feed on pests in the surrounding fields and functionas natural pest controllers. The trees help in holding back moisture and reduce surface run-off rates thus helping in the control of soil erosion.

10.8.2 Components of Wasteland Reclamation

The first major task is the identification of the problem at the micro level. For this it is necessary to have district, village and plot level surveys of the wasteland. A profile of the maps indicating the detailed distribution and information on the wasteland is essential. With the help of local government institutions such as the village panchayats, along with block development officers, revenue department functionaries, a plan based on the community needs must be produced. This must be done through a participatory exercise that involves all the different take holders in the community. A think-tank of administrators, ecologists, and local NGOs must also be involved in the process. The next step is to identify the factors that are responsible for the formation of wastelands. Based on these factors the wasteland is classified into: marginally, partially or severely deteriorated lands. Local specific strategies for reclaiming the wasteland must be worked out. Government officials along with the local NGOs must assist the farmers by demonstrating improved methods of cultivation, arranging for loans for the small, marginal and landless farmers and the people from the weaker sections of the society. Involving local women has proved to be of great value. Another essential component of the program is to organize publicity campaigns, integrated with training farmers and front line government and forest department staff on the various aspects of wasteland utilization. Environmental scientists can help by suggesting the necessary changes in cropping patterns particularly for drought-prone areas. Other tasks that should be addressed include the selection of appropriate crops for fodder and trees that provide local people with non-wooded forest products according to the nature of the wasteland. Testing soil in laboratories provides guidance to the farmers on the proper land management to be used. Irrigation and other expertise needed for improving productivity without creating unsustainable patterns of development provide the local people with newer technological advances. Guidelines regarding control of waterlogging must be provided. Appropriate technologies must be made available to people belonging to the weaker sections and landless farmers. Collective efforts have to be made to check soil losses through water and wind erosion to prevent the collapse of the irrigation system through siltation. Plans concerning wasteland reclamation and utilization prepared at various stages must be properly integrated for a successful long-term outcome.

CASE STUDY (Tehri, Uttar Pradesh)

Nagchaund village in Tehri District of Uttar Pradesh was once an eroded and deforested land. When Soban Singh Bhandari returned to his village after retirement from the army in 1987 he was struck by this degradation. After six months he became the pradhan of the village and decided to implement various village development schemes differently. Through the Jawahar Rozgar Yojana, he gained immense community support. In1990 the Forest Department selected a 30-hectare barren piece of community land fora micro-watershed development program. The villagers controlled grazing in the area, undertook plantations for fuel and fodder. Bhandari helped the village raise money by selling the fodder from the area to a neighboring village and the money was used for development and maintenance work. This community effort has had a great impact on the ecology of the area. The moisture content of the area increased and the water sources of the villages were recharged. Local people now have access to all the natural resources they need for their daily lives. The demands of our increasing human population for environmental goods and services have imposed severe pressures on the available land resources especially on the forests and green cover. This is closely linked to the well-being of the rural population which constitutes a large per cent of the population which depends on local natural resources for their survival. The increasing demand for food, fodder, fuelwood, etc., has led to activities that are responsible for increasing environmental degradation. This is responsible for the extension of wastelands. Thus, the development of agro forestry based agriculture and forestry has become the prime prerequisite for an overall development of the economy in the country. The pressure on land is already very high and the only hope of increasing productivity lies in bringing appropriate improvement in the various categories of wasteland spread over the country.

10.8.3 Consumerism and Waste Products

Modern societies that are based on using large amounts of goods, especially those that are manufactured for one time use, are extremely wasteful. The increasing consumption of natural resources has led to serious environmental problems around the world. Current consumption patterns are depleting non-renewable resources, poisoning and degrading ecosystems, and altering the natural processes on which life depends. The present pattern of consumption, especially in affluent societies, is mainly responsible for the high level of utilization of resources. People in the industrialized countries make up 20% of the world population but consume 80% of the world's resources and produce 80% of wastes. This is due to a pattern of economic development that ensures that people go on consuming even more than they actually need. India is rapidly moving into this unsustainable pattern of economic growth and development.

The rich in such a society get richer often at the cost of the poor whose lives are not improved by the process of development. It is seen that today's consumption patterns are depleting natural resources at a rapid rate and widening the inequalities in consumption in different societies. Consumerism causes wasteful use of energy and material far beyond that needed for everyday living at a comfortable level. Money is not the only way to measure the cost of an item that we use. When one adds up all the raw material and energy that goes into the manufacture of goods or the services provided by nature that one uses during a day's activities, the toll on the environment is large. When this cost is multiplied over a lifespan, the amount is staggering. If one considered the over utilization in each family, city or a country, the impacts are incredibly high. For example: two hundred billion cans, bottles, plastic cartons and paper cups, are thrown away each year in the "developed" world. "Disposable" items greatly increase this waste. Rather than compete on quality or reliability, many industrial consumer products are made for one-time use. Buying quality product that are warranted against failure or wearing out, learning about the raw materials that things are made of, and an appreciation of their origin from nature's storehouse, as well as knowing the conditions of the workers that make them, are some ways of resisting consumerism and decreasing waste.

Himachal Pradesh was the first State in India to regulate the manufacture and use of plastics. The State proposed a ban on all types of polythene packing. While there may be some new appliances and cars that are more productive and energy efficient, discarding the old often leads to an almost total waste of the energy and material already invested in these products. This alone may more than nullify the energy savings of the new product. This is a tricky problem. Consumerism is related to the constant purchasing of new goods, with little attention to their true need, durability, product origin, or the environmental consequences of their manufacture and disposal. Consumerism is driven by huge sums spent on advertising designed to create both a desire to follow trends, and a personal feeling of satisfaction based on acquisition. Materialism is one of the end results of consumerism. Consumerism interferes with the sustainable use of resources in a society by replacing the normal common sense desire for an adequate supply of life's necessities, with an insatiable quest for things that are purchased by larger and larger incomes to buy them. There is little regard for the true utility of what is bought. An intended consequence of this strategy which is promoted by those who profit from consumerism is to accelerate the discarding of the old, either because of lack of durability or a change in fashion. Especially in developed countries, landfills are being rapidly filled with cheap discarded products that fail to work within a short time and cannot be repaired. In many cases consumer products are made psychologically obsolete by the advertising industry long before they actually wear out. The inordinate amount of waste that is generated by consumer-oriented societies around the world is now a serious environmental issue. Most human activities are related to production and consumption cycles which produce excessive amounts of waste in the form of solid, liquid and gaseous waste products. The problems of waste management in the urban and rural sectors are different. Rural communities that were smaller once had a limited amount of waste which was recycled as the communities used them effectively. With the advent of an industrial civilization the highly complex technological processes for production of goods has rapidly increased problems due to inadequate waste disposal. This creates a heavy burden on natural resources, degrades the environment and creates health hazards. With the rapid increase in population, the amount of waste in terms of quantity and quality has increased waste management pressures manifold in recent years. If the high quantities of waste generated continue, mankind will be drowned under heaps of garbage, and streams of sewage. The problems of waste disposal must be changed to the careful utilization of resources, recovery of used material by waste recycling. Therefore, reuse of goods and waste utilization should become apart of the production-consumption cycle. Utilizing various forms of waste must be made apart of the planning and development process. Current patterns in the industrial sector have led to the disposal of waste in a careless uneconomical manner. Burning or dumping wastes into streams and oceans, or creating more landfills damages the environment. Large quantities of solid, liquid and gaseous waste is produced by urban industrial communities in the form of plastic, paper, leather, tin cans, bottles, mineral refuse, and pathological waste from hospitals. Dead animals, agricultural wastes, fertilizer and pesticide overuse, and human and animal excreta are essentially rural concerns. The waste is either discharged into the atmosphere, into water sources, or buried underground. These wastes are not considered to have any economic value. This attitude towards waste has led to disastrous effects on the environment besides overexploiting natural resources.

10.8.4 Reduce, Reuse, and Recycle

Reduce, reuse, recycle, or the 3Rs principle, is the new concept in waste management. But what does actually mean? Although some waste is inevitable in any society, we must minimize the generation of waste at the source by using minimal resources. Do not use what you do not need. The goal of every society should be to reach a low-waste or no waste society, e.g., fancy packaging of consumer products into or three layers is not necessary.

Use your own reusable cloth/ jute bags instead of plastic bags. The residual waste can be converted into a useable resource. In developed countries waste is used to produce energy. Several technological breakthroughs have recently been made to recover material from industrial waste such as heavy metals and chemicals such as mercury and nitric acid. Thus, the waste does not remain a waste product anymore, but becomes a useful resource, e.g., using kitchen wet waste to make compost that can be used as an organic fertilizer. Using sewage in a biogas plant to make fuel. One industry's waste could be a valuable resource for another industry, e.g., cloth rags from the textile industry are bought and used by paper and other industries. Bagasse, a waste product of the sugar industry, is used in the paper, ply industries. The material left over after extraction of oil from seeds is used as cattle feed.

CASE STUDY: Plastic to Oil

The Indian Oil Corporation Limited and the Department of Science and Technology are expected to establish India's first plant to convert waste plastic into petrol, diesel and LPG. The generated waste or discarded material that cannot be used again in its original form can be sent back to the industry to be broken down and used as a resource to be made into a new product of the same type or into something entirely different, e.g., plastic items are recycled into new plastic products. Metal scrap and broken glass is used to make new products.

Finally, the waste material generated which can neither be reused nor recycled, must be disposed of in a proper manner with minimum impact to the environment.

- Non-toxic solid waste should be properly segregated and disposed of in landfills that are properly sealed to avoid leakage and contamination of surrounding land and groundwater.
- Toxic wastes should be treated or disposed of separately in a proper manner.
- Sewage and industrial wastewater should be adequately treated and raw materials recovered from it where possible before it is released into our rivers and waterways. The 3Rs principle of Reduce, Reuse, Recycle, should be followed in that order.
- Reduction is the best option. If we reduce at source, there is a smaller chance of waste generation and the pressure on our already stretched natural resources is reduced. Reuse is the next best option, as the products reused in its current form without any energy expended to convert it into a new item. Recycling is the last option, as although it converts a waste into a resource, it uses energy to transform that resource into a new useable product. Thus, by following the 'Reduce, Reuse, Recycle'principle, i.e. by reducing use at source, by reusing and recycling whatever possible and finally by proper disposal of residual waste, we can cut down or the waste generated and ensure that the minimal residual waste does not harm our environment. This principle can be followed by everyone, from an individual or an industry to a whole country.

What can I do? You can follow the 3Rs principle in the following ways:

1. Use only as much as you need, be it any resource– water, food, paper, etc.
2. Next time you throw away something, think about whether it is really a waste. If it is of no use to you, could someone else use it? Reuse rinses water to water your garden, etc. Donate old clothes to the needy, instead of throwing them away.
3. If you are sure the item is not usable in its present form, can it be recycled? Paper, plastics, glass, metal can all be recycled.
4. Segregate your waste into wet and dry garbage. Wet garbage includes most kitchen wastes. Most of this can be used for composting. Most dry garbage is recyclable. The amount of dry waste generated in your household is an indicator of how well you are following the 3Rs principle. A lot of dry waste means you should go back to the 'Reduce and Reuse' principles and try to follow them better.
5. Avoid the use of non-biodegradable materials such as Styrofoam and certain types of plastics.

   Although most plastics are recyclable, recycling still takes up energy, which is another precious resource not to be wasted. If thrown away as waste, Styrofoam and plastics can take hundreds of years to decompose.

6. Do not litter or throw garbage in public places. Garbage and litter is a visual contaminant and can cause diseases and health problems. Proper disposal of garbage is an important part of waste management.
7. Be a conscious consumer and do not buy products that are over-packaged. Try choosing products that are made from recycled material or are organically grown.

10.8.5 Suggestions for Better Waste Management

(1) Every country must survey all the different forms of waste generation along with its sources. They must set up priorities concerning waste utilization. Most waste can be converted to resources which can enhance the economy of the country.

(2) Plans should be prepared for controlling waste at the source. This must include segregation of wet and dry waste, where the wet waste can be converted to compost and used and the dry waste is recycled.

(3) Research and developmental programmes to find innovative methods of waste recycling must be encouraged. Recycling should be a part of conservation and environmental protection programs. Private and public organizations for waste recycling and management should beset up.

(4) Uneconomical methods of waste disposal like land filling, or incineration must be reduced to a minimum. Plans for appropriate disposal of non-utilizable hazardous waste from chemical industries must be implemented and strictly monitored.

(5) Every community should organize extensive programmes on education and demonstration on the reduction of waste, and the proper disposal and effective reutilization of waste material. People should be informed of the need for waste management to protect the quality of the environment. This should be included in the curriculum at school and college level.

(6) Every society should make efforts to design peoples' lifestyles and cultural patterns based on low waste production. The goal of every society should be to reach a low-waste or no waste society.

Resources must be conserved by proper selection, production technologies, recovering and recycling what is usable and reducing unnecessary demands for consumption and inventing technologies which would make it possible for reusing the waste resources so as to reduce over exploiting of our existing resources.

10.9 ACTS AND LEGISLATIONS

Several acts and legislations have been introduced by government to comprehend sustainable development.

10.9.1 The Environment (Protection) Act

The Environment (Protection) Act, 1986 not only has important constitutional implications but also an international background. The spirit of the proclamation adopted by the United Nations Conference on Human Environment which took place in Stockholm in June 1972, was implemented by the Government of India by creating this Act. Although there were several existing laws that dealt directly or indirectly with environmental issues it was necessary to have a general legislation for environmental protection because the existing laws focused on very specific types of pollution, or specific categories of hazardous substances or were indirectly related to the environment through laws that control land use, protect our national parks and sanctuaries and our wildlife. However there were no overarching legislation and certain areas of environmental hazards were not covered. There were also gaps in areas that were potential environmental hazards and there were several inadequate linkages in handling matters of industrial and environmental safety. This was essentially related to the multiplicity of regulatory agencies. Thus, there was a need for an authority which could assume the lead role for studying, planning and implementing long-term requirements of environmental safety and give directions to, as well as coordinate a system of speedy and adequate response to emergency situations threatening the environment. This Act was thus passed to protect the environment, as there was a growing concern over the deteriorating state of the environment. As impacts grew considerably environmental protection became a national priority in the 1970s. The decline in the environmental quality was evidenced by increasing pollution, loss of forest cover and an increasing threat to biodiversity. The presence of excessive concentrations of harmful chemicals in the atmosphere and aquatic ecosystems leads to the disruption of food chains and a loss of species. These are symptoms of a rapidly deteriorating environment. The growing risks of environmental accidents and threats to life support systems now loom threateningly over our civilization. The decision taken at the conference in Stockholm strongly voiced these environmental concerns and several measures were made possible for environmental protection. While the need for a wider general legislation to protect our environment is now in place, it has become increasingly evident that our environmental situation continues to deteriorate. We need to implement this Act much more aggressively if our environment is to be protected. Public concern and support is crucial for implementing the EPA. This must be supported by an enlightened media, good administrators, highly aware policy makers, informed judiciary and trained technocrats who together can influence and prevent further degradation of our environment. Each of us has a responsibility to make this happen.

10.9.2 The Air (Prevention and Control of Pollution) Act

The Government passed this Act in 1981 to clean up our air by controlling pollution. Sources of air pollution such as industry, vehicles, power plants, etc. are not permitted to release particulate matter, lead, carbon monoxide, sulfur dioxide, nitrogen oxide, volatile organic compounds (VOCs) or other toxic substances beyond a prescribed level. To ensure this, pollution control boards (PCBs) have been set up by Government to measure pollution levels in the atmosphere and at certain sources by testing the air. This is measured in parts per million or in milligrams or micrograms per cubic metre. The particulate matter and gases that are released by industry and by cars, buses and two wheelers is measured by using air sampling equipment. However, the most important aspect is for people themselves to appreciate the dangers of air pollution and reduce their own potential as polluters by seeing that their own vehicles or the industry they work in reduces levels of emissions. This Act is created to take appropriate steps for the preservation of the natural resources of the earth which among other things includes the preservation of high quality air and ensures controlling the level of air pollution. The main objectives of the Act are as follows:

(a) To provide for the prevention, control and abatement of air pollution.

(b) To provide for the establishment of central and state boards with a view to implement the Act.

(c) To confer on the boards the powers to implement the provisions of the Act and assign to the boards functions relating to pollution. Air pollution is more acute in heavily industrialized and urbanized areas, which are also densely populated. The presence of pollution beyond certain limits due to various pollutants discharged through industrial emission is monitored by the pollution control boards set up in every state.

Powers and Functions of the Boards

Central Board: The main function of the Central Board is to implement legislation created to improve the quality of air and to prevent and control air pollution in the country. The Board advises the Central Government on matters concerning the improvement of air quality and also coordinates activities, provides technical assistance and guidance to state boards and lays down standards for the quality of air. It collects and disseminates information in respect of matters relating to air pollution and performs functions as prescribed in the Act.

State Pollution Control Boards: The State Boards have the power to advice the State Government on any matter concerning the prevention and control of air pollution. They have the right to inspect at all reasonable times any control equipment, industrial plant, or manufacturing process and give orders to take the necessary steps to control pollution. They are expected to inspect air pollution control areas at intervals or whenever necessary. They are empowered to provide standards for emissions to be laid down for different industrial plants with regard to quantity and composition of emission of air pollutants into the atmosphere. A State Board may establish or recognize a laboratory to perform this function. The state governments have been given powers to declare air pollution control areas after consulting with the State Board and also give instructions for ensuring standards of emission from automobiles and restriction on use of certain industrial plants.

Penalties: Persons managing industry are to be penalized if they produce emissions of air pollutants in excess of the standards laid down by the State Board. The board also makes applications to the court for restraining persons causing air pollution. Whoever contravenes any of the provision of the Act or any order or direction issued is punishable with imprisonment for a term which may extend to three months or with a fine of Rs. 10,000 or with both ,and in case of continuing offence with an additional fine which may extend to Rs. 5,000 for every day during which such contravention continues after conviction for the first contravention.

What can an individual do to control air pollution?

(1) When you see a polluting vehicle takes down the number and send a letter to the Road Transport Office (RTO) and the Pollution Control Board (PCB).

(2) If you observe an industry polluting air, inform the Pollution Control Board in writing and ascertain the action is taken.

(3) Use cars only when absolutely necessary. Walk or cycle as much as possible instead of using fossil fuel powered vehicles.

(4) Use public transport as far as possible, as more people can travel in a single large vehicle rather than using multiple small vehicles which add to pollution.

(5) Share a vehicle space with relatives and friends. Car pools minimize the use of fossil fuels.

(6) Do not use air-fresheners and other aerosols and sprays which contain CFCs that deplete the ozone layer.

(7) Do not smoke in a public place. It is illegal and endangers not only your own health but also that of others.

(8) Coughing can spread bacteria and viruses. Use a handkerchief to prevent droplet infection which is air borne. It endangers the health of other people. It is a citizen's duty to report

to the local authorities such as the Collector or the Pollution Control Board, and the press about offences made by a polluter so that action can be taken against the offender. It is equally important to prevent and report to the authorities on cutting down of trees, as this reduces nature's ability to maintain the carbon dioxide and oxygen levels. Preventing air pollution and preserving the quality of our air is a responsibility that each individual must support so that we can breathe air that will not destroy our health.

10.9.3 The Water (Prevention and Control of Pollution) Act

The Government has formulated this Act in 1974 to be able to prevent pollution of water by industrial, agricultural and household wastewater that can contaminate our water sources. Wastewaters with high levels of pollutants that enter wetlands, rivers, lakes, wells as well as the sea are serious health hazards. Controlling the point sources by monitoring levels of different pollutants is one way to prevent pollution by giving a punishment to a polluter. However, it is also the responsibility of people in general to inform the relevant authority when they see a likely source of pollution. Individuals can also do several things to reduce water pollution such as using biodegradable chemicals for household use, reducing use of pesticides in gardens, and identifying polluting sources at workplaces and in industrial units where oil or other petroleum products and heavy metals are used. Excessive organic matter, sediments and infecting organisms from hospital wastes can also pollute our water. Citizens need to develop a watchdog force to inform authorities to take appropriate actions against different types of water pollution.

A polluter must pay for his actions. However, preventing pollution is better than trying to cure the problems it has created, or punishing offenders. The main objectives of the Water Act are to provide for prevention, control and abatement of water pollution and the maintenance or restoration of the wholesomeness of water. It is designed to assess pollution levels and punish polluters. The central government and state governments have set up Pollution Control Boards that monitor water pollution.

Functions of the Pollution Control Boards: The Government has given the necessary powers to the PCBs to deal with the problems of water pollution in the country. The Government has also suggested penalties for violation of the provisions of the Act. Central and State water testing laboratories have been set up to enable the boards to assess the extent of water pollution and standards have been laid down to establish guilt and default. The Central and State Boards are entitled to certain powers and functions which are as follows:

Central Board: It has the power to advise the Central Government on any matters concerning the prevention and control of water pollution. The board coordinates the activities of the state boards and also resolves disputes. The Central Board can provide technical assistance and guidelines to state boards to carry out investigations and research relating to water pollution, and organizes training for people involved in the process. The board organizes a comprehensive awareness programme on water pollution through mass media and also publishes data regarding water pollution. The board lays down or modifies the rules in consultation with the state boards on standards of disposal of waste.

The main function of the Central Board is to promote the cleanliness of rivers lakes streams and wells in the country.

State Boards: They have the power to advise the state government on any matters concerning water pollution. It plans a comprehensive programme for the prevention of water pollution. It collects and disseminates information on water pollution and participates in research in collaboration with the Central Board in organizing training of people involved in the process. The Board inspects sewage or trade effluents, treatment plants, purification plants and the systems of disposal and also evolves economical and reliable methods of treatment of sewage and other effluents. It plans the utilization of sewage water for agriculture. It ensures that if effluents are to be discharged on land the waste is diluted. The State Board advises state governments with respect to location of industries. Laboratories have been established to enable the Board to perform its functions. The state boards have the power to obtain information from officers empowered by it who make surveys, keep records of flow, volume, and other characteristics of the water. They are given the power to take samples of effluents and suggest the procedures to be followed in connection with the samples. The concerned board analyst is expected to analyze the sample sent to him and submit a report of the result to the concerned board. The board is required to send a copy of the result to the respective industry. The board also has the power of inspecting any plant record, register, document or any material object, and can conduct a search in any place in which there is reason to believe that an offence has been conducted under the Act. Penalties are charged for acts that have caused pollution. This includes failing to furnish information required by the board, or failing to inform the occurrence of any accident or other unforeseen act. An individual or organization that fails to comply with the directions given in the subsections of the law can be convicted or punished with imprisonment for a term of three months or with a fine of Rs 10,000 or both and in case failure continues an additional fine of Rs. 5,000 every day. If a person who has already been convicted for any offence is found guilty of the same offence again, he/she after the second and every subsequent conviction would be punishable with imprisonment for a term not less than two years but which may extend to seven years with fine.

What can individuals do to prevent water pollution?

1. Inform the Pollution Control Board of any offender who is polluting water and ensure that appropriate action is taken. One can also write to the press.
2. Do not dump wastes into a household or industrial drain which can directly enter any water body, such as a stream, river, pond,lake or the sea.
3. Do not use toilets for flushing down waste items as they do not disappear but reappear at other places and cause water pollution.
4. Use compost instead of chemical fertilizers in gardens.
5. Avoid use of pesticides at home like DDT, Melathion, Aldrin, and use alternative methods like paste of boric acid mixed with gram flour to kill cockroaches and other insects. Use dried Neem leaves to help keep away insects.

10.9.4 The Wildlife Protection Act

This Act passed in 1972, deals with the declaration of national parks and wildlife sanctuaries and their notification. It establishes the structure of the state's wildlife management and the posts designated for Wildlife Management. It provides for setting up Wildlife Advisory Boards. It prohibits hunting of all animals specified in Schedules I to IV of the Act. These are notified in order of their endangeredness. Plants that are protected are included in schedule VI. The Amendment to the Wildlife Protection Act in 2002 is more stringent and prevents the commercial use of resources by local people. It has brought in new concepts such as the creation of Community Reserves. It has also altered several definitions. For instance in animals, fish are now included. Forest produce has been redefined to ensure protection of ecosystems. While there are several changes, the new Act still has serious issues concerned with its implementation. Laws are only as good as the ones that can be complied with. The Act is expected to deter people from breaking the law. However, there are serious problems due to poaching. One cannot expect to use the Act to reduce this without increasing forest staff, providing weapons, jeeps, radio equipment, etc., for establishinga strong deterrent force.

Penalties: A person who breaks any of the conditions of any license or permit granted under this Act shall be guilty of an offence against this Act. The offence is punishable with imprisonment for a term which may extend to three years or with a fine of Rs 25,000 or with both. An offence committed in relation to any animal specified in Schedule I, or Part II of Schedule II, like the use of meat of any such animal, or animal articles like a trophy, shall be punishable with imprisonment for a term not less than one year and may extend to six years and a fine of Rs 25,000. In the case of a second or subsequent offence of the same nature mentioned in this sub-section, the term of imprisonment may extend to six years and not less than two years with a penalty of Rs.10,000.

What can an individual do?

(1) If you observe an act of poaching, or see a poached animal, inform the local forest department official at the highest possible level. One can also report the event through the press. Follow up to check that action is taken by the concerned authority. If no actionis taken, one must take it up to the Chief Wildlife Warden of the State.

(2) Say 'no' to the use of wildlife products and also try to convince other people not to buy them.

(3) Reduce the use of wood and wood products wherever possible.

(4) Avoid misuse of paper because it is made from bamboo and wood, which destroys wildlife habitat. Paper and envelopes canal may be reused.

(5) Create a pressure group and ask government to ensure that the biodiversity of our country is conserved.

(6) Do not harm animals. Stop others from inflicting cruelty to animals.

(7) Do not disturb birds' nests and fledglings.

(8) When you visit the zoo do not tease the animals by throwing stones or feeding them, and prevent others from doing so.

(9) If you come across an injured animal do what you can to help it.

(10) If the animal needs medical care and expert attention contact the Society for the Prevention of Cruelty to Animals in your city.

(11) Create awareness about biodiversity conservation in your own way to family and friends.

(12) Join organizations, which are concerned with protection of biodiversity, such as Worldwide Fund For Nature –India (WWFI),Bombay Natural History Society (BNHS), or a local conservation NGO.

10.9.5 Forest Conservation Act

To appreciate the importance of the Forest Conservation Act of 1980, which was amended in1988, it is essential to understand its historical background. The Indian Forest Act of 1927 consolidated all the previous laws regarding forests that were passed before the 1920's. The Act gave the Government and Forest Department the power to create Reserved Forests, and the right to use Reserved Forests for Government use alone. It also created Protected Forests, in which the use of resources by local people was controlled. Some forests were also to be controlled by a village community, and these were called Village Forests. The Act remained in force till the 1980s when it was realized that protecting forests for timber production alone was not acceptable. The other values of protecting the services that forests provide and its valuable assets such as biodiversity began to overshadow the importance of their revenue earnings from timber. Thus a new Act was essential. This led to the Forest Conservation Act of 1980 and its amendment in 1988. India's first Forest Policy was enunciated in 1952. Between 1952 and 1988, the extent of deforestation was so great that it became evident that there was a need to formulate a new policy on forests and their utilization. Large tracts of forestland had already been diverted to other uses. The earlier forest policies had focused attention on revenue generation only. In the 1980s it became clear that forests must be protected for their other functions such as maintenance of soil and water regimes centred on ecological concerns. It also provided for the use of goods and services of the forest for its local inhabitants. The new policy framework made conversion of forests into other uses much less possible. Conservation of the forests as a natural heritage findsa place in the new policy, which includes the preservation of its biological diversity and genetic resources. It also values meeting the needs of local people for food, fuelwood, fodder and non-wood forest products that they subsist on. It gives priority to maintaining environmental stability and ecological balance. It expressly states that the network of Protected Areasshould be strengthened and extended.

In 1992, the 73rd and 74th Amendments to the Constitution furthered governance through panchayats. It gives states the ability to provide power to the local panchayats to manage local forest resources. The Forest Conservation Act of 1980 was enacted to control deforestation. It ensured that forest lands could not be de-reserved without prior approval of the Central Government. This was created as States had begun to De-reserve the Reserved Forests for non-forest use. States had regularized encroachments and resettled 'Project Affected People' from development projects such as dams in these de-reserved areas. The need for a new legislation became urgent. The Act made it possible to retain a greater control over the frightening level of deforestation in the country and specified penalties for offenders.

Penalties for offences in Reserved Forests: No person is allowed to make clearings or set fire to a reserved forest. Cattle are not permitted to trespass into the Reserved Forest. Felling, collecting of timber, bark or leaves, quarries or collecting any forest product is punishable with imprisonment for a term of six months, or with a fine which may extend to Rs. 500, or both.

Penalties for offences in Protected Forests: A person who commits any of the following offences like felling of trees, or strips off the bark or leaves from any tree or sets fire to such forests, or kindles a fire without taking precautions to prevent its spreading to any tree mentioned in the Act, whether standing or felled, or fells any tree, drags timber, or permits cattle to damage any tree, shall be punishable with imprisonment for a term which may extend to six month or with a fine which may extend to Rs. 500, or both. When there is a reason to believe that a forest offence has been committed pertaining to any forest produce, the produce together with all tools used in committing such offences may be seized by any forest officer or police officer. Every officer seizing any property under this sections hall put on the property a mark indicating the seizure and report the seizure to the magistrate who has the jurisdiction to try the offence. Any forest officer, even without an order from the magistrate or a warrant, can arrest any person against whom a reasonable suspicion exists.

What can an individual do to support the Act?

(1) Be alert to destructive activities in your local green areas such as Reserved Forests and Protected Forests, and in Protected Areas (National Parks and Wildlife Sanctuaries).

Report any such act to the Forest Department as well as the Press. Report of violations can be made to the Conservator of Forest, District Forest Officer, Range Forest Officer, Forest Guard or the District Commissioner,or local civic body.

(2) Acquaint yourself with the laws, detailed rules and orders issued by the Government.

(3) Be in touch with concerned local NGOs and associations. Organize one with other like minded people if none exist in your area.

(4) Create awareness about the existence and value of national parks and sanctuaries and build up a public opinion against illegal activities in the forest or disturbance to wildlife.

(5) Pressurize the authorities to implement the forest and wildlife laws and rules to protect green areas.

(6) Take legal action if necessary and if possible through a Public Interest Litigation (PIL) against the offending party. Use the help of NGOs who can undertake legal action.

(7) Help to create public pressure to change rules laws and procedures when necessary.

(8) Use better, ecologically sensitive public transport and bicycle tracks. Do not litter in a forest area.

(9) Participate in preservation of greenery, by planting, watering and caring for plants. Whom should forest offences be reported to? If you as a citizen come across anyone felling trees, encroaching on forest land, dumping garbage, cutting green wood, lighting a fire, or creating a clearing in Reserved Forests, Protected Forests, National Park, Sanctuary or other forest areas, you must report it to the forest / wildlife officers concerned. For urgent action one can contact the police. In fact you should file an FIR in any case because it serves as an important proof that you have made the report.

10.10 ISSUES INVOLVED IN ENFORCEMENT OF ENVIRONMENTAL LEGISLATION

Environmental legislation is evolved to protect our environment as a whole, our health, and the earth's resources. The presence of a legislation to protect air, water, soil, etc. does not necessarily mean that the problem is addressed. Once legislation is made at the global, national or state level, it has to be implemented. For a successful environmental legislation to be implemented there has to be an effective agency to collect relevant data, process it and pass it on to a law enforcement agency. If the law or rule is broken by an individual or institution, this has to be punished through the legal process. Information to law enforcement officials must also come from concerned individuals. In most situations, if no cognizance is given, the interested concerned individual must file a Public Interest Litigation (PIL) for the protection of the environment. There are several NGOs in the country such as WWF-I, BEAG and the BNHS which take these matters to court in the interest of conservation. Anyone can request them to help in such matters. There are also legal experts such as MC Mehta who have successfully fought cases in the courts to support environmental causes. A related issue is the fact that there are several irregular practices for which a bribe to an unscrupulous official is used to cover up an offence. Thus the general public must actas a watch dog not only to inform concerned authorities, but also to see that actions are taken against offenders.

10.10.1 Environment Impact Assessment (EIA)

For all development projects, whether government or private, the MoEF requires an impact assessment done by a competent organization. The EIA must look into physical, biological and social parameters. EIAs are expected to indicate what the likely impacts could be if the project is passed. The Ministry of Environment and Forests (MoEF) has identified a large number of projects that need clearance on environmental grounds. The EIA must define what impact it would have on water, soil and air. It also requires that a list of flora and fauna identified in the region is documented and to specify if there are any endangered species whose habitat or life could be adversely affected. Most development projects such as industries, roads, railways and dams may also affect the lives of local people. This must be addressed in the EIA. There are 30 different industries listed byMoEF that require a clearance before they are set up. Impacts created by each type of industry differ and the proposed sites also vary in their sensitivity to impacts. Some areas are more fragile than others. Some have unique ecosystems. Others are the habitats of wildlife and some may be the home of endangered species of plants or animals. All these aspects require evaluation before a development project or an industry siteis cleared. New projects are called 'green field projects'where no development has been done. Projects that already exist but require expansion must also apply for clearance. These are called 'brownfield projects'. After the Environmental Protection Act of 1986 was passed, an EIA to get an environmental clearance for a project became mandatory. Project proponents are expected to select a competent agency to undertake an EIA. Projects can be classified into those with a mild impact, a moderate impact or a serious impact. Some may have temporary major impacts, during the construction phase, which could later become less damaging, or be mitigated by a variety of measures. In other situations the impact may continue and even increase, for example, where toxic solid waste will be constantly generated. Some projects could thus cause temporary reversible damage while others can have irreversible or even permanent impacts. To get an environmental clearance the proposer of the project is expected to apply to the State Pollution Control Board. The PCB checks and confirms that the EIA can be initiated. The Agency that does the assessment submits a report to the proposer. This may take several months. A Report of the Environmental Statement is forwarded to the MoEF, which is the impact assessment authority. After 1997, the MoEF has stipulated that a public hearing should be done at the local level. The Pollution Control Board puts an advertisement about the hearing in the local vernacular press. An Environmental Impact Statement which is an Executive Summary of the EIA is kept for the public to read. The venue and time of the Public Hearing is declared. Once the hearing is held and opinions have been expressed, both for and against the project, the minutes of the meeting are sent to the MoEF. Though this is done, it is evident that the voices of project affected people are still not heard. In some cases NGOs have taken up the cause of local people. Until educational levels and environmental awareness becomes a part of public thinking and is objectively based on the facts of the case, these hearings will remain an inadequate tool to control possible impacts of new development projects. Experience shows that a large number of EIAs are inadequately researched and frequently biased as they are funded by the proposer of the project. While most EIAs are adequate for studies on the possibilities of air, water and soil pollution, they generally deal inadequately with issues such as preservation of biodiversity and the social issues that may arise from future environmental impacts. Biodiversity concerns frequently are sketchily considered and mostly consist of a listing of species without population assessments, or census figures of wildlife, or a study of the effects on the ecosystem as a whole. Changes in land use patterns effect whole communities of living organisms. This is rarely taken into account, as such issues are difficult to assess in quantifiable terms.

Issues related to equity of resources that are inevitably altered by development related projects are also not fully addressed. These cryptic concerns must be dealt with more seriously in environmental assessments and the public at large should know and appreciate these inadequacies. It is not sufficient to say that an EIA has been done. It is the quality and sincerity of the EIA that is of importance.

An EIA is not intended to stop all types of development. The sight of an industry can be selected carefully and if it is likely to damage a fragile area an alternate less sensitive area must be selected. In some cases it is essential to drop projects altogether if the anticipated impacts are likely tube very severe. In other cases it is necessary for the project to counter balance its effects by mitigating the ill effects on the environment. This means compensating for the environmental damage by forestation or creating a protected area in the neighborhood at the cost of the project.

Rehabilitation and resettlement of project affected people is a key concern which should be given adequate funds and done after a consent is clearly obtained from the people living in the area. In most cases it is advisable to avoid resettlement altogether. If an area's vegetationis being affected project costs must include the cost of compensatory forestation and other protective measures.

10.10.2 Environmental Management Systems (EMS)

EMS stands for Environmental Management System. It is a system that enables any organization irrespective of its size, type and setting to:

- Manage environmental impacts arising out of its activities, product and services
- Ensures compliance to regulations
- Bring continual improvements
- Demonstrate high environmental performance to others by conforming to policy, objectives and targets.

The reasons why many companies are now adopting an Environmental Management System include the following:

- Improve compliance with legislative and regulatory requirements.
- Facilitate access to overseas markets.
- Reduce liability/risks.
- Helps in attracting a high-quality workforce.
- Prevent pollution and reduce waste.
- Generate a desire to profit in the market for "green" products.
- Improve the internal management methods.
- Serves the interests of the shareholder groups.
- Create community goodwill.
- Helps in getting insurance coverage.

ISO 14001?

ISO 14001 is an Environment Management System (EMS) standard published by International Organization for Standardization in year 1996 and later updated in the year 2004. It provides a highly effective, globally accepted framework for establishing and continually improving the effectiveness of environmental management. Implementation of ISO 14001 may bring with it both reductions in environmental risk and environmental costs.

For ISO 14001 certification, the assessment consists of two-stage assessment. Both the assessments are conducted at site.

Stage I: In the first stage, in addition to the system document assessment the following issues are assessed:

1. A brief overview of the client's Environmental Management System
2. Identification of environmental aspects and impacts
3. Environmental regulations and laws are observed
4. Internal environmental audit and management review

Stage II: On recommendation of the assessment team, the assessment moves to the state II of the audit. All non - conformities raised at stage I should be closed before going for Stage II Audit. The aspects assessed in State II are:

1. Operation
2. Implementation
3. Effectiveness of the environmental management system in the organization.

Compliance to ISO 14001 standard is demonstrated by registering an organization's environment management system through a third party certification scheme. The process involves an assessment of the supplier's documentation and on-site audit performed by a certification body. Periodic surveillance assures the maintenance of the organization's Environment Management Systems.

10.10.3 Environmental Information Systems (ENVIS)

The Environmental Information System was implemented by the Ministry by end of 6th Five Year Plan as a Plan Scheme for environmental information collection, collation, storage, retrieval and dissemination to policy-planners, decision-makers, scientists and environmentalists, researchers, academicians and other stakeholders. It is a decentralized computerized network database system consisting of the focal point located in the Ministry and a chain of network partners, known as ENVIS centres located in the potential organizations/institutions throughout the country. In order to develop ENVIS network as a comprehensive distributed environmental information network system, the ambit of ENVIS was extended to cover all the States/UTs of the country. Necessary guidelines and policies were provided to revise the earlier EFC of the scheme and the new concept for covering the state government departments were completed through a revised SFC.

Presently, the ENVIS network consist of 76 ENVIS centres apart from the focal point, out of which, 29 ENVIS centres are on state government departments dealing with the status of environment and related issues of the concerned state government and the remaining 47 have been set up on various environmental disciplines covering from air pollution, water pollution, noise pollution, biodiversity, solid waste management, ecology and ecosystems, environmental education, NGOs, media and even environmental parliament, coastal ecosystem, clean technology, etc.

The broad objectives of ENVIS are:

Long-term objectives:

1. To build up a repository and dissemination centre in Environmental Science and Engineering;
2. To gear up the modern technologies of acquisition, processing, storage, retrieval and dissemination of information of environmental nature; and
3. To support and promote research, development and innovation in environmental information technology.

Short-term objectives

1. To provide national environmental information service relevant to present needs and capable of development to meet the future needs of the users, originators, processors and disseminators of information;
2. To build up storage, retrieval and dissemination capabilities with the ultimate objectives of disseminating information speedily to the users;
3. To promote, national and international cooperation and liaison for exchange of environment related information;
4. To promote, support and assist education and personnel training programmers designed to enhance environmental information processing and utilization capabilities;
5. To promote exchange of information amongst developing countries.

10.10.4 Citizen's Actions and Action Groups

Citizens must learn to act as watchdogs to protect their own environment from the consequences of unsustainable projects around them. Well informed citizens not only have rights but also have a duty to perform in this regard. They can join action groups to develop a lobby to strengthen the environmental movements in the country, their state, town or village. Individuals can take one or several possible actions when they observe offenders who for their own self interest damage the environment for others living in the area. An individual has the right to bring an environmental offence or nuisance to the attention of concerned authorities. This ranges from government line agencies such as the police, the forest department, the collector or commissioner of the area as the case may be. At times the concerned officials may not be able to easily appreciate complex environmental concerns and the individual may have to learn how to communicate these issues in a way in which it becomes essential for the concerned officer to act in a pro-environmental fashion. If this does not work a citizen can seek legal redressed under relevant statutes of law. The Environment Protection Act and the Wildlife Protection Act are the most frequently used legal instruments for these purposes. It is possible to move courts by a Public Interest Litigation, and take this up to the Hon. Apex Court – the Supreme Court of India, which in the recent past has given several highly enlightened pro-conservation judgments. Citizen groups can resort to alternate means of pressure such as 'rastarokos', 'dharnas', etc. to draw attention to important environmental concerns. They can also elicit public support through the press and electronic media.

CASE STUDY: The Narmada Issue

The controversy over the plan to build several dams on the Narmada River and its tributaries symbolizes the struggle for a just and equitable society in India. The construction of these dams displaces many poor and underprivileged communities, destroying their relatively self-sufficient environmentally sound economy and culture and reducing a proud people to the status of refugees or slum dwellers. The Narmada Bachao Andolan (Save the Narmada Movement) is one of the most dynamic people's movements fighting for the rights of these underprivileged people who are being robbed of their homes, livelihoods and way of living in the name of 'national interest'.One such dam, the Sardar Sarovar Dam,when completed will drown 37,000 hectares of fertile land and displace 200,000 people and cause incomprehensible loss to the ecology.

CASE STUDY: Silent Valley

The proposed hydel project at Silent Valley a unique pocket of tropical biodiversity in South India, in the 1970s was stopped and the area declared a National Park in 1984.This was achieved by several dedicated individuals, groups and organizations lobbying to save the area from being submerged and protect its rich biodiversity. Among the many environmental battles that have been fought in this country some have been won while many others have been lost. These projects have led to serious environmental degradation in spite of the laws intended to control such damage.

10.11 ROLE OF NON-GOVERNMENTAL ORGANIZATIONS (NGOs)

Voluntary organizations can help by advising the government about some local environmental issues and at the same time interacting at the grassroot levels. They can act as effective and viable link between the two. They can act both as an action group or a pressure group. They can be very effective organizing public movements for the protection of environment through creation of awareness. The Chipko Movement for conservation of trees by Dasholi Gram Swarajya Mandal in Gopeshwar or the Narmada Bachao Andolan organized by Kalpvariksh are some of the instances where NGOs have played a landmark role in the society for conservation of environment.

The Bombay Natural History Society (BNHS), the World Wide Fund for Nature- India (WWF-India), Kerala Sastra Sahitya Parishad, Centre for Science and Environment (CSE) and many others are playing a significant role in creating environmental awareness through research as well as extension work. The recent report by CSE on more than permissible limits of pesticides in the cola drinks sensitized the people all over the country. Before we can all take up the task of environmental protection and conservation, we have to be environmentally educated and aware. It can therefore say "If you want to act green, first think green."

10.11.1 Enforcement of Environmental Legislation: Major Issues

We have seen that there are a number of important environmental laws in the form of Acts for safeguarding our environmental quality. But in spite of these we find that we are not able to achieve the target of bringing 33% of our land cover under forests. Still we are losing our wildlife. The rivers have been turned into open sewers in many places and the air in our big cities is badly polluted. The status of environment shows that there are drawbacks in environmental legislations and problems in their effective implementation.

10.11.2 Public Environmental Awareness

Public awareness about environment is at a stage of infancy. Off late, some awareness has taken place related to environmental degradation, pollution etc. but incomplete knowledge information and ignorance about many aspects has often led to misconceptions. Development has paved the path for rise in the levels or standards of living but it has simultaneously led to serious environmental disasters. Issues related to environment have been often been branded as anti-development. The wisdom lies in maintaining a balance between our needs and supplies so that the delicate ecological balance is not disrupted. Some of the main reasons responsible for widespread environmental ignorance are detailed below:

1. Our courses in science, technology, economics etc. have so far failed to integrate the knowledge in environmental aspects as an essential component of the curriculum.
2. Our planners, decision-makers, politicians and administrators have not been trained so as to consider the environmental aspects associated with their plans.
3. In a zeal to go ahead with some ambitious development projects, quite often there is a purposeful concealment of information about environmental aspects.
4. There is greater consideration of economic gains and issues related to eliminating poverty by providing employment that overshadows the basic environmental issues.

10.11.3 Methods to Propagate Environmental Awareness

There is immense need for environmental awareness. It is to be created through formal and informal education to all sections of the society. Everyone needs to understand it because 'environment belongs to all' and 'every individual matters' when it comes to conservation and protection of environment. Various stages and methods that can be useful for raising environmental awareness in different sections of the society are given below:

1. Among students through education: Such education should be imparted to the students right from the childhood age. These studies are now being incorporated at all stages in schools, colleges as per the directives of the Supreme Court.
2. Among the masses through mass-media: Media can play an important role to educate the masses through articles, rallies, campaigns, street plays, TV serials etc. This will appeal all age groups at the same time.
3. Among the planners, decision makers and leaders: It is very important to give these classes of people necessary orientation and training through specially organized workshops and training programmes.

Learning Outcomes

The main learning outcomes from this chapter should be fundamental understanding of: Unsustainable to Sustainable development; Urban problems related to energy; Water conservation; rain water harvesting; watershed management; Resettlement and rehabilitation of people- its problems and concerns, Case Studies; Environmental ethics: Issues and possible solutions; Climate change; global warming; acid rain; ozone layer depletion; nuclear accidents and holocaust, Case studies; Wasteland reclamation; Consumerism and waste products, Environment Protection Act ; Air (Prevention and Control of Pollution). Act, Water (Prevention and Control-of Pollution) Act; Wildlife Protection Act; Forest Conservation Act; Role of Government, Legal aspects, Initiatives by Non-governmental Organizations (NGO), Environmental Education, Issues involved in enforcement of environmental legislation; and Public awareness.

Readers are also advised to carry out following exercise:

Exercise

1. Write summary of case study on Tehri Dam project with suitable example and government legislations.

Chapter 11

Global Warming and Carbon Sequestration

11.1 INTRODUCTION

Sunlight is one of the major sources of energy for living organisms. Sunlight also helps in keeping the atmosphere warm. When sunrays strike the surface of the earth, they are partly absorbed, and partly reflected back into the atmosphere. The reflected rays are known as infrared radiation, are captured by greenhouse gases which are the main reason behind keeping our atmosphere warm. As a result we can say that greenhouses gases acts like a blanket. Infrared light also causes the heat you feel from ordinary light bulbs. If greenhouse gases were not available for trapping heat in the atmosphere, the earth would be a very cold place. The process of keeping earth atmosphere warm with the help of greenhouse gases is called the greenhouse effect. Carbon dioxide, methane and water-vapour are some of the greenhouse gases. Around only 1% of all these gases form our atmosphere, they are extremely vital in maintaining the ecological balance and sustaining life on this planet. The temperature of the earth would have been lesser than 30°C,which means that survival of living species wouldn't be possible. So, it is quite clear that the presence of greenhouse gases is important for the survival of living species. Nature maintains balance in the atmosphere through greenhouse effect. In past hundred years, human activity has caused an increase in the percentage of greenhouse gases in the gases. So in a simple way we can say that greenhouse effect is an increase in the average temperature of the earth. From the Industrial Revolution, humans are adding greenhouse gases in excessive amount to the atmosphere, which is enhancing greenhouse effect. As greenhouse gases are responsible for absorbing the heat, the more greenhouse gases is in the atmosphere, the warmer the air will be. The balance of life will be disrupted if air gets too hot and this may also results in the death of many plants and animal species. The food chain could be upset which would cause many serious problems in the world. So, this increase in greenhouse gases has the potential to cause catastrophic problems for earth and its inhabitants.

11.1.1 Greenhouse Gases

The most important GHGs directly emitted by humans include CO_2, CH_4, nitrous oxide (N_2O), and several others. The sources and recent trends of these gases are detailed below.

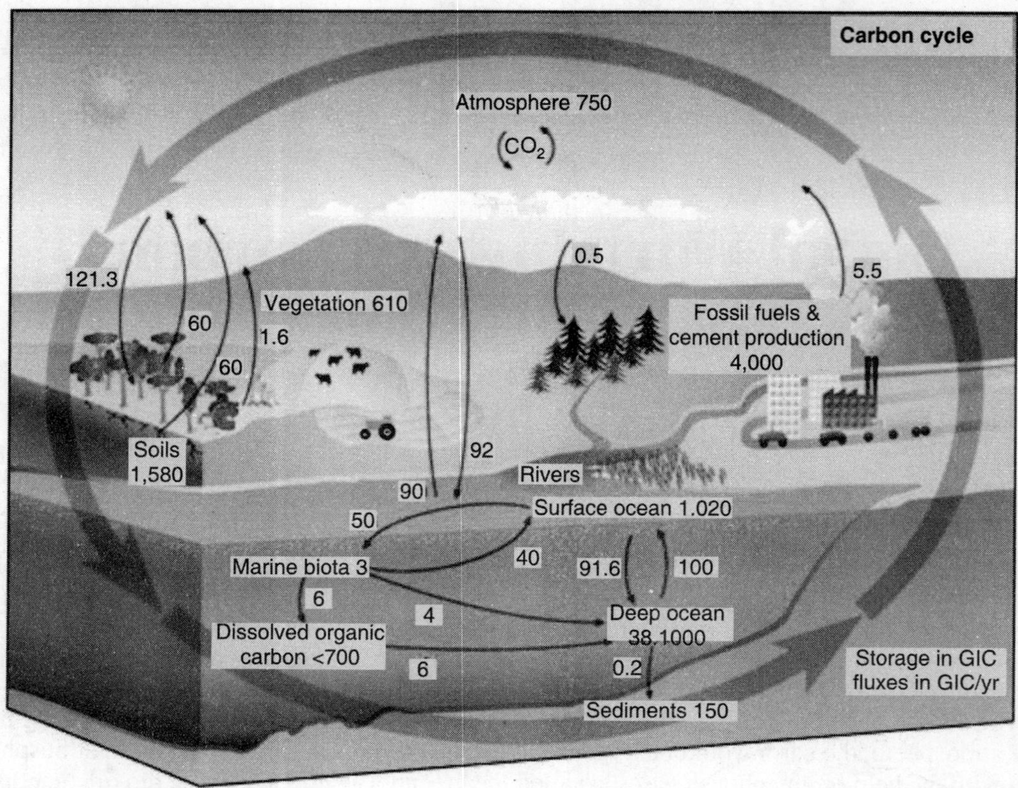

Fig. 11.1 Carbon cycle (Image source: NASA. (Note, values shown represent Carbon Gigatons being absorbed and released).

(i) Carbon dioxide: Carbondioxide is the primary greenhouse gas that is contributing to recent climate change. CO_2 is absorbed and emitted naturally as part of the carbon cycle, through animal and plant respiration, volcanic eruptions and ocean-atmosphere exchange. Human activities, such as the burning of fossil fuels and changes in land use, release large amount of carbon to the atmosphere, causing carbon dioxide concentrations in the atmosphere to rise. In 2010, the carbon dioxide concentration has risen from pre-industrial levels of 280 parts per million by volume (ppmv) to about 390 ppmv and concentrations have risen by 75 ppmv since 1958. The level of carbon dioxide currently is higher than it has been in at least 8,00,000 years. In the distant past volcanic eruptions released large quantities of carbon dioxide. Human activities emit now more than 135 times carbon dioxide, as volcanoes each year. Every year over 30 billion tonnes of carbon dioxide is released into the atmosphere by human activities.

(ii) Methane: Methane is more abundant in the earth's atmosphere now than at any time in at least the past 650,000 years. During most of the 20th century, methane concentrations increased sharply due to human activities and are now more than two-and-a-half times pre-industrial levels. Through both natural and human activities methane is produced. For example, agricultural activities, natural wetlands and fossil fuel extraction and transport all emit methane.

(iii) Nitrous oxide: Mainly through agricultural activities and natural biological processes nitrous oxide is produced. About 18% of nitrous oxide have risen since, the start of the Industrial Revolution. Through natural and human activities nitrous oxide is produced. By the end of the 20th century nitrous oxide level increased. Before the onset of the industrial period, the atmospheric concentration of nitrous oxide varied only slightly for a period of 11,500 years.

(iv) Water Vapour: Water vapour is the most abundant greenhouse gas despite having a short atmospheric lifetime. In terms of its contribution to the natural greenhouse effect water vapour is most important gas. Some human activities can influence local water vapour levels. The concentration of water vapour is controlled by temperature on a global scale. It influences overall rates of evaporation and precipitation. By direct human emissions, the global concentration of water vapour is not substantially affected.

(v) Ozone (O_3): Ozone is a potent greenhouse gas which also has a short atmospheric life. In the presence of sunlight, chemical reactions create ozone from emission of nitrogen oxides and volatile organic compounds from automobiles, power plants, and other industrial and commercial sources. Ozone is a pollutant that can cause respiratory health problems and damage crops and ecosystems.

Chlorofluorocarbons, hydrochlorofluorocarbons, hydrofluorocarbons, perfluorocarbons and sulfur hexafluoride together called F-gases. These F-gases are often used in coolants, foaming agents, fire extinguishers, solvents, pesticides, and aerosol propellants. These F-gases have long atmospheric life and some of these emissions affect the climate for many decades or centuries.

11.2 CONTRIBUTORS OF GREENHOUSE GAS

Those gas molecules which have three or more atoms in the earth's atmosphere are called greenhouse gases, because they can capture outgoing infrared energy from the earth, thereby warming the earth. As we all know that greenhouse gases include water vapour, ozone, carbon dioxide, methane, etc., at the same time the trace quantities of chlorofluorocarbons can have a disproportionately large effects.

(i) Deforestation: Deforestation is one of the man-made causes of the greenhouse effect. This results in increase amount of carbon dioxide in the atmosphere, as due to absence of trees photosynthesis cannot be possible. The levels of deforestation have increased about 9% in recent times. Due to burden of our needs deforestation is rampant today. Burning of wood also releases more carbon dioxide into the atmosphere.

(ii) Burning of Fossils: Greenhouse gases are released into the atmosphere due to the burning of fossil fuels, oil, coal and gas. These materials are used increasingly and rampantly in industries. Many gas which last for longer time in the atmosphere are produced by most of the factories. These gases are naturally available in the atmosphere. These gases contribute to the greenhouse effect and also increase the global warming on the earth. Therefore, industries are also a major cause of the greenhouse effects.

(iii) Electrical Appliances: One of the man-made causes of the increase in the greenhouse effect is the emission of greenhouse gases by electrical appliances. Refrigerators emit gases which contribute to greenhouse effect. These gases are called chlorofluorocarbons

Fig. 11.2 Natural and human enhanced greenhouse effect.

and are used in refrigerators, aerosol cans, some foaming agents in the packaging industry, fire extinguishers, chemicals and cleaners used in the electronic industry. Some processes of the cement manufacturing industries also act as a cause towards the greenhouse effect.

(iv) Population Growth: Population growth is an indirect contributor and one of the major causes of the greenhouse effect. The needs and wants of people increase with the increase in population, which as a result increases the manufacturing and the industry process. This in turn increase the release of industrial gases which catalyze the greenhouse effect.

11.3 IMPACT OF GREENHOUSE GAS

(i) Sea Level Rise: Sea level rise is the most dangerous aspect. The world's oceans have already risen 4-8 inches. This leads to erosion of some islands. As a result, people living in the low lying coastal area have to relocate to higher ground. Further sea level rise could cause great sufferings. About 15 million people living with in 1 metre of sea level in Bangladesh alone and another 8 million in a similar circumstance in India. If sea levels continue to rise, inhabitated land could be inundated.

(ii) Droughts and floods: Changes in the climate due to excess greenhouse gases are causing both increased drought and increased flooding. Violent storm will increase and at the same time more water evaporates from the oceans, as temperatures rise. This includes more powerful hurricanes, pacific typhoons and an increased frequency of severe localized storms and tornadoes. In coastal areas insurance companies struggle to cover escalating costs, as these storms often result in flooding and property damage. Many dry areas are experiencing more severe droughts. Since 1970, the amount of land on the earth suffering from drought condition has doubled.

(iii) The human price of climate change: Excess of greenhouse gases results drought and it is the driving current increases in food prices around the world, in combination with increased use of grains for fuel. Until the late 1990s, the number of malnourished people decreased globally. The disease carriers will expand their territory, either by moving to higher elevations in mountainous area or by expanding their territory further from equator. Millions of humans are exposed to deadly infectious diseases due to this expansion. About 150,000 annual deaths worldwide have been tied to climate change, according to a 2005 World Health Organization report. Industrialized countries may be sheltered from the current impacts of climate change. Floods and more powerful storms linked to climate change as well as heat waves and droughts are responsible for these deaths.

(iv) Approaching a Slippery Slope: Due to temperature increase the vast Artic tundra is melting and thus releasing enormous volumes of both carbon dioxide and methane into the atmosphere. Due to more carbon dioxide and methane that are released from the Artic tundra, the greenhouse effect will be enhanced, which creates the possibility of a self-reinforcing loop of climate change. Because of rising water temperatures the world's oceans are losing their ability to absorb carbon. This is significant as, the world's oceans hold 50 times more carbon than do the world's forests and grasslands.

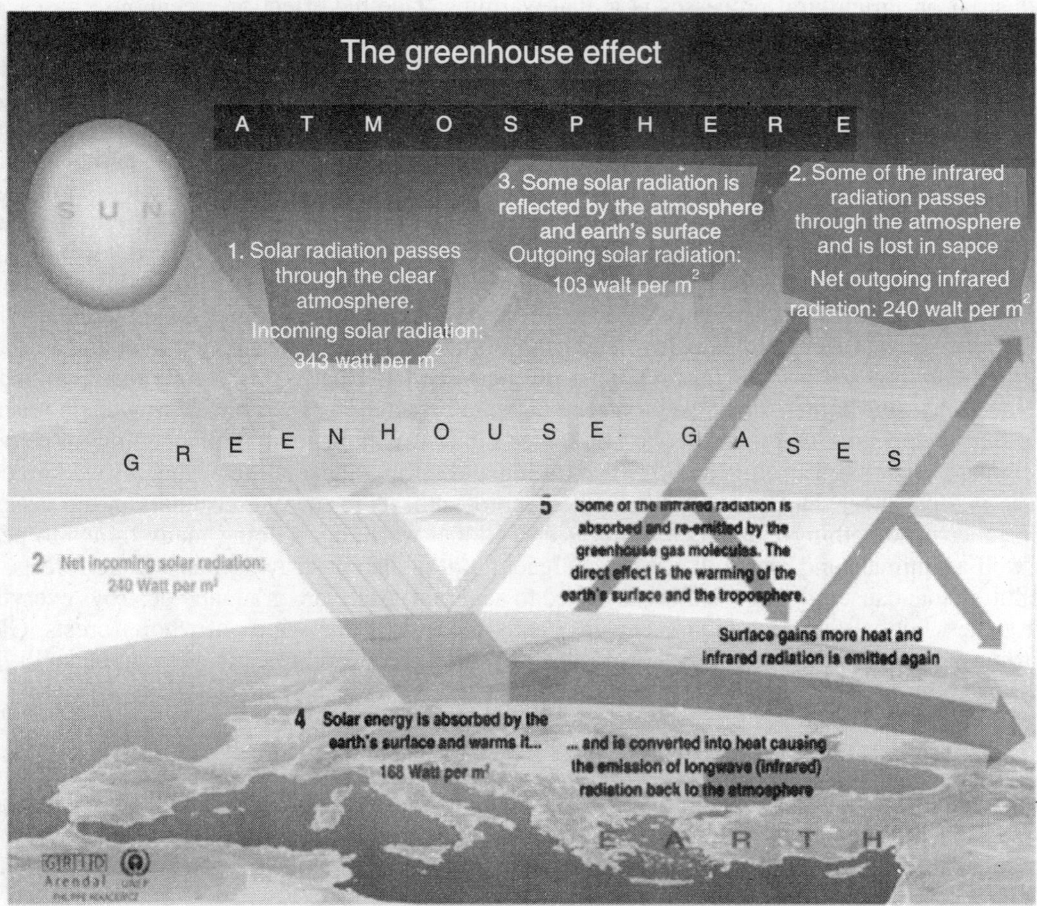

Fig. 11.3 Greenhouse process in out atmosphere.

(v) **Rapid Climate Shifts:** The past cycles of climate change on the earth have been slow and incremental. Climate change happens suddenly and violently due to human-created climate change. For many thousands of years the earth's climate exists in a stable state. The pressures for change builds from increases or decreases in carbon levels as well as changes in solar radiation. All over the world, weather patterns may be disrupted. We must not gamble with all of life on the earth. It is very much necessary to take decisive action to avoid devastating climate change.

There are some of the factors that cause an increase in the greenhouse effect. It is not the greenhouse effect that cause global warming. But it is the unrestrained human activity that has caused an increase in greenhouse gases, which in turn have led to global warming. It is necessary now to spare a thought on the damage that we might be causing to the environment. We may not live long enough to witness the repercussions of our actions, but our future generations might have to pay for our actions.

11.4 GLOBAL WARMING

An increase in the earth's temperature, which can be caused by the use of fossil fuels and industrial or agricultural processes is global warming. Due the effect of greenhouse gases, such as carbon dioxide emission from burning fossil fuels or from deforestation, which trap heat that would otherwise escape from the earth. These harmful processes have been leading to a large increase in the buildup of greenhouse gases, which are carbon dioxide, methane, nitrous oxide, chlorofluorocarbons and water vapour. Today the world is facing an extremely severe problem of global warming. The effects and causes of global warming have been on the rise and people need to do something about it before the problem gets any worse. Until industrial revolution the world did not have such a terrifying problem of harmful gases in the atmosphere. With the increasing population growth, increased fossil fuel burning, and deforestation the world faces a major problems of the combination. Many people, animals and plants cannot accept this change and finally they die. Many parts of the world are affected by global warming, for example sea level rise and as a result water covers many low land islands. Plants die due to excess of water and as a result the animals lose a source of food, along with their habitat. Due to this people lose two sources of food and their homes too. This would be called a break in the food chain, or a chain reaction, one thing happening that leads to another and so on. Many things that are happening to the ocean are linked to global warming, and this is harming and killing algae in the ocean. As we know that algae is a producer that makes food through photosynthesis to many consumers in the ocean. A consumer is something that eats the producers. Global warming is doing many things to people as well as animals and plants. It is killing algae, but it is also destroying many huge forests. The pollution that causes global warming is linked to acid rain. Acid rain gradually destroys everything it touches. Global warming is also causing many more fires that wipe out whole forests. Global warming can make the earth very hot. In forests, some plants and trees leaves can be so dry that they catch on fire.

11.4.1 Causes of Global Warming

All of us believe that the greenhouse effect is the main cause of global warming. Excessive use of fossil fuels has increased the levels of greenhouse gases which retain the heat radiated by the sun, consequently increasing the temperature of earth's atmosphere. Reflecting ultraviolet radiation and trapping infrared heat are the jobs of the earth's atmosphere. Greenhouse gases heat the atmosphere more than the need which ultimately causes global warming. Strictly speaking greenhouse gas does not cause global warming and greenhouse effect is a natural process which keeps our earth warm. So it is quite clear that the enhanced greenhouse effect is what causes global warming as these processes increase the levels of greenhouse gases in the atmosphere. Thus the increase in concentration of greenhouse gases is the main problem. People have started using glass houses to grow plants in winter which will trap the heat from the sun and these are called greenhouses. Similarly, greenhouse gases in the earth atmosphere work like the glass and prevent the heat from escaping.

Sunlight is absorbed by land, water and air. Due to increase in carbon dioxide emissions in the last couple of decades, more and more heat is being absorbed in the atmosphere. The sunlight enters piercing through the cloud of greenhouse gases and when it reaches the earth's surface, it is absorbed by land, water, and air. After absorption the same energy is sent back into the atmosphere. Some of it remains trapped in the atmosphere due to the greenhouse gases and this energy makes the earth warmer. Since there has been an increase in CO_2 emissions in the last couple of decades, more and more heat is being absorbed in the atmosphere. This is increasing the temperature of the earth gradually and causing global warming.

(i) **Carbon dioxide emissions from burning gasoline for transportation:** About 33% of our modern car culture and appetite for globally sourced goods are responsible for emission. With our population growing at an alarming rate, the demand for more cars and consumer goods means that we are increasing the use of fossil fuels for transportation and manufacturing. Our consumption is outpacing our discoveries of ways to mitigate the effects, with no end in sight to our massive consumer culture.

(ii) **Methane emissions from animals, agriculture such as rice paddies, and from Arctic sea beds:** Ranking right behind carbon dioxide, methane is another extremely potent greenhouse gas. Methane is produced when organic matter is broken down by bacteria under oxygen-starved conditions as in rice paddies.This process also takes place in the intestines of herbivorous animals, and with the increase in the amount of concentrated live stock production, the levels of methane released into atmosphere. Methane clathrate is another source of methane,which is a compound containing large amounts of methane trapped in the crystal structure of ice. The rate of global warming increases significantly as methane escapes from Artic seabed.

(iii) **Increase in usage of chemical fertilizers on cropland:** The use of chemical fertilizers has risen dramatically in the last half of the 20th century. The high rate of application of nitrogen-rich fertilizers has effects on the heat storage of cropland and the run-off of excess fertilizers creates dead-zones in our oceans. This ultimately affects the high nitrate levels in groundwater due to over-fertilization are causes for concern for human health.

(iv) **Electrical Pollution:** Many things cause global warming. Electrical pollution also cause global warming. Fossil fuels are burnt to create electricity. Fossil fuels are made of dead plants and animals. Many pollutants are sent into the air when fossil fuels are burnt. Some of these are greenhouse gases. We use these sources of energy much more than the sources that give off less pollution. One source of energy is petroleum, which is used a lot for transportation, making electricity and making many other things.Although this source of energy gives off a lot of pollution.

The cause of global warming is due to an increase of carbon dioxide in the atmosphere. The levels of carbon dioxide has increased from 280 parts per million to 379 parts per million over the last 100 years. This increase in carbon dioxide is strongly correlates with the increase of global surface temperatures. Increases in carbon dioxide is mostly due to burning of fossil fuels such as coal, oil and natural gas. Some believe that global warming is simply a short-term trend caused by the natural rhythms of the planet, or our sun.

11.4.2 Effects of Global Warming

(i) **Rise in sea levels worldwide:** An increase in sea levels worldwide due to the melting of two massive ice sheets in Antarctica and Greenland is one of the effects of global warming. However, many nations around the world will experience the effects of rising sea levels, which could displace millions of people.

(ii) **More killer storms:** The severity of storms such as hurricanes and cyclones is increasing. Global warming will significantly increase the intensity of the most extreme storms worldwide. The maximum wind speeds of the strongest tropical cyclones have increased significantly since 1981.

(iii) **Massive crop failures:** Almost 90% chance that 3 billion people worldwide will have to choose between moving their families to milder climes and going hungry due to climate change within 100 years. Climate change is expected to have the most severe impact on water supplies. Shortages in future are likely to threaten food production, reduce sanitation, hinder economic development and damage.

(iv) **Widespread extinction of species:** Rising temperatures could lead to the extinction of more than a million species. Existing without a diverse population of species on the earth is scary news for humans. Climate change led to a great threat to the number of species surviving on the earth as habitat destruction and modification.

(v) **Disappearance of coral reefs:** A report on coral reefs says that, coral populations will collapse by 2100 due to increased temperatures and ocean acidification. The 'bleaching' of corals from small but prolonged rises in sea temperature is a severe danger for ocean ecosystems, and many other species in the oceans rely on coral reefs for their survival. For reefs, warming waters and acidification are like a pair of jaws that threaten to make them the first global ecosystem to disappear.

Fig. 11.4 Consequences of global warming.

11.4.3 Consequences of Global Warming

- An increase in the frequency of heat waves.
- A reduction in the surface area and thickness of glaciers.
- The melting of ice.
- Loss of mass of polar ice sheets in the Antarctic and Greenland.
- Increased average ocean levels.
- Disrupted water cycle, which may lead to very heavy rainfall (flooding) or prolonged droughts.
- Changes in vegetation (flowering dates, harvest times).
- Migration of animal species(e.g. insects).

Entire population affected by global warming has effects on public health that are difficult to quantify. Nevertheless, public health is affected by climate:

- Heat waves have consequences on particularly vulnerable populations (older people, newborns, and people with respiratory problems,particularly asthma).
- Air quality is altered, which also affects vulnerable people and those suffering from cardio-respiratory problems.
- Changes in pollen production are one of the reasons why allergies are on the increase in most countries.
- Movements of insects carrying epidemics (malaria, dengue) can lead to the extension of the areas at risk of these epidemics.
- Climate changes also affect farm production and fishing resources, which in turn have consequences on people's food supply.

11.5 CLIMATE CHANGE

Climate is the long term statistical expression of short term weather. Climate can be defined as expected weather. When changes in the expected weather occur, we call these climate changes. They can be defined by the differences between average weather conditions at two separate times. Climate may change in different ways over different time scales and at different geographical scales. Climate change is defined as changes in the earth's weather, including changes in temperature, wind patterns and rainfall, especially the increase in the temperature of the earth's atmosphere that is caused by the increase of particular gases, especially carbon dioxide. The overall state of the global climate is determined by the amount of energy stored by the climate system, and in particular the balance between energy the earth receives from the sun and the energy which the earth releases back to space, called the global energy balance. Major causes of climate involve any process that can alter the global energy balance and the energy flows within the climate system. Causes of climate change include changes in the earth's orbit around the sun, changes in the amount of energy coming from the sun, changes in ocean circulation or changes in the composition of the atmosphere. Large volcanic eruptions can affect the global climate over only a few years. By contrast, the movement of continents around the world over hundreds of millions of years can also affect global climate. Throughout the earth's history climate has fluctuated between periods of relative warmth and relative cold. During this period, the amount of greenhouse gases in the atmosphere has increased, largely as a result of the burning of fossil fuels for energy and transportation, and land use changes, for food by mankind. Global warming is now considered most probably to be due to the man-made increases in greenhouse gas emissions. It is the major, overriding environmental issue of our time, and the single greatest challenge facing environmental regulators. It is a growing crisis with economic, health and safety, food production, security, and other dimensions. Whilst other natural causes of climate change, including changes in the amount of energy coming from the sun and shifting patterns of ocean circulation, can cause global climate to change over similar periods of time, the balance of evidence now indicates that there is a discernible human influence on the global climate. Shifting weather patterns, for example, threaten food production through increased unpredictability of precipitation, rising sea levels contaminate coastal freshwater reserves and increase the risk of catastrophic flooding, and a warming atmosphere aids the pole-ward spread of pests and diseases once limited to the tropics. Combined with thermal expansion warm water occupies more volume than cold. The melting of ice sheets and glaciers around the world is contributing to rates and an ultimate extent of sea-level rise that could far outstrip those anticipated in the most recent global scientific assessment. Mountain glaciers are in alarming retreat and the downstream effects of reduced water supply in the dry months. They will have repercussions that transcend generations. Climate feedback systems and environmental cumulative effects are building across earth systems demonstrating behaviours. The potential for runaway greenhouse warming is real and has never been more present. Climate change is changing our economy, health and communities in diverse ways. Climate change is a long-term change in the statistical distribution of weather patterns over periods of time that range from decades to millions of years. Climate change may be limited to a specific region, or may occur across the whole earth.

11.5.1 Causes of Climate Change

The causes of climate change can be divided into two categories, human and natural causes. The climatic changes occurring today have been speeded up because of man's activities. The natural variability and the climate fluctuations of the climate system have always been the part of the earth's history. The atmosphere growing at an unprecedented rate and magnitude due to changes in the concentration of greenhouse gases.

(i) **Natural cause:** The climate of earth's atmosphere is influenced and changed through natural causes like volcanic eruption, ocean current, the earth's orbital changes and solar variations.

(ii) **Volcanic eruptions:** When a volcano erupts it throws out large volumes of sulfur dioxide (SO_2), water vapour, dust, and ash into the atmosphere. Large volumes of gases and ash can influence climatic patterns for years by increasing planetary reflectivity causing atmospheric cooling. Tiny particles called aerosols are produced by volcanoes. They have a cooling effect on the world because they reflect solar energy back into space. The greenhouse gas, carbon dioxide is also produced. The CO_2 produced volcano is insignificant when compared to emissions created by humans which .is also responsible for climate change.

(iii) **Ocean current:** The oceans are a major component of the climate system. Ocean currents move vast amounts of heat across the planet. Deep ocean circulation of cold water from the poles towards the equator and movement of warm water from the equator back towards the poles. Winds push horizontally against the sea surface and drive ocean current patterns. Without this movement the poles would be colder and the equator

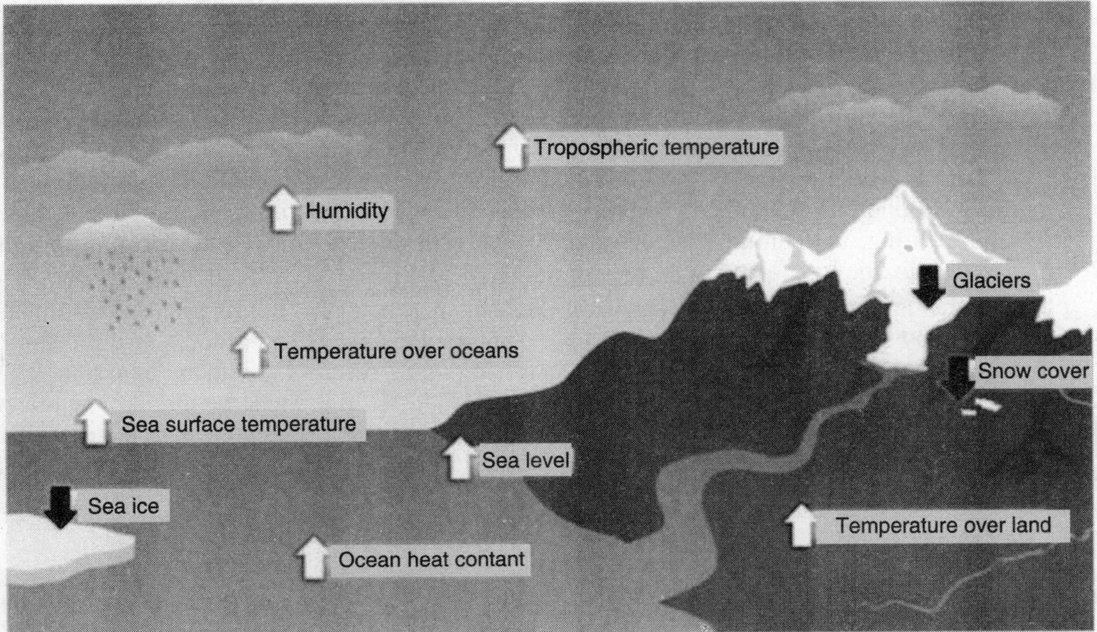

Fig. 11.5 Ten indicators for a warming world, Past Decade Warmest on Record According to Scientists in 48 Countries, NOAA, July 28, 2010.

warmer. Changes in ocean circulation may affect the climate through the movement of CO_2 into or out of the atmosphere.The oceans play an important role in determining the atmospheric concentration of CO_2.

(iv) Earth orbital changes: The earth makes one full orbit around the sun each year. It is tilted at an angle of 23.5°C to the perpendicular plane of its orbital path. Climatically important changes in the strength of the seasons are due to changes in the tilt of the earth, more tilt means warmer summers and colder winters; less tilt means cooler summers and milder winters. Slow changes in the earth's orbit lead to small but climatically important changes in the strength of the seasons over tens of thousands of years. Climate feedbacks amplify these small changes, thereby producing ice ages.

(v) Solar variations: The sun is the source of energy for the earth's climate system. Although from an everyday point of view, sun's energy output appears constant and small changes over an extended period of time can lead to climate changes. It would be reasonable to assume that changes in the sun's energy output would cause the climate to change, as the sun is the fundamental source of energy that is instrumental in our climate system. Solar variations have performed a role in past climate changes. Current global warming however cannot be explained by solar variations. Cooling in the upper atmosphere, a warming at the surface and in the lower parts of the atmosphere was observed, which is due to greenhouse gasses capturing heat in the lower atmosphere.

(vi) Human cause: It has been demonstrated that the climate is changing due to man-made greenhouse gases. Over the next 30 years, we are already committed to future substantial change and change is likely to accelerate over the rest of the 21st century. Over the last half-century the warming of the earth has been caused largely by human activity, such as the burning of fossil fuels and changes in land use, including agriculture and deforestation. The Industrial Revolution in the 19th century saw the large-scale use of fossil fuels for industrial activities. Fossil fuels such as oil, coal and natural gas supply most of the energy needed to run vehicles, generate electricity for industries and households. About 3/4 of the carbon dioxide emissions, 1/5 of the methane emissions and a large quantity of nitrous oxide, the energy sector is responsible. The most important greenhouse gas in the atmosphere is undoubtedly carbon dioxide. Changes in land use pattern, deforestation, land clearing, agriculture, and other activities have all led to a rise in the emission of carbon dioxide.

The another important greenhouse gas in the atmosphere is methane. It is released from animals such as dairy cows, goats, pigs, buffaloes, camels, horses and sheep. The process of oil drilling, coal mining, leaking gas pipelines, landfills and waste dumps results in methane emission. The certainty of global warming can be seen through some of the natural phenomenon like the effect on crops and extreme weather conditions around the world. It is especially clear in the dramatic change of the polar caps, i.e., the Arctic ice cap is shrinking and the Antarctica ice shelf is melting.

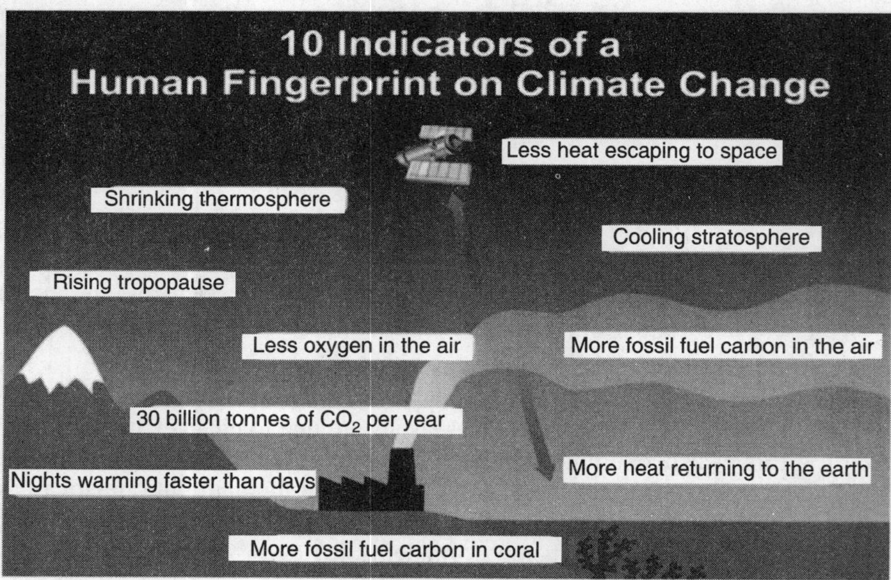

Fig. 11.6 John Cook, 10 Indicators of a Human Fingerprint on Climate Change, Skeptical Science July 30, 2010.

11.5.2 Main Contributors and Cause of Climate Change

(i) **Agriculture as a contributor to the causes of climate change:** Fossil fuels, land use, and agriculture are the three main causes of the increase in greenhouse gases observed over the past 250 years. According to the intergovernmental panel on climate change. Through the production and release of greenhouse gases such as carbon dioxide, methane, and nitrous oxide agriculture has been shown to produce significant effects on climate change. Another contributing cause of climate change is when agriculture alters the earth's land cover, which can change its ability to absorb or reflect heat and light. Together deforestation and desertification, with use of fossil fuels, are the major anthropogenic sources of carbon dioxide.

(ii) Deforestation as a contributor to the causes of climate change: Resource rainforests play important role in our world. They form part of a delicate ecosystem that has taken millions of years to evolve. Almost 20% of man-made CO_2 emissions is absorbed by rainforest and therefore deforestation can be classed as a major contributor to the causes of climate change. More deforestation means more CO_2 build up in the atmosphere. Cutting down rainforests faster than they can be replaced has a devastating effect on the carbon emission cycle producing an extra 17% of greenhouse gases. Deforestation by means of cutting down and burning these tropical rainforests usually pave the way for agriculture and industry which often produce even more CO_2. Remember trees absorb CO_2. Both climate change and forests are interlinked. One of the main causes of climate change. The increased destruction of the rainforest forming a precious cooling band around the earth's equator. Forests trap and store carbon dioxide, playing a major role in mitigating climate change. On the flip side of the coin, forests when destroyed or over-harvested and burned become the sources of the greenhouse gas, carbon dioxide.

Both climate change and forests are interlinked. The increased destruction of the rainforest forming a precious cooling band around the earth's equator. Forests trap and store carbon dioxide, playing a major role in mitigating climate change. On the flip side of the coin, forests when destroyed or over-harvested and burnt become the sources of the greenhouse gas, carbon dioxide. Both climate change and forests are interlinked. One of the main causes of climate change.The increased destruction of the rainforest forming a precious cooling band around the earth's equator. Forests trap and store carbon dioxide, playing a major role in mitigating climate change. On the flip side of the coin, forests when destroyed or over-harvested and burnt become the sources of the greenhouse gas, carbon dioxide. Forests if not harmed ensure that they are enabling to continue to produce the benefits; to mitigate the effects of a changing climate; and to compensate for fossil fuel emissions through carbon storage; and to enhance ecosystem health, sustainability, and resilience. Forests reduce greenhouse gas emissions to combat global warming. 20% of global greenhouse gas emissions result from deforestation and degradation of forest, more than all the world's cars, trucks, ships and planes combined. Fossil fuels contribute to global warming and climate change due to release of carbon dioxide into the atmosphere. Forest alleviates this change by converting carbon dioxide to carbon during photosynthesis. Trees possess about 20 per cent carbon by weight and biomass of forest acts as a "carbon sink." The soil organic matter also acts as a carbon store which is produced by the decomposition of dead plant material. Consequently forests store enormous amount of carbon.

(iii) Nitrous oxide no laughing matter: Nitrous oxide which is popularly known as laughing gas is very much overlooked when it comes to greenhouse gases. It is believed to be the third most important gas that contributes to global warming. As a by-product of a microbial process working on nitrogen, nitrous oxide is produced along side, as a harmless gas dinitrogen. As human activities such as intensive agriculture and the burning of fossil fuels hugely increased nitrous oxide. When compared to CO_2, nitrous oxide molecule,

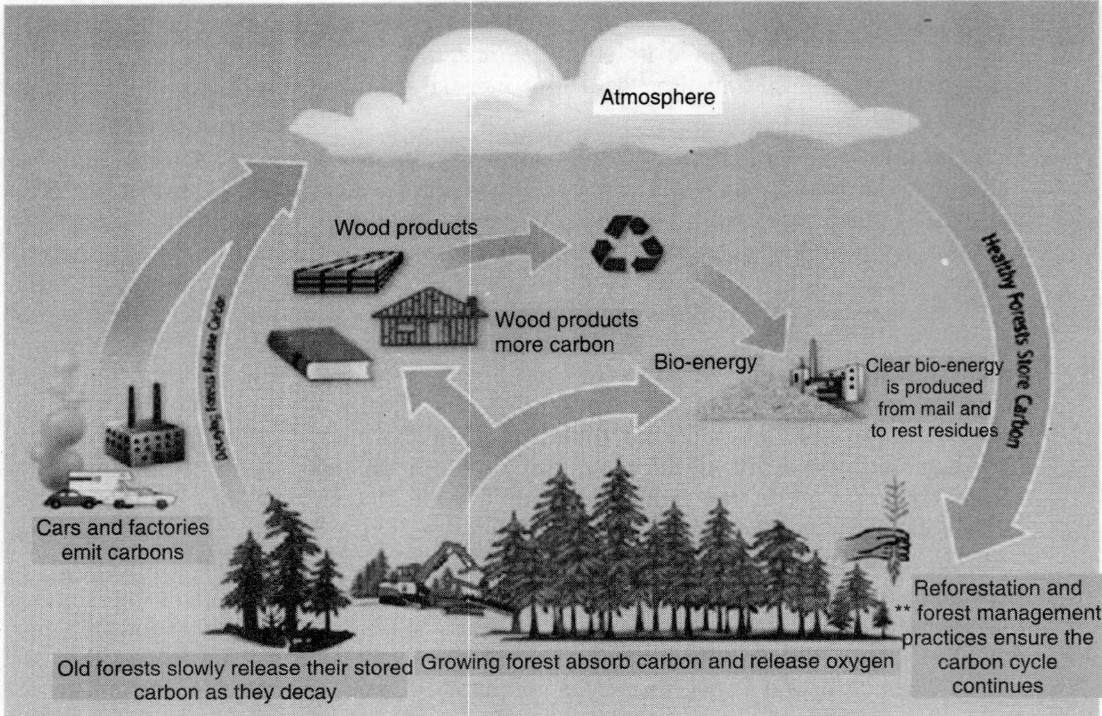

Fig. 11.7 Forests, through the process of photosynthesis, absorb carbon dioxide from the atmosphere.

is 300 times more potent when it comes to warming up the earth. Although it may be classed as a smaller contributor to climate change as it is emitted less than CO_2. It is increasing by over 0.2 per cent each year. It is also believed to be the leading remaining threat to the ozone layer. Ozone layer plays an essential part in protecting the human race and all life on this planet from the sun's highly dangerous ultra violet radiation.

(iv) Scientific Basis of Climate Change and Causes

- Of all parts of the climate system an enormous body of observations was analyzed.
- Catalogues increasing concentrations of atmospheric greenhouse gases.
- In order to govern the climate system our understanding of the processes and feedbacks should be assessed.
- Using a wide range of models of future emissions of greenhouse gases and aerosols scenarios of future climate change should be observed.
- Whether causes by a human influence on climate can be identified and studied in detail.
- Suggests gaps in information and understanding that remain in our knowledge of climate change and how these might be addressed.

(v) Livestock and Climate Change Causes: In the production of greenhouse gases the life cycle and supply chain of animals raised for food has been an area greatly underestimated. They account for at least half (50%) of human caused greenhouse emissions. To find better alternatives by replacing livestock with other food sources is a better strategy in reversing climate change. They have far more of a rapid effect on greenhouse emissions and their atmospheric concentrations therefore the rate on which the climate is warming and the unnecessary urgency to replace fossil fuels with renewable energy. It is widely accepted that livestock are known to contribute to greenhouse gases. About18 per cent of annual worldwide greenhouse gases, are attributable to cattle, buffalo, sheep, goats, camels, horses, pigs, and poultry.

11.5.3 Effects of Climate Change

People worldwide began burning more coal and oil for homes, factories, and transportation over 100 years ago,. Burning these fossil fuels releases carbon dioxide and other greenhouse gases into the atmosphere. These added greenhouses gases have caused earth to warm more quickly than it has in the past.

(i) **Sea level is rising:** Sea level rose about 15 cm (6 inches) due to melting glacier ice and expansion of warmer seawater during the 20th century. Sea level may rise as much as 59 cm (23 inches) during the 21st century, threatening coastal communities, wetlands, and coral reefs. Arctic sea ice is melting. Melting ice may lead to changes in ocean circulation. Plus melting sea ice is speeding up warming in the Arctic.

(ii) **Glaciers and permafrost are melting:** Mountain glaciers in all areas of the world have decreased in size over the past 100 years and so has the amount of permafrost in the Arctic. Greenland's ice sheet is melting faster too.

(iii) **Sea-surface temperatures are warming:** About a quarter of the world's coral reefs in the last few decades undergoes death due to the warmer waters in the shallow oceans. Many of the coral animals died after weakened by bleaching, a process tied to warmed waters.

(iv) **The temperatures of large lakes are warming:** The temperatures of large lakes world-wide have risen dramatically. Temperature rises have increased algal blooms in lakes, favour invasive species, increase stratification in lakes and lower lake levels.

(v) **Heavier rainfall cause flooding in many regions:** Warmer temperatures have led to more intense rainfall events in some areas. Heavy rainfall can cause flooding.

(vi) **Extreme drought is increasing:** Higher temperatures cause a higher rate of evaporation and more drought in some areas of the world.

(vii) **Crops are withering:** Crop productivity around the world is declined due to increased temperatures and extreme drought. Decreased crop productivity can mean food shortages which have many social implications.

(viii) **Ecosystems are changing:** As temperatures warm, species may either move to a cooler habitat or die. Species that are particularly vulnerable include endangered species, coral reefs, and polar animals. Changes in the timing of spring events and the length of the growing season is caused by warming.

(ix) **Hurricanes have changed in frequency and strength:** There is evidence that the number of intense hurricanes has increased in the Atlantic since 1970. More frequent heat waves. It is likely that heat waves have become more common in more areas of the world.

(x) **Warmer temperatures affect human health:** Due to heat waves there have been more deaths and more allergy attacks as the pollen season grows longer. There have also been some changes in the range of animals that carry disease like mosquitoes.

(xi) **Seawater is becoming more acidic:** Carbon dioxide dissolving into the oceans, is making seawater more acidic which may show their impacts on coral reefs and other marine life.

11.5.4 Physical Impact of Climate Change

(i) **Global average temperature:** One of the most-cited indicators of global climate change is global average temperature. The global surface temperature is based on air temperature data over land and sea-surface temperatures. There is a clear long-term global warming trend. Each individual year does not always show a temperature increase relative to the previous year, and some years show greater changes than others. Due to natural process shows temperatures fluctuations year-to-year.

(ii) **Diurnal temperature:** There has been a general, but not global, tendency toward reduced diurnal temperature range. It is the difference between daily high or maximum and daily low or minimum temperatures. A variety of factors likely contribute to this change in DTR, particularly on a regional and local basis, including changes in cloud cover, atmospheric water vapour, land use and urban effects.

(iii) **Indirect indicators of warming:** Borehole temperatures, snow cover, and glacier recession data are indirect indicators of warming, and are in substantial agreement with the more direct indicators of recent warmth. Evidence such as changes in glacial mass balance is useful since it not only provides qualitative support for existing meteorological data, but glaciers often exist in places too remote to support meteorological stations.

(iv) **Effects on weather:** Increasing temperature is likely to lead to increasing precipitation but the effects on storms are less clear. Extra tropical storms, which is predicted to weaken in the northern hemisphere as the polar region warms more than the rest of the hemisphere, partly depend on the temperature gradient.

(v) **Fire:** Fire is a major agent for conversion of biomass and soil organic matter to CO_2. There is a large potential for future alteration in the terrestrial carbon balance through altered fire regimes. An increase in global mean temperature of about 0 to 2°C by 2100. Relative to the 1990–2000 period results in increased fire frequency and intensity in many areas. An increase in the region of 2°C or above would lead to increased frequency and intensity of fires.

(vi) **Extreme Weather:** Storm strength leading to extreme weather is increasing, such as the power dissipation index of hurricane intensity Hurricane power dissipation is highly correlated with temperature, reflecting global warming. The increase in power dissipation in recent decades cannot be completely attributed to global warming. Hurricane stimulated under warmer, high-CO_2 conditions, are more intense, however, hurricane frequency will be reduced. The warming induced by greenhouse gas may lead to increasing occurrence of highly destructive category storms. A substantially higher risk of extreme weather does not necessarily mean a noticeably greater risk of slightly-above-average weather. It is clear that severe weather and moderate rainfall are also increasing. Increases in temperature are expected to produce more intense convection over land and a higher frequency of the most severe storms.

(vii) Increased Evaporation: Evaporation rates have reduced worldwide; this is thought by many to be explained by global dimming. The causes of global dimming are reduced, evaporation will increase due to warmer oceans as the climate grows warmer. Because the world is a closed system this will cause heavier rainfall, with more erosion. This erosion, in turn, leads to desertification in vulnerable tropical areas and on the other hand, increased rainfall leads to growth of forests in dry desert areas. Increased evaporation could result in more extreme weather as global warming progresses. Global average water vapour concentration and precipitation are projected to increase during the 21st century.

(viii) Increased Freshwater Flow: Increase in the flow of freshwater into the world's oceans, partly from melting ice and partly from increased precipitation driven by an increase in global ocean evaporation. The increase in global freshwater flow, was about 18%. One effect, as perhaps experienced in the 2010 Pakistan floods, is to overwhelm flood control infrastructure.

11.5.5 Regional Climate Change

It was likely that human influences had significantly contributed to surface temperature increases in every continent except Antarctica. The magazine *Scientific American* reported on December 23, 2008, that the 10 places most affected by climate change were Darfur, the Gulf Coast, Italy, northern Europe, the Great Barrier Reef, island nations, Washington, D.C., the Northwest Passage, the Alps, and Uganda.

(i) Northern hemisphere: Arctic dipole: Over the last 50 years, in the northern hemisphere, the southern part of the Arctic region has experienced a temperature rise of 1°C to 3°C. Canada, Alaska and Russia are experiencing initial melting of permafrost. By increasing bacterial activity in the soil which leads to these areas becoming carbon sources instead of carbon sinks may disrupt ecosystems. A study of changes, western Siberia is at the initial stage where melting permafrost is creating new lakes, which will eventually start disappearing as in the east. Furthermore, permafrost melting will eventually cause methane release from melting permafrost peat bogs.

Polar regions: Model projections showed that Arctic terrestrial ecosystems and the active layer (the top layer of soil or rock in permafrost that is subjected to seasonal freezing and thawing) would be a small sink for carbon. Increased emissions of carbon from thawing of permafrost could occur. This would lead to an amplification of warming.

(ii) Atmosphere: Noctilucent cloud: Temperature trends in the lower stratosphere, mid to upper troposphere, lower troposphere, and surface. In higher parts of the atmosphere an enhanced greenhouse effect is expected to cause cooling. Due to increased carbon dioxide concentrations, the strongest cooling and contraction occurring in that layer during solar minimum a contraction of the thermosphere has been observed. The most recent contraction in 2008–2009 was the largest such since at least 1967.

Water vapour Feedback: Due to both greenhouse gases and ozone-depleting substances the warming and evaporation temperatures in the lower stratosphere near the tropopause declined, reducing water vapour levels and removing its warming effect. The water vapour in the stratosphere arrives through tall thunderstorms. About 15% of this vapour is delivered by tropical cyclones, and through chemical breakdown of methane into water vapour and carbon dioxide, both of which are greenhouse gases. The vapour is frozen out of the stratosphere as more of the water is subjected to temperatures that freeze it out of the stratosphere in increases in tropopause cold point temperatures.

(iii) Gas hydrates: Clathrate gun hypothesis *and* Arctic methane release

Gas hydrates are ice-like deposits containing a mixture of water and gas, the most common gas of which is methane. Gas hydrates are found underneath the oceans and permafrost regions and are stable under high pressures and at relatively low temperatures. Future warming will tend to destabilize gas hydrates resulting in the release of large quantities of methane. On the other hand, projected rapid sea level rise in the coming centuries associated with global warming will tend to stabilize marine gas hydrate deposits.

(iv) Carbon cycle: Carbon cycle gets affected by climate change. The efficiency of the land and ocean carbon cycle to absorb human-induced CO_2 is reduced by climate change. A larger fraction of human-induced CO_2 will stay air borne if climate change controls the carbon cycle.

(v) Cryosphere: In recent decades northern hemisphere average annual snow cover has declined. This pattern is consistent with warmer global temperatures. In the spring and summer months some of the largest declines have been observed.

(vi) Sea ice: Snow cover and sea ice extent decrease as climate warms. Sea ice in the Antarctic has shown very little trend over the same period, or even a slight increase since 1979. Though extending the Antarctic sea-ice record back in time is more difficult due to the lack of direct observations in this part of the world. Reduction of sea ice in both the Arctic and Antarctic was found. The range of model responses was large. Summer sea ice cover in the Arctic would disappear entirely by the latter part of the 21st century.

(vii) Glacier retreat and disappearance

Retreat of glaciers since 1850: Warming temperatures lead to the melting of glaciers and ice sheets. Mountain glaciers and snow cover had decreased in both the northern and southern hemispheres. Glacier retreat has become increasingly rapid and ubiquitous, and has threatened the existence of many of the glaciers of the world. This widespread decrease in glaciers and ice caps has contributed to observed sea level rise. The total volume of glaciers on the earth is declining sharply. Glaciers have been retreating worldwide for at least the last century; the rate of retreat has increased in the past decade.

IPCC made a number of projections related to future changes in glaciers:

- Mountainous areas in Europe will face glacier retreat.
- In Latin America, changes in precipitation patterns and the disappearance of glaciers will significantly affect water availability for human consumption, agriculture, and energy production.
- In Polar regions, there will be reductions in glacier extent and the thickness of glaciers.

Glaciers around the world retreated as the climate warmed. Glacier retreat declined and reversed in many cases as a slight global cooling occurred. Currently glacier retreat rates and mass balance losses have been increasing in the Andes, Alps, Pyrenees, Himalayas, Rocky Mountains and North Cascades The loss of glaciers not only directly causes landslides, flash floods and glacial lake overflow, but also increases annual variation in water flows in rivers. In India alone, the Ganges provides water for drinking and farming for more than 500 million people. Lots of people live in the drainage basin of the Himalayan rivers. India, China, Pakistan, Bangladesh, Nepal and Myanmar could experience floods followed by droughts in coming decades. The Himalayan glaciers that are the sources of Asia's biggest rivers Ganges, Indus, Brahmaputra, Yangtze, Mekong, Salween and Yellow could diminish as temperatures rise.

Retreat of the Helheim Glacier, Greenland: The mountain and valley glaciers of temperate latitudes amount to a small fraction of glacial ice on the earth despite their proximity and importance to human populations. Glacier retreat has been observed in these outlet glaciers, resulting in an increase of the ice flow rate. In Greenland the period since the year 2000 has brought retreat to several very large glaciers that had long been stable. In western Greenland JakobshavnIsbræ had been moving at speeds of over 24 m (79 ft)/day with a stable terminus. Helheim, JakobshavnIsbræand Kangerdlugssuaq Glaciers, jointly drain more than 16% of the Greenland Ice Sheet.

(viii) Oceans: The role of the oceans in global warming is a complex one. The oceans serve as a sink for carbon dioxide, but increased levels of CO_2 have led to ocean acidification. As the temperature of the oceans increases, they become less able to absorb excess CO_2. Global warming is projected to have a number of effects on the oceans. Rising sea levels due to thermal expansion and melting of glaciers and ice sheets, and warming of the ocean surface are ongoing effects.

Current sea level rise - Amundsen Sea

Global average sea level had risen at an average rate of 1.8 mm/yr. With increasing average global temperature, the water in the oceans expands in volume, and additional water enters them which had previously been locked up on land in glaciers and ice sheets. Sea level rise due to the collapse of an ice sheet would be distributed non-uniformly across the globe. The loss of mass in the region around the ice sheet would decrease the gravitational potential there, reducing the amount of local sea level rise or even causing local sea level fall. The loss of the localized mass would also change the moment of inertia of the earth. This change in the moment of inertia results in true polar wander, in which the earth's rotational axis remains fixed with respect to the sun. This changes the location of the equatorial bulge of the Earth and further affects the geoid, or global potential field.

(ix) Temperature rise and ocean heat content: Time series of seasonal (red dots) and annual average (black line) global upper ocean heat content for the 0-700m layer between 1955 and 2008. The graph shows that ocean heat content has increased over this time period. Increasing heat content in the ocean is also consistent with sea level rise, which is occurring mostly as a result of thermal expansion of the ocean water as it warms.

Ocean acidification: Ocean acidification is an effect of rising concentrations of CO2 in the atmosphere, and is not a direct consequence of global warming.Oceans currently absorb about one tonne of CO_2 per person per year. The oceans soak up much of the CO_2 produced by living organisms, either as dissolved gas, or in the skeletons of tiny marine creatures that fall to the bottom to become chalk or limestone. In water, CO_2 becomes a weak carbonic acid. There are concerns that increasing acidification could have a particularly detrimental effect on coralsand other marine organisms with calcium carbonate shells.

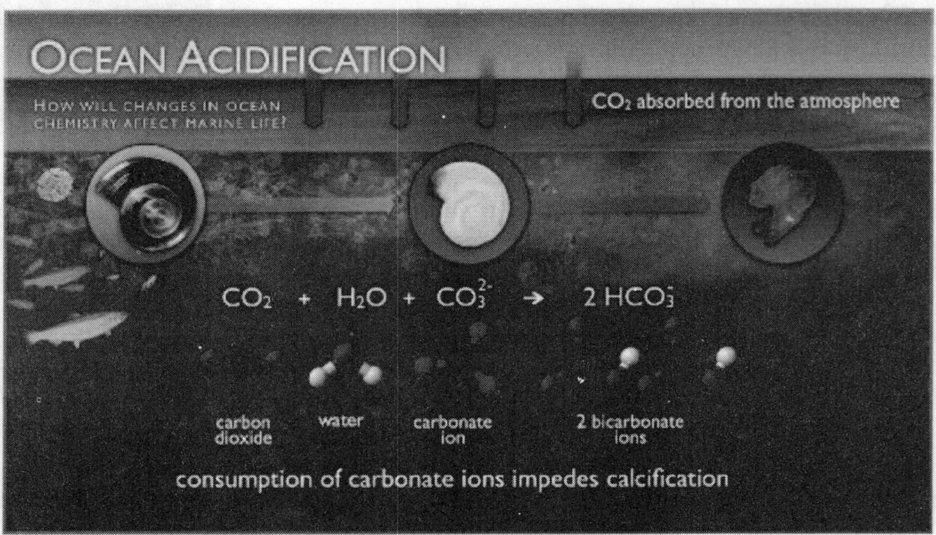

Fig. 11.8 Ocean acidification; consumption of carbonate ions impede calcification.
Source: Pacific Marine Environment Laboratory, NOAA.

(x) Oxygen depletion: The amount of oxygen dissolved in the oceans may decline, with adverse consequences for ocean life.

(xi) Sulfur aerosols: The layer of the atmosphere in which they exist is known simply the stratospheric aerosol layer. Stratospheric sulfur aerosols are sulfur-rich particles which exist in the stratosphere region of the earth's atmosphere. Sulfur aerosols, especially stratospheric sulfur aerosols have a significant effect on climate. These particles consist of a mixture of sulfuric acid and water. They are created naturally, such as by photochemical decomposition of sulfur-containing gases, e.g. carbonyl sulfide. When present in high levels by reflecting sunlight, and by modifying clouds as they fall out of the stratosphere they produce a cooling effect. Disruption to the oceans as a result of ocean acidification or disruptions to the thermohaline circulation may result in disruption of the sulfur cycle, thus reducing its cooling effect on the planet through the creation of stratospheric sulfur aerosols.

(xii) Shutdown of thermohaline circulation: Global warming could, trigger localized cooling in the North Atlantic and lead to cooling, or lesser warming, in that regionvia a shutdown or slowdown of the thermohaline circulation, This would affect in particular areas like Scandinavia and Britain that are warmed by the North Atlantic drift. The chances of this near-term collapse of the circulation are unclear.

11.5.6 Geology

(i) Volcanoes: The retreat of glaciers and ice caps can cause increased volcanism. Reduction in ice cover reduces the confining pressure exerted on the volcano, increasing deviatoric stresses and potentially causing the volcano to erupt. This reduction of pressure can also cause decompression melting of material in the mantle, resulting in the generation of more magma. The effects of current sea level rise could include increased crustal stress at the base of coastal volcanoes. Current melting of ice is predicted to increase the size and frequency of volcanic eruptions. In particular, lateral collapse events at strato volcanoes are likely to increase, and there are potential positive feedbacks between the removal of ice and magmatism.

(ii) Earthquakes: A numerical modeling study has demonstrated that seismicity increases during unloading, such as that due to the removal of ice.

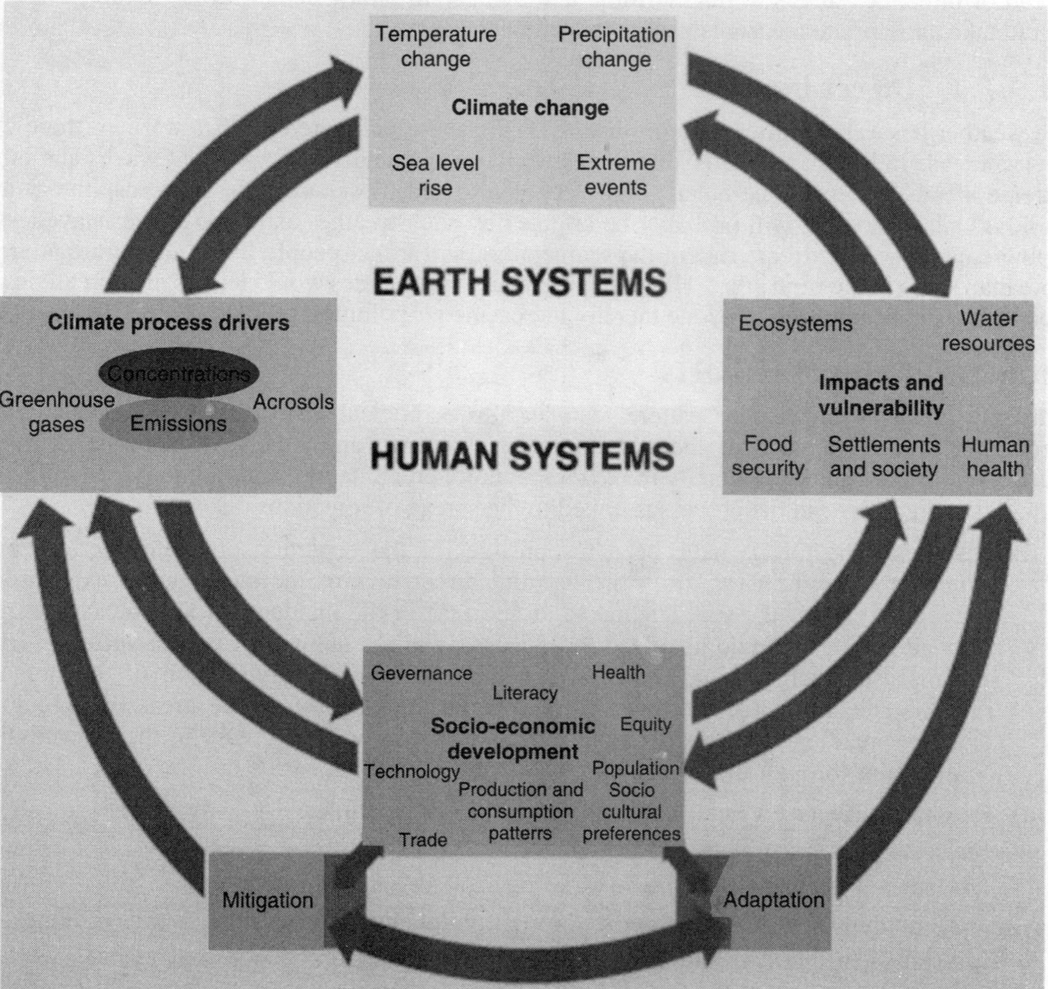

Fig. 11.9 Schematic framework of anthropogenic climate change drivers, impacts and responses.

11.5.7 Health Impacts of Climate Change

Climate change is a major problem caused by the increase of human activities leading to several direct and indirect impacts on health. The combustion of fossil fuels, increasing number of industries, and large-scale deforestation are some of the causes for the accumulation of GHGs (greenhouse gases) in the atmosphere. GHGs, like methane, ozone, nitrous oxide, and chlorofluorocarbons, in the atmosphere lead to changes in rainfall and snowfall, more intense or frequent droughts, floods, and storms, as well as a rise in sea level. These climatic changes will have wide-ranging harmful effects including increase in heat-related mortality, dehydration, spread of infectious diseases, malnutrition, and damage to public health infrastructure. Thus we should take appropriate measures to stop this climate change.

11.5.7.1 *Direct impacts*

The weather has a direct impact on our health. If the overall climate becomes warmer, there will be an increase in health problems, due to greater frequency and severity of heat waves and other extreme weather events. The elderly, the very young and those suffering from respiratory and cardiovascular disorders will probably be affected by such weather extremes as they have lesser coping capacity. An extreme rise in the temperature will affect people living in the urban areas more than those in the rural areas. Higher temperatures in the cities would lead to an increase in the ground-level concentration of ozone thereby increasing air pollution problems.

11.5.7.2 *Indirect impacts*

Indirectly, changes in weather pattern, can lead to ecological disturbances, changes in food production levels, increase in the distribution of malaria, and other vector-borne diseases. Fluctuation in the climate especially in the temperature, precipitation, and humidity can influence biological organisms and the processes linked to the spread of infectious diseases.

(i) **Diseases:** The GHGs have been responsible for the depletion of stratospheric ozone, which protects the earth from the harmful direct rays of the sun. Higher exposure to ultraviolet rays of the sun, leading to an increase in the incidents of skin cancer in light skinned people. It could also lead to an increase in the number of people suffering from eye diseases such as cataract. It is also thought to cause suppression of the immune system. Due to global warming there will be an increase in the areas of habitat of disease-spreading insects such as the mosquito, causing an increase in the transmission of infection through these carriers.

(ii) **Potential effects on health due to sea level rise include:**
- Death and injury due to flooding.
- Reduced availability of fresh water due to saltwater intrusion.
- Contamination of water supply through pollutants from submerged waste dumps.
- Change in the distribution of disease-spreading insects.
- Effect on the nutrition due to a loss in agriculture land and changes in fish catch.
- Health impacts associated with population displacement.

11.5.8 Preventive Measures

- Increased use of renewable sources and reduction in the use of non-renewable sources of energy will undoubtedly decrease the emission of GHGs substantially. This decrease in the GHGs will have a positive effect on the health and well-being of the people.

- Furthermore, switching to energy-efficient technologies and cleaner fuels will reduce local pollutants and therefore, have an added beneficial impact on health.

(i) **Some of the green myths that contribute to the causes of climate change:** Paper bags cause more global warming than plastic. They need much more space to store so require extra energy to transport them from manufacturers to shops. Burning wood for fuel is better for the environment than recycling it. Organic dairy cows are worse for the climate. Traditional nappies (diapers) are as bad as disposables and while throwaway nappies make up 0.1 per cent of landfill waste. They produce less milk so their methane emissions per litre are higher. Trees, regarded as shields against global warming because they absorb carbon.

(ii) **Contributors and cause according to UK government**

According to the UK Government the main contributors of man-made causes of climate change in the UK are:

- 4% of carbon emissions come from industrial processes.
- 7% come from agriculture, for example, methane emissions from livestock and manure, and nitrous oxide emissions from chemical fertilizers.
- 21% carbon emissions from transport.
- 65% come from the use of fuel to generate energy (excluding transport).

About 40% of carbon emissions in the UK are the result of decisions taken directly by individuals. The biggest sources of emissions for most people are likely to be:

- Energy use in the home (the main use is heating).
- Driving a car.
- Air travel.

(iii) **Global climate change: recent impact**

- Cold days, cold nights and frost less frequent over land areas.
- More frequent hot days and night.
- Heat waves more frequent over most land areas.
- Increased incidence of extreme high sea level.
- Global area affected by drought has increased.
- Increase in intense tropical cyclone activity in North Atlantic.
- Above points excludes the tsunamis, which are not due to climate change but considered more natural disaster however may have linkage of climate change.
- Global climate change future trends.
- Contraction of snow cover areas, increased thaw in permafrost regions, decrease in sea ice extent.
- Increased frequency of hot extremes, heat waves and heavy precipitation.
- Increased in tropical cyclone intensity.
- Precipitation increases in high latitudes.
- Precipitation decreases in subtropical land regions.
- Decreased water resources in many semi-arid.

11.6 CARBON SEQUESTRATION

Efforts to mitigate the greenhouse gas problem have traditionally focused on avoiding the production of carbon dioxide (CO_2) by reducing fossil fuel use. One alternative to CO_2 abatement would be to capture CO_2 emissions and sequester them in carbon reservoirs such as deep aquifers, deep oceans, or minerals. Carbon Sequestration is capturing and securely storing carbon dioxide emitted from the global energy system. One economic problem is that CO_2 sequestration is not a perfect substitute for avoiding CO_2 production as sequestered CO_2 may leak back into the atmosphere and impose future climate damages. In contrast, avoiding CO_2 production would not impose the legacy of CO_2 leakage. **Human and industrial development causing a dramatic increase in atmospheric CO_2concentration over the past hundred years has led to a huge increase in fossil fuel consumption and CO_2emissions, This increased CO_2 is believed to be responsible for a significant rise in global temperature over the past several decades. Increased atmospheric CO_2 also leads to ocean acidification, which will have drastic consequences for marine ecosystems.**

So, carbon sequestration is the process of capture and long-term storage of atmospheric carbon dioxide (CO_2) and may refer specifically to:

- "The process of removing carbon from the atmosphere and depositing it in a reservoir." When carried out deliberately, this may also be referred to as carbon dioxide removal, which is a form of geo-engineering.
- The process of carbon capture and storage, where carbon dioxide is removed from flue gases, such as on power stations, before being stored in underground reservoirs.
- Natural biogeochemical cycling of carbon between the atmosphere and reservoirs, such as by chemical weathering of rocks.

Carbon sequestration describes long-term storage of carbon dioxide or other forms of carbon to either mitigate or defer global warming and avoid dangerous climate change. This is a way to slow the atmospheric and marine accumulation of greenhouse gases, which are released by burning fossil fuels. Carbon dioxide is naturally captured from the atmosphere through biological, chemical or physical processes. Some anthropogenic sequestration techniques exploit these natural processes. CO_2 sequestration includes the storage part of carbon capture and storage, which refers to large-scale, permanent artificial capture and sequestration of industrially produced CO_2 using subsurface saline aquifers, reservoirs, ocean water, aging oil fields, or other carbon sinks. In general, carbon sequestration is enhanced by:

- Minimizing soil disturbance and erosion.
- Maximizing the return of crop residues to soils.
- Maximizing water and nutrient use efficiencies in crop production. Growing plants with a large capacity to store carbon in above-ground and below-ground biomass.
- Implementing reduced impact logging techniques to minimize forest disturbance.
- Aforestation (planting new forests) on degraded and marginal agricultural soils.

11.6.1 Types of Sequestration

There are number of technologies for sequestering carbon from the atmosphere. Its main categories are:

(i) **Ocean sequestration:** Carbon stored in oceans through direct injection or fertilization. The ocean represents the largest potential sink for anthropogenic CO_2. The amount of carbon that would cause a doubling of the atmospheric concentration would change the ocean concentration by less than 2%. A large fraction of the emitted CO_2 will naturally be absorbed by the ocean. The CO_2 carrying capacity of the ocean is determined by its carbonate alkalinity. To form lakes of CO_2 at the bottom of the ocean is the another approach.

Clathrates are solids that can form at temperatures slightly higher than the melting point of water. The most significant factor determining the acceptability of ocean storage is the environmental impact. Environmental concerns in ocean sequestration focus on long term chronic issues of changing the ocean chemistry as well as on the local effects of low pH and its effect on marine organisms. To enhance the net oceanic uptake from the atmosphere, through iron fertilization is an alternative approach to ocean sequestration of CO_2.

(ii) Geologic sequestration: For long-term carbon dioxide natural pore spaces in geologic formations serve as reservoirs. For near-term application geologic sequestration is thought to have the largest potential. A major option for disposing of carbon dioxide is underground storage in geologic formation which is already practised. Active oil fields, coal beds, depleted oil and gas reservoirs, deep saline aquifers, mineral salt domes and rock caverns are potential sites for geological storage. Displacing methane tied up in deep un-mineable coal adds another small carbon sink to the portfolio of options.

Either by physical or chemical absorption most of the injected carbon dioxide will be immobilized on the coal surface, which is one of the attractive methods. The injected carbon dioxide will likely exist as a supercritical phase without being fully dissolved for thousands of years in some of the other reservoirs such as aquifers and that creates a potential problem of a higher leakage rate. So, geologic sequestration is the process of injecting carbon dioxide, captured from an industrial or energy- related source, into deep subsurface rock formation for long-term storage. This is a part of a process frequently referred to as "carbon capture and storage" or CCS.

(iii) Terrestrial sequestration: Increasing carbon fixation through photosynthesis, slowing down or reducing decomposition of organic matter, and changing land use practices can enhance carbon uptake in these natural sinks. A large amount of carbon is stored in soils and vegetation, which are our natural carbon sinks. Terrestrial sequestration involves the transfer of carbon dioxide from atmosphere into soils and vegetation. It includes both the removal of carbon dioxide from the atmosphere and limiting carbon dioxide emissions from terrestrial ecosystems into the atmosphere. The process of photosynthesis uses sunlight as the energy which is necessary to convert carbon in the atmosphere into the organic compounds that constitute trees, grasses and other plants, and agricultural crops.

Carbon dioxide is released back into the atmosphere when trees are cut down, crops harvested or decayed, and when the soil is disturbed or eroded. As a result of deforestation and traditional agricultural practices a considerable amount of carbon originally contained in soils and vegetation has been released. To reduce the amount of CO_2 that is released while enhancing the storage capability of soils, grazing and crop lands, and trees through changes in management practices is the goal of terrestrial sequestration. In the next three to four decades, terrestrial sequestration can play an important role in stabilizing CO_2 concentrations. The terrestrial carbon pool is the third largest (after oceans and geologic processes), and contains about four times as much carbon as is held currently in the atmosphere.

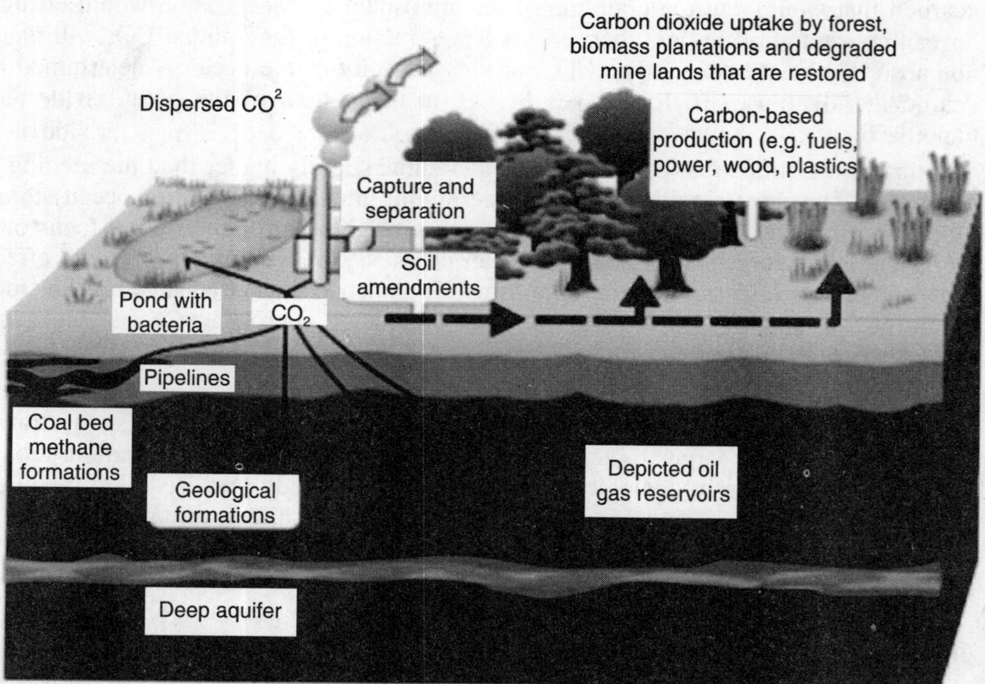

Fig. 11.10 Schematic showing both terrestrial and geological sequestration of carbon dioxide emissions from a coal-fired plant.

Terrestrial sequestration can be implemented rapidly, allowing time for the development of a more robust suite of carbon management technologies. Adoption of recommended management practices provides many ancillary, value-added environmental benefits by reducing soil erosion, improving soil fertility, creating new or improving existing habitats that encourage biodiversity, enhancing water retention and transmission, reducing fire risk and reclaiming degraded lands.

(iv) Biological sequestration: One of the largest carbon flows through the environment is driven by photosynthesis. Plants take up carbon dioxide and water, and turn it into reduced carbon compounds such as starch and cellulose. About 100 Gt, photosynthesis fixes carbon per year. By respiration and decomposition most of this is returned, but it seems that over the last decade mid latitude forests have been net sinks for carbon dioxide. During the growth periods trees and other plants sequester carbon dioxide. Thus, forestation and agricultural fixation of carbon, either in biomass or soil carbon, play a role in carbon sequestration. Biological sequestration could make up for past deforestation, which is responsible for a small fraction of the increased atmospheric carbon dioxide. It is most useful when it is the by-product of other societal goals.

(v) Mineral sequestration: As the amount of carbon dioxide stored increases, it becomes progressively more difficult to guarantee a physical barrier that prevents carbon dioxide from returning to the atmosphere. Carbon dioxide is the anhydrous form of carbonic acid and therefore cab be used to displace weaker acids. Thermodynamically, carbon dioxide can be bound as a carbonate. So, mineral sequestration would provide a means of storing carbon dioxide and this is a new and less studied method of sequestration. This sequestration process offers a safe and permanent method of carbon dioxide disposal, since there is almost no possibility of accidental release of carbon dioxide from the disposal site.

Mineral sequestration can proceed in underground or industrial processes. In the first process, CO_2 could be injected into selected underground mineral deposits for carbonation (in situ carbonation). This process envisions pumping CO_2 directly into an underground deposit of porous magnesium or calcium bearing rock. This process would result in chemically stable carbonates. The reaction would be aided by naturally high pressures and could proceed more rapidly than mineral weathering on the earth's surface.

11.6.2 Process of CO_2 Sequestration

(i) **Biological process:** Bio-sequestration or carbon sequestration through biological processes affects the global carbon cycle which include major climatic fluctuations. By manipulating such processes, geo-engineers seek to enhance sequestration.

(ii) **Peat production:** Peat bogs are a very important carbon store. Carbon can be sequestered by creating new bogs, or enhancing existing ones.

(iii) **Reforestation:** The replanting of trees on marginal crop and pasture lands to incorporate carbon from atmospheric CO_2 into biomass is reforestation. Reforestation with long-lived trees (>100 years) will sequester carbon for a more graduated release, minimizing impact during the "carbon crisis" of the 21st century. For this process to succeed the carbon must not return to the atmosphere from burning or rotting when the trees die. The trees must grow in perpetuity or the wood from them must itself be sequestered, e.g., via biochar, bio-energy with carbon storage or landfill.

(iv) **Wetland restoration:** Wetland soil is an important carbon sink; 14.5% of the world's soil carbon is found in wetlands, while only 6% of the world's land is composed of wetlands.

(v) **Agriculture:** Globally, soils are estimated to contain more organic carbon, than the amount in vegetation and the atmosphere. Modification of agricultural practices is a recognized method of carbon sequestration as soil can act as an effective carbon sink. Reducing and/or displacing emissions and enhancing carbon removal are the two categories of carbon emission reduction methods in agriculture. Some of these reductions involve increasing the efficiency of farm operations while some involve interruptions in the natural carbon cycle. Also, some effective techniques can negatively impact other environmental concerns.

(vi) **Reducing emissions:** Increasing yields and efficiency generally reduces emissions. Techniques include more accurate use of fertilizers, less soil disturbance, better irrigation, and crop strains bred for locally beneficial traits and increased yields. Replacing more energy intensive farming operations can also reduce emissions, more likely to release its CO_2 to the atmosphere as it decays, reducing the net carbon reduction.

(vii) **Enhancing carbon removal:** All crops absorb CO_2 during growth and release it after harvest. To use the crop and its relation to the carbon cycle to permanently sequester carbon within the soil is the goal of agricultural carbon removal. This is done by selecting farming methods that return biomass to the soil and enhance the conditions in which the carbon within the plants will be reduced to its elemental nature and stored in a stable state. Methods for accomplishing this include:

- Use cover crops such as grasses and weeds as temporary cover between planting seasons.
- Concentrate livestock in small paddocks for days at a time so they graze lightly but evenly.
- This encourages roots to grow deeper into the soil.
- Cover bare paddocks with hay or dead vegetation. This protects soil from the sun and allows the soil to hold more water and be more attractive to carbon-capturing microbes.
- Restore degraded land, which slows carbon release while returning the land to agriculture or other use.

Agricultural sequestration practices may have positive effects on soil, air, and water quality, be beneficial to wildlife, and expand food production.The effects of soil sequestration can be reversed. If the soil is disrupted or tillage practices are abandoned, the soil becomes a net source of greenhouse gases. Many factors affect the costs of carbon sequestration including soil quality, transaction costs and various externalities such as leakage and unforeseen environmental damage. Farmers can be reluctant to adopt more expensive agricultural techniques when there is not a clear crop, soil, or economic benefit because reduction of atmospheric CO_2 is a long-term concern.

(viii) Ocean-related iron fertilization: Ocean iron fertilization is an example of a geo-engineering technique. This technique is controversial due to limited understanding its complete effects on the marine ecosystem. Iron fertilization attempts to encourage phytoplankton growth, which removes carbon from the atmosphere for at least a period of time. Such effects include release of nitrogen oxides, and disruption of the ocean's nutrient balance.

(ix) Urea fertilization-ocean nourishment: To encourage phytoplankton growth a nitrogen rich substance urea is used to fertilize the ocean. Urea into the ocean to boost CO_2-absorbing phytoplankton growth as a way to combat climate change.

(x) Biomass-related: Bio-energy with carbon capture and storage (BECCS) refers to biomass in power stations and boilers that use carbon capture and storage. The carbon sequestered by the biomass would be captured and stored, thus removing carbon dioxide from the atmosphere.

(xi) Burial: Burying biomass directly, mimics the natural processes that created fossil fuels. Landfills also represent a physical method of sequestration.

Biochar burial: Biochar is charcoal created by pyrolysis of biomass waste. The resulting material is added to a landfill or used as a soil improver. Biogenic carbon is recycled naturally in the carbon cycle. Pyrolysing it to biochar renders the carbon relatively inert so that it remains sequestered in soil. The mechanisms related to biochar are referred to as bio-energy with carbon storage. Further, the soil encourages bulking with new organic matter, which gives additional sequestration benefit. In the soil, the carbon is unavailable for oxidation to CO_2 and consequential atmospheric release.

Fig. 11.11 Biochar burial.

Biochar can be land filled, used as a soil improver or burned using carbon capture and storage.

Mixing layers: Encouraging various ocean layers to mix can move nutrients. By placing large vertical pipes in the oceans to pump nutrient-rich water to the surface mixing may be achieved. The triggering blooms of algae, store carbon when they grow and export carbon when they die. This produces results somewhat similar to iron fertilization. One side-effect is a short-term rise in CO_2, which limits its attractiveness.

(xii) Ocean storage: Transporting material such as crop waste out to sea and allowing it to sink exploits this idea to increase carbon storage. River mouths bring large quantities of nutrients and dead material from upriver into the ocean as part of the process that eventually produces fossil fuels.

(a) Subterranean injection: Carbon capture and storage: Carbon dioxide can be injected into depleted oil and gas reservoirs and other geological features, or can be injected into the deep ocean.The gas comes in part from anthropogenic sources, but is principally from large naturally occurring geologic formations of CO_2. It is transported to the oil-producing fields through a large network.

(b) Physical process: Physical activation using the internal grinding unit was preferred because when a fluidized bed reactor is used. This method could provide physical agitation without significant energy requirements. The bombardments between the larger and harder grinding media and fine serpentine particles in the slurry would chip away the layer of SiO_2 that inhibits further dissolution.

(c) Removal of SiO_2 layers from the surface of the serpentine particles: It was found that removing the diffusion limiting SiO_2 layer to expose the inner Mg layer is important to promote the further dissolution of the mineral. Several methods were proposed including internal grinding, ultrasound, acoustic, and microwave.

11.6.3 Effect of Internal (in Situ) Grinding Unit on Serpentine

Dissolution: An internal (in-situ) attrition grinding system was chosen to investigate its effect on serpentine dissolution with and without adding any chemical additives from various proposed

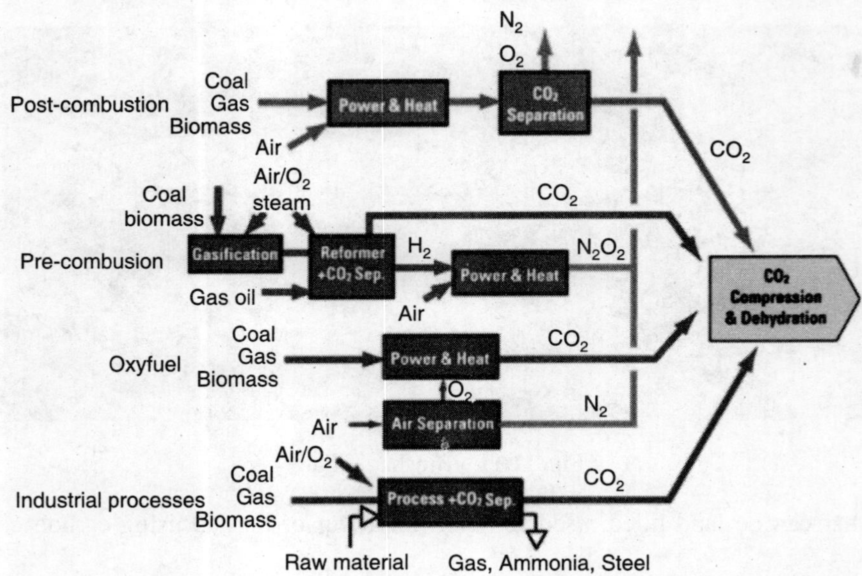

Fig. 11.12 Schematic diagram illustrates the primary routes for carbon capture.

methods of physical activation. Air was used as the fluidizing gas to study the effect of internal grinding on serpentine dissolution independently from the mineral dissolution and the carbonation caused by dissolved CO_2.

pH swing process: Based on the findings in the dissolution studies, aqueous carbonation of serpentine was investigated using ground serpentine and the prepared solvent.

11.6.4 Methods of Carbon Sequestration

11.6.4.1 MEA scrubbing process

Carbon dioxide is scrubbed from the flue gases of the combustion process which is a commercial process. A chemical absorption process is used with a mono-ethanolamine (MEA) solvent. The process allows the MEA solution to come into contact with the flue gases and mix in the absorber. The carbon dioxide rich MEA solution is then sent to a stripper where it is reheated to release almost pure carbon dioxide. This carbon dioxide can be used for industrial processes or to supply the needs of a photo-bioreactor system The MEA solution is then recycled to the absorber. This process is generally deemed uneconomic as it results in large equipment sizes and high regeneration energy requirements.

11.6.4.2 Membrane technology

To assist the carbon dioxide capture process in coal combusting power plants the principle behind membrane technology could be used. The benefit of using membrane technology was to increase the mass transfer area in a given volume, and to avoid the problems associated with vapour–liquid contact surfaces. In this application, the membrane did not serve the function of separation of the CO_2 from the flue gas. For most membrane types, the amine either destroyed or wetted the surface, causing blockage. Only the Pdytetra-fluoro-ethylene (PTFE) membrane type was successful.

11.6.4.3 Molecular sieve

A molecular sieve is a specially designed sieve that separates molecules based on their molecular weight. The arguments in favour of this type of separation technology are that they are cost effective, produce minimal waste and can be adapted to a variety of carbon sequestration schemes. Several demonstrations of separations have been performed, notably the separation of CO_2 from CH_4, CO_2 from air and CO_2, CO, H_2S and H_2O from a mixture of gases. This type of sieve has an open structure that allows fluid to flow freely through the material.

11.6.4.4 Desiccant adsorption

A process referred to as pressure and temperature swing adsorption (PTSA) that may be applied to electric power plant flue gas. Using zeolite as desiccant, carbon dioxide can be adsorbed at near normal pressure. As the absorbent is heated, the carbon dioxide can be regenerated under depressurization. Further regenerative capacity and power reductions can be achieved with a moderate temperature swing. The desorbed gas was then fed back into the inlet of the pilot plant for an improvement in terms of system configuration. This resulted in an overall reduction in power consumption of 3%. In the overall test results, the target removal efficiencies of 90–99% purity of CO_2 removed were achieved.

11.6.4.5 Disposal options by direct injection

Direct injection of carbon dioxide refers to the process whereby the gas to be sequestered is pumped into a sink usually capable of holding many mega-tonnes of gas over a period of time. It is recognized that most sinks will leak sequestered gas over a period of time, but it is still necessary to remove carbon dioxide from the atmosphere to reduce the otherwise inevitable temperature rises due to the greenhouse effect.

11.6.4.6 Geologic injection

Geologic injection may be superior to oceanic injection due to its higher expected retention rate. Even longer residence times could be achieved if the carbon dioxide reacted underground to form carbonate minerals, thus reducing the possibility of escape into the atmosphere. Expected residence times are at least thousands of years compared to that of oceanic injection of only hundreds of years. The consideration of carbon credits should be made on the retention ability of the geologic reservoir. The amount of carbon dioxide that leaks into the atmosphere should be considered as the difference of the amount sequestered in the geologic formation versus the actual quantity remaining.

11.6.4.7 Enhanced oil recovery (EOR)

When carbon dioxide is injected into an oil reservoir, the gas becomes miscible with the oil and promotes movement of the oil through the rock reservoir and out of the wells. As the carbon dioxide is pushed out with this oil, it is recaptured and reinjected until as much oil as is feasible has been extracted. Generally speaking, the economic viability of EOR projects depends on the price of oil and the proximity of a relatively pure carbon dioxide flood source. A major reason for the lack of economic viability of EOR is the high cost of the CO_2 removal process.

11.6.4.8 Coal seams

Another form of geologic injection that could be considered is the use of abandoned, uneconomic coal seams. Carbon dioxide diffuses through the pore structure of coal and is physically adsorbed, thus retention on a permanent basis is possible. Coal beds often contain large amounts of methane. The extraction of this methane could represent a value added process.

11.6.4.9 Oceanic injection

The ocean contains an estimated 40,000 Gt of carbon, whereas the atmosphere and the terrestrial biosphere contain an estimated 750 and 2200 Gt, respectively. A doubling of the carbon concentration levels in the atmosphere, therefore, represents only enough carbon to increase the ocean's concentration levels by about 2%. It is estimated that around 80% of today's anthropogenic emissions of carbon dioxide will be absorbed by the ocean. By direct injection into the ocean, this natural process is effectively enhanced, thus reducing peak atmospheric carbon dioxide concentrations and their rate of increase. However, using this method, it is estimated that around 15–20% of the carbon dioxide injected into the ocean will leach back into the atmosphere over hundreds of year.

There have been a number of injection techniques suggested in order to dissolve the carbon dioxide into the ocean. These are:

1. **Droplet plume**—liquid CO_2 injected from a manifold below 1000 m, forming a rising plume.
2. **Dense plume**—a dense CO_2 seawater mix that sinks, injected at a depth between 500 and 1000 m.
3. **Dry ice**—dropped off a boat and allowed to sink and diffuse.
4. **Towed pipe**—injected from a boat at a depth of 1000 m, forming a rising plume.
5. **CO_2 lake**—injection at a depth of around 4000 m to form a stable ''deep lake''.

In the short term, the droplet plume and towed pipe methods are probably the most viable due to technological and economic reasons.

11.6.4.10 Oceanic fertilization

Oceanic injection can be further enhanced through a method of marine fertilization, thus enhancing the biological carbon uptake. Carbon is absorbed through photosynthetic organisms where organic matter is broken down by bacteria into inorganic compounds in a process of remineralisation. This method of sequestration has limited potential, high costs and a high level of uncertainty. There are drawbacks to these fertilization methods. Changes in plankton structures can have long-term effects on the ocean ecosystem. Large-scale iron fertilization as large quantities of sinking organic mass can trigger the production of methane gas. This would counteract the beneficial effect of carbon fixation.

11.6.4.11 A photosynthetic solution of carbon dioxide fixation

The concept of photosynthetic conversion to fix carbon dioxide using bacteria or micro-algae under a controlled environment is examined in this section. Cyanobacteria or micro-algae have been suggested to perform the role of photosynthesis. In order to promote uniform growth of the organisms, the distribution of photosynthetic photon flux light in the wavelength range of 400–700 nm needs to be delivered to the bioreactor. The efficient distribution of light throughout the photobioreactor will affect the carbon dioxide uptake rates. Although the central role of technology limitations has not yet been properly addressed, a number of facilities capable of mitigating over a million tonnes of CO_2 would have a significant impact on the reduction of greenhouse gas emissions.

Learning Outcomes

This chapter on global warming and carbon sequestration will provide readers comprehensive understanding of:

Global warming, greenhouse gases, contribution of greenhouse gases in global warming, impact of greenhouse gasses, causes, effect and consequences of global warming, climate change and its causes, contributors, effects and impacts, carbon sequestration, types and methods of sequestration.

Readers are advised to carry out following exercise.

Exercise

1. Write all possible implications of climate change on humans.
2. Scientific note on the present the requirement of carbon sequestration and recent advances in the field of technology.
3. Write a 100-word note on how climate change problem can be targeted through sustainable development.

Chapter **12**

Financial Promotion for Sustainable Development: Carbon Credits

12.1 INTRODUCTION

Carbon credit is indeed a great step towards environment protection, yet it must be modified from time to time to meet the existing needs of environment protection. One thing is certain that carbon credit is not just a term in the field of environmental practices but is also a driver in promoting sustainable development and its importance for the mankind. The concept of "carbon credits" came into existence as a result of increasing awareness of the need for controlling emissions of several greenhouse gases. It is defined as a permit that allows the holder to emit one tonne of carbon dioxide. Also a carbon credit (often called a carbon offset) is a financial instrument that represents a tonne of CO_2 (carbon dioxide) or CO_2 (carbon dioxide equivalent gases) removed or reduced from the atmosphere from an emission reduction project. Credits are awarded to countries or groups that have reduced their greenhouse gases below their emission quota. The IPCC (Intergovernmental Panel on Climate Change) observed that, the policies that can provide a real or implicit price of carbon could create significant benefits for producers and consumers to invest in low-GHG products, technologies and processes. Such policies may include economic instruments, government funding and regulation. Carbon credits are an element used to aid in regulation of the amount of gases that are being released into the air. This is part of a larger international plan which has been created in an effort to reduce global warming and its effects. These credits are often traded between businesses. However, they also are bought and sold in international markets at whatever the determined market value for them is. There are also times when these credits are used to fund carbon reduction plans between trading partners.

Carbon projects may focus on such areas as reforestation, fuel-switching, carbon capture and storage, and energy efficiency. As previously stated, the central idea is that using these credits will eventually mitigate the effects of global warming. Some detractors claim, however, that the plan does more to make money for its developers than it does to save the earth from itself. The goal of the group is to meet emission reduction targets. Over 170 countries make up part of this agreement. There are a number of countries that work with people who are interested in decreasing the size of their carbon footprint in an effort to help save the environment. These people are not mandated to use carbon credits, but they are adamant about using them in the right way. Many major examples deal with renewable energy options. Solar panels, solar lighting, and solar heating are also important alternative energy areas.

The main reason why this whole project has come about is because of the burning of fossil fuels by industries across the world. Proponents to the programme claim that using carbon credits will make companies more responsible and will even make some companies more money if they manage the credits properly. Fuels produce gaseous by-products such as carbon dioxide, which eventually makes its way into the ozone layer and contributes to global warming. While it's impractical for many people to stop driving or using electricity, consumers can offset their greenhouse gas contributions by purchasing carbon credits. Carbon credits originate from a range of emission reduction activities associated with the removal of existing emissions from the atmosphere and the reduction of future emissions. These are commonly called "methodologies."

The key means by which existing emissions can be removed from the atmosphere are aforestation and reforestation. Carbon credits originated through these emission reduction activities can be created under a variety of voluntary and compliance market mechanisms, schemes and standards.Kyoto targets and others provide avenues for voluntary offsetting purposes. Around the world some schemes clearly deliver more environmental benefits than others. Developing parts of the world produce the most carbon credits.

12.1.1 Function

One carbon credit equals 1 tonne of carbon dioxide emissions. Through a variety of environmental organizations consumers purchase carbon credits. Organizations perform an action that will remove 1 tonne of carbon from the atmospheres when anyone purchases a carbon credit. Some groups plant trees, which use carbon dioxide during photosynthesis. Others contract with farmers to plant a specific amount of plants on unused fields which consume the carbon and store it within their cells, preventing it from reaching the earth's ozone layer.

12.1.2 Types

Carbon credits can be categorized as voluntary or compliant. Some federal and local governments require industries to purchase carbon credits to offset a portion of their emissions. Agriculture, forestry, waste management and coal mining industries, which are all major greenhouse gas producers, face only voluntary carbon credit standards. Individual consumers can choose to purchase carbon credits to offset air travel or even everyday lifestyle choices.

12.1.3 Considerations

Despite their name, carbon credits can be used to offset all varieties of greenhouse gases, not just carbon. These gases include methane, nitrous oxide and fluorocarbons from refrigerants. The cost and the value of each credit is determined by how much each gas contributes to global warming. Methane gas has more than 20 times the environmental effect as the same amount of carbon.

12.1.4 Benefits

Carbon credits offer environmental benefits to people and wildlife all over the world. As of 2007, carbon credits were reduced greenhouse gas emissions by 23 million metric tonnes each year. This reduction in emissions reduces air and water pollution, which protects both short and long-term health.

12.1.5 Drawbacks

One of the major obstacles to the use of carbon credits has been the large number of scams and lack of regulation for this industry. Despite the benefits offered by carbon credits, consumers in the U.S. have been slow to embrace offsets. Even buyers who believe in the principles of carbon offset enough to pay for it might be discouraged by the number of illegitimate firms engaged in this practice. For example, European consumers offset 1.6 billion tonnes of carbon credits in 2007, only 23 million tonnes were offset in the United States.

12.2 COMPLIANCE V VOLUNTARY MARKETS

The compliance market comprises several legally-binding mandatory emission-trading schemes largely established under the Kyoto Protocol linked to the United Nations Framework on Climate Change (UNFCCC). It also includes some regional compliance markets in the USA and Australia. The Voluntary Carbon Offset Market functions outside of the compliance market and enables companies and individuals to purchase carbon credits on a voluntary basis to satisfy personal or Corporate Social Responsibility (CSR) objectives. This basic conclusion has been endorsed by almost all the scientific societies and academies of science all over the world including every national academies of science of the major industrialized countries. The average temperature of earth has increased during the last century.

The Intergovernmental Panel on Climate Change concludes that increasing industrial activities and human activities have given rise to increasing greenhouse gas emissions which has increased the surface temperature of the earth. Because of the large heat capacity of the oceans and the long life time of carbon dioxide in the atmosphere, warming is expected to continue. As we know that, global temperature rise will have catastrophic effects which include sea level rise, expansion of subtropical deserts, continuing retreat of glaciers, permafrost, increase in the intensity of extreme weather, species extinctions and reduce agricultural yields. Governments are working out on various mechanisms to curb the rising carbon dioxide levels and have come out with two models which can be used effectively to reduce and slow down the process of carbon dioxide emissions.

One mechanism is taxing carbon and the other is capping the emission of carbon. The compliance market comprises several legally-binding mandatory emission-trading schemes. The quality of the credits is based in on the validation process and sophistication of the fund or development company that acted as the sponsor to the carbon project. So, carbon credits is a mechanism adopted by national and international governments to mitigate the effects of greenhouse gases (GHGs). Capped greenhouse gases are traded in the market to regulate the emissions from the sources.The idea is to allow market mechanisms to drive industrial and commercial processes in the direction of low greenhouse gases(GHGs). These mitigation projects generate credits, which can be traded in the international markets for monetary benefits. On a voluntary basis, there are also many companies that sell carbon credits to commercial and individual customers who are interested in lowering their carbon footprint. These carbon credits are bought from an investment fund or a carbon development company that has aggregated the credits from individual projects.

Voluntary units typically have less value than the units sold through the rigorously-validated Clean Development Mechanism. Fossil fuels are the major source of greenhouse gas emissions. industries such as power, textile, fertilizer use fossil fuels for their high volumes of operations. The major greenhouse gases emitted by these industries are carbon dioxide, methane, nitrous oxide, hydrofluorocarbons (HFCs), etc., all of which can increase the temperature of earth's atmosphere by trapping infrared radiations from the Sun. The increasing awareness about the environmental degradation gave rise to the concept called carbon credit. The Intergovernmental Panel on Climate Change (IPCC) has observed that: mechanism/ policies that provide a price of carbon could create incentives for producers and consumers to significantly invest in low-greenhouse gas products, technologies and processes. Such policies could include financial instruments, government funding and regulation.

12.2.1 Emission Allowances

Both developed and developing countries subsequently set quotas on the emissions of installations run by local business, industries and other organizations. The protocol sets up 'caps'(quotas) on the maximum amount of greenhouse gases for developed and developing countries. The industries which come under the purview of the government regulated environmental policies are termed 'Operators'. Countries manage this through 'registries' that are set up, which are required to be validated and monitored for compliance by the UNFCCC. Each operator has an allowance of credits, where each allowance gives the owner the right to emit one metric tonne of carbon dioxide or other equivalent greenhouse gas. Operators can sell their unused allowances as carbon credits if they have not used up their quotas. Businesses that are about to exceed their quotas can buy the extra allowances as credits on the open market. As demand for energy grows over time, the total emissions must still stay within the cap. However, it allows industry some flexibility and predictability in its planning to accommodate this so that emissions remain under control. The allowances are bought and sold in carbon markets in such a way that an operator can seek the most effective way of reducing emissions.

12.2.2 Kyoto's 'Flexible Mechanisms'

A credit is an emissions allowance which was allocated originally or auctioned by the national administrators of a cap-and-trade programme. It can also be an offset of emissions. In any developing country which has ratified the Kyoto Protocol, the activities of offsetting and mitigating can occur. Post Approval, these units are termed Certified Emission Reductions, or CERs. The protocol allows these projects to be constructed and credited in advance of the Kyoto trading period. The Kyoto Protocol can be used in three ways that can enable countries or operators in developed countries to acquire greenhouse gas credits:

- Under Joint Implementation (JI), a project in another developed country would set by a developed country which incurs high cost in the reduction of GHGs.
- Under the Clean Development Mechanism (CDM) a developed country can fund the projects in developing country where the cost of funding is relatively. For carrying out projects which would reduce emissions and help in meeting its emission reduction targets, the developed country would be given credits, while the developing country would benefit from the capital investment. .
- Under International Emissions Trading (IET) to cover their shortfall in allowances, countries can trade in the international carbon credit market. Countries can sell the remaining credits in the markets which have under-used the credits and countries which have fully used their credits can buy the credits from other players in the market. These carbon projects can be created by a national government or by an operator within the country.

12.2.3 Emission Markets

Allowances can be traded privately or in the international market at the prevailing market price. One allowance or carbon emission reductions (CER) is equivalent to one metric tonne of carbon dioxide emissions. Each international transfer is validated by the UNFCCC. To discover the current market prices and maintain liquidity exchanges have been established to provide market for allowances. Carbon prices are measured in Euros per tonne of carbon dioxide. Other GHGs can also be traded. Many companies now engage in emissions reductions, offsetting, and sequestration programmes so that credits are generated which can then be sold in international markets.

12.2.4 Market Price for Carbon

The number of companies needing to buy credits will increase over the period of time. Supply-demand for credits will determine the price of the carbon which will in turn encourage companies to go cleaner. Energy usage and emissions should be kept under constant check. Because of the lack of a mature secondary market for CERs and a lack of homogeneity between projects which causes difficulty in pricing, an allowance, such as a Kyoto Assigned Amount Unit (AAU) or its near-equivalent European Union Allowance (EUA), may have a different value in the market to an offset such as a carbon emission reduction (CER). Moreover offsets generated by a carbon project under clean development mechanism are under limited in value mainly because the operators are restricted as to what percentage of allowances can be generated.

12.2.5 Raising Price of Carbon Achieves Four Goals

- First, the consumers will get informed about the goods which are high in carbon or which utilized high carbon.
- Second, it would drive producers to use inputs which are less in carbon content which results in companies using low carbon substitutes.
- Third, it would foster innovation in coming up with technologies which are cleaner and efficient.
- Fourth it would incentivize everyone who are involved in reduction of greenhouse gases.

12.2.6 Criticisms

Environmental restrictions and activities have been imposed on businesses through regulation. The only internationally agreed mechanism for Carbon Credit Regulation is Kyoto-Protocol and its supporting organization, the UNFCCC, is the only organization with a global say on the overall effectiveness of emission control systems. Business investment runs over large number of years, this adds uncertainity to their plans. Businesses in capped countries may see themselves to be working at a competitive. A key concept behind the cap and trade system is that national quotas should be chosen to represent fairly correct, genuine and meaningful reductions in national output of emissions.This would ensure reduction in overall emissions along with the facts that costs of trading is carried across to all the trading parties. A question has been raised over deviating from the rules of allowances. Countries within the European Union have granted their incumbent businesses most or all of their allowances for free. This can sometimes be perceived as a protectionist obstacle to new entrants into their markets. There have also been accusations of power generators getting a heavy profit by passing on these emissions 'charges' to their customers. Establishing a meaningful offset project is complex, voluntary offsetting activities outside the mechanism are unregulated and there have been criticisms of offsetting in these unregulated activities. Concerns have also been raised over the validation of credits. One is to the accurate assessment of additionality. Others relate to the complexities involved to get a project approved.

12.3 CARBON TAX

Carbon tax is a form of pollution tax. Carbon tax is based on the economic principle of negative externalities. Externalities are costs or benefits generated by the production of goods and services. Negative externalities are costs that are not paid for. The government sets a price per tonne on carbon. The fee is determined by assessment of the amount of carbon released into the atmosphere on combustion. When utilities, businesses or homeowners consume fossil fuels, they create pollution that has a societal cost; everyone suffers from the effects of pollution. Proponents of a carbon tax believe that the price of fossil fuels should account for these societal costs.

12.3.1 Benefits of Carbon Tax

The primary purpose of carbon tax is to lower greenhouse-gas emissions. This would encourage businesses and individuals to reduce the consumption of carbon rich substances. Carbon tax also encourages alternative energy by making it cost-competitive with cheaper fuels. Carbon tax is stable. The tax is based on the carbon content of the fuel that was involved in combustion. As the tax get imposed, a number of steps that can be taken to reduce costs can be envisaged. The prices of carbon under cap and trade policy may vary due to weather and economic conditions. This is because cap-and-trade schemes set a definite limit on emissions, not a definite price on carbon. For example: People may switch to fluorescent lamps from incandescent bulbs, a business may try to install newer appliances which are cleaner and efficient. And since carbon tax sets a definite price on carbon, there is a guaranteed return on expensive efficiency investments. A tax on a carbon rich and inexpensive fuel such as coal raises its per British Thermal Unit (Btu) price to one comparable with cleaner forms of power. The money that is raised by taxes can be used to subsidize environment friendly projects or can be used to rebate taxes. Many fans of carbon tax believe in progressive tax-shifting. This would mean that some of the tax burden would be takes off from federal taxes and sales taxes. Since carbon tax is simple, it is easily comprehendible.

12.3.2 Price Determination

Tax based on the carbon content would fairly reflect the prices and hence carbon content has to be the basis for price determination instead of weights and volumes. The carbon content of oil, coal and gas varies. Proponents of a carbon tax want to encourage the use of efficient fuels. Each fuel variety also has its own carbon content. Bituminous coal, for instance, contains considerably more carbon than lignite coal. Residual fuel oil contains more carbon than gasoline. Every fuel variety needs to have its own rate based on its Btu heat content.

Carbon tax can be applied at different points of production and consumption. Some taxes target the top end of the supply chain — the transaction between producers like coal mines and oil well heads and suppliers like coal shippers and oil refiners, whereas some taxes charge distributors – the oil companies and utilities. And other taxes charge consumers directly through electric bills. Different carbon taxes, both real and theoretical, support varying points of implementation. Even if the tax is charged for the companies, they will pass on some of the cost to consumers by increasing the cost of energy. Consumption can be charged easily than production without majority affecting the economy of a country. Taxation on production may also result in higher domestic energy prices which would encourage foreign imports. Hence it is consumers who are targeted for carbon taxes. Carbon tax is not a widely accepted mechanism around the world.

12.3.3 Carbon Tax Vs Carbon Credit

Carbon tax is a better alternative than carbon credit mainly because of the following six reason:

- Energy prices are easily predictable by the mechanism of tax than by the mechanism of cap and trade. The high volatility of the carbon credits that are generated by the mechanism of cap and trade has consistently discouraged energy efficient schemes.
- Tax system can be quickly implemented than cap and trade. Since the environment is getting polluted at a faster rate, it is high time that necessary actions are taken quickly and efficiently.
- Carbon taxes are transparent and easily understandable, making them more likely to garner public support than complex cap and trade.
- Carbon taxes cannot be easily manipulated and hence cannot be easily exploited whereas the complexity of cap and trade always provides room for exploitation for special interests.
- Carbon taxes address emissions of carbon from every sector, whereas some cap-and-trade systems discussed to date have only targeted the electricity industry.
- Carbon tax revenues would most likely be returned to the public through dividends or progressive tax-shifting, while the costs of cap-and-trade systems are likely to become a hidden tax as dollars flow to market participants, lawyers and consultants.

12.3.4 Carbon Taxes Will Lend Predictability to Energy Prices

Carbon taxes being ramped up through a multi-year phase-in it makes the prediction of future energy and power prices easy. Once the prediction of prices become easy, it will give rise to millions of energy decisions that could have a major impact on how we use energy and design systems that use energy. Cap and trade on the other hand will worsen the volatility of the prices as the price of carbon allowances varies with the fluctuations in weather and economic factors. These factors affect the demand for energy.

12.3.5 Carbon Taxes Will Provide Quicker Results

Taxes are simple and are designed to be simple and to be easily adopted. On the other hand, cap-and-trade systems are highly complex and will take years to develop and implement. Disruptive issues must be addressed intellectually and resolved politically; the appropriate level of the cap, timing, allowance allocations, certification procedures, regional conflicts, penalties, standards for use of offsets, the inevitable requests for exceptions by affected parties and a several other complex issues.

12.3.6 Carbon Taxes Are Transparent and are Easier to Understand than Cap-and-Trade

A carbon tax is simple, transparent and easy to comprehend. The government simply imposes a tax on a tonnes of carbon emitted, which is easily translated into a tax per kWh of electricity, gallon of gasoline or natural gas. The prices which are set for carbon credits under cap and trade systems with high market fluctuations and hence difficult to predict.

12.3.7 A Carbon Tax's Simplicity Prevents it Against Incentives and Potential for Vested Interests that Will Accompany Cap-and-Trade

Carbon Tax is simple and hence not many negotiations are required for the implementation of carbon tax. Carbon tax is simple and straight forward. By contrast the lengthy negotiations required for the implementation of Cap and trade system will always provide opportunities for the industry.

3.8 Carbon Taxes Encompasses all Sectors and Activities Producing Carbon Emissions

Carbon taxes are applicable to all the sectors – energy, industry, transportation whereas most of the cap and trade is applicable to the electric industry. Industries apart from power sector contribute more than 50% of the emissions and hence it is necessary that these sectors are targeted for the significant reduction in carbon emissions.

Learning Outcomes

This chapter provides knowledge transfer on; introduction to carbon credits, potential benefits and drawback, carbon tax, difference between carbon credit and carbon tax.

Readers are also advised to carry out following exercise:

Exercise

1. Write case study for carbon credits earned from Wind turbine of 1000 kW during its life time.

Appendix

The Batteries (Management and Handling) Rules, 2001*

Notification New Delhi, the 16th May, 2001

S.O. 432(E) Whereas a notification of the Government of India in the Ministry of Environment and Forests was published in the Gazette of India, Extraordinary, Part II-section 3, sub-section (ii) vide No. S.O. 491 (E) dated 24h May, 2000 and corrigendum published in the Gazette of India Extraordinary Part-II section 3, sub-section (ii) vide No. SO 593 (E) dated 23d June, 2000 under powers conferred by sections 6, 8 and 25 of the Environment (Protection) Act, 1986 (29 of 1986), inviting objections from persons likely to be affected, within a period of sixty days from the date of publication of the said notification with regard to the Government's intention to notify the Battery (Management and Handling) Rules, 2000. And, whereas all objections received have been duly considered by the Central Government;

Now, therefore, in exercise of the powers conferred by sections 6, 8 and 25 of the Environment (Protection) Act, 1986 (29 of 1986), the Central Government hereby notifies the Batteries (Management and Handling) Rules, 2001.

1. **Short Title and Commencement**

   (1) These rules may be called the **Batteries (Management and Handling) Rules, 2001.**

   (2) They shall come into force on the date of their publication in the Official Gazette.

2. **Application**

   These rules shall apply to every manufacturer, importer, re-conditioner, assembler, dealer, recycler, auctioneer, consumer and bulk consumer involved in manufacture, processing, sale, purchase and use of batteries or components thereof.

3. **Definitions**

   In these rules, unless the context otherwise requires,

   (a) 'Act' -means the Environment (Protection) Act, 1986 (29 of 1986);

   (b) 'Assembler' – means a person who manufactures lead acid batteries by assembling various components;

   (c) 'Auction' – means bulk sale of used lead acid batteries or component (s) thereof by invitation of tenders or auction, contract or negotiation by individual(s), companies or Government Departments;

   (d) 'Auctioneer' – means a person (s) who auctions used lead acid batteries or components, thereof,

*Ministry of Environment and Forests

(e) **'Battery'** – means lead acid battery which is a source of electrical energy and contains lead metal;

(f) **'Bulk consumer'** – means a consumer such as Central or State Government Departments of Railways, Defence, Telecom, Posts & Telegraph, State Road Transport Undertakings, State Electricity Boards and others who purchase batteries through central 'rate' or running contract centrally placed on behalf of individual departments or user units under their jurisdiction.

(g) **'Components'** – means lead bearing components of a lead acid battery;

(h) **'Consumer'** – means a person using lead acid batteries excluding bulk consumers;

(i) **'Dealer'** – means a person who sells and receives lead acid batteries or components thereof to and from the consumers or other dealers or retailers on behalf of the manufacturers, importers, assemblers and reconditioners or otherwise;

(j) **'Designated collection centre'** – means a collection centre established, individually or jointly by one or more manufacturers or importers, assemblers and re-conditioners in pursuance of their responsibilities under rule- 4 of these rules.

(k) **'Importer'** – means a person who imports new lead acid batteries or components containing lead thereof for the purpose of sale;

(i) **'Manufacturer'** – in relation to any factory manufacturing lead acid batteries or components thereof means a person or Chief Executive Officer (CEO) of the company who has control over the affairs of the factory or the premises for sale and collection of lead acid batteries or components thereof,

(m) **'Original equipment manufacturer'** – means manufacturer of equipment or product using lead acid batteries as a component;

(n) **'Reconditioned'** – means a person involved in repairing of lead acid batteries for selling the same in the market;

(o) **'Recycler'** – means an occupier who processes used lead acid batteries or components thereof for recovering lead

(p) **'Registered recycler'** – means a recycler registered with the ; Ministry of Environment and Forests or an agency designated by it for reprocessing used lead acid batteries or components thereof;

(q) **'State Board'** – means the concerned State Pollution Control Board or the Pollution Control Committee as the case may be;

(r) **'Used batteries'** – means used, damaged and old lead acid batteries or components thereof; and

(s) The words not defined in these rules will have the same meaning as defined in the Environment (Protection). Act, 1986 and the rules framed there under.

4. **Responsibilities of manufacturer, importer, assembler and re-conditioner:** It shall be the responsibility of a manufacturer, importer, assembler and re-conditioner to

(i) Ensure that the used batteries are collected back as per the Schedule against new batteries sold excluding those sold to original equipment manufacturer and bulk consumer(s);

(ii) Ensure that used batteries collected back are of similar type and specifications as that of the new batteries sold;

(iii) File a half-yearly return of their sales and buy-back to the State Board in Form- I latest by 3 June and 30 December of every year;

(iv) Set up collection centres either individually or jointly at various places for collection of used batteries from consumers or dealers;

(v) Ensure that used batteries collected are sent only to the registered recyclers,

(vi) Ensure that necessary arrangements are made with dealers for safe transportation from collection centres to the premises of registered recyclers;

(vii) Ensure that no damage to the environment occurs during transportation;

(viii) Create public awareness through advertisements, publications, and posters or by other means with regard to the following

(a) Hazards of lead;

(b) Responsibility of consumers to return their used batteries only to the dealers or deliver at designated collection centres; and

(c) Addresses of dealers and designated collection centres.

(ix) Use the international recycling sign on the batteries;

(x) Buy recycled lead only from registered recyclers; and

(xi) Bring to the notice of the State Board or the Ministry of Environment and Forests any violation by the dealers.

5. **Registration of Importers:** The importer shall get himself registered with the Ministry of Environment and Forests or an agency designated by it by submitting details in Form-11.

6. **Customs clearance of imports of new lead acid batteries:** Customs clearance of imports shall be contingent upon. Valid registration with the Reserve Bank of India (with Importer's Code Number); one time registration with the Ministry of Environment and Forests or an agency designated by it in Form-11; Undertaking in Form-111; and (iv) a copy of the latest half-yearly return in Form-IV.

7. **Responsibilities of dealer:** It shall be the responsibility of a dealer to

(i) Ensure that the used batteries are collected back as per the Schedule against new batteries sold;

(ii) Give appropriate discount for every used battery returned by the consumer;

(iii) Ensure that used batteries collected back are of similar type and specifications as that of the new batteries sold;

(iv) File half-yearly returns of the sale of new batteries and buy-back of old batteries to the manufacturer in Form V by 31st May and 30th November of every year;

(v) Ensure safe transportation of collected batteries to the designated collection centres or to the registered recyclers; and

(vi) Ensure that no damage is caused to the environment during storage and transportation of used batteries.

8. **Responsibilities of recycler:** Each recycler shall –

(i) Apply for registration to the Ministry of Environment and Forests or an agency designated by it if not applied already, by submitting information in Form VI;

(ii) Ensure strict compliance of the terms and conditions of registration; however, those already registered with the Ministry of Environment and Forests or all agency designated by it for reprocessing used batteries would be bound by the terms and conditions of such registration;

(iii) Submit annual returns as per Form VII to the State Board;

(iv) Make available all records to the State Board for inspection;

(v) Mark 'Recycled' on lead recovered by reprocessing; and

(vi) Create public awareness through advertisements, publications, posters or others with regard to the following:

(a) Hazards of lead; and

(b) Obligation of consumers to return used batteries only to the registered dealers or deliver at the designated collection centres.

9. Procedure for registration/renewal of registration of recyclers

(1) Every recycler of used lead acid batteries shall make an application in Form VI along with the following documents to the Joint Secretary, Ministry of Environment and Forests or any officer designated by the Ministry or an agency designated by it for grant of registration or renewal;

(a) A copy of the valid consents under Water (Prevention and Control of Pollution) Act, 1974, as amended and Air (Prevention and Control of Pollution) Act, 198 1, as amended;

(b) A copy of the valid authorization under Hazardous Wastes (I Management and Handling Rules, 1989 as amended;

(c) A copy of valid certificate of registration with District Industries Centre ; and

(d) A copy of the proof of installed capacity issued by either State Pollution Control Board/District Industries Centre.

(2) The Joint Secretary, Ministry of Environment and Forests or any officer designated by the Ministry or an agency designated by it shall ensure that the recyclers possess appropriate facilities, technical capabilities, and equipment to recycle used batteries and dispose of hazardous waste generated;

(3) The Joint Secretary, Ministry of Environment and Forests or any officer designated by the Ministry or an agency designated by it shall take decision on application for registration within 120 days of receipt of application form with complete details;

(4) The registration granted under this rule shall be in force for a period of two years from the date of issue or from the date of renewal unless suspended or cancelled earlier;

(5) An application for the renewal of registration shall be made in Form VI at least six months before its expiry. The Joint Secretary, Ministry of Environment and Forests or any officer designated by the Ministry or an agency designated by it shall renew the registration of the recycler granted under sub rule(4) of this rule, after examining each case on merit;

(6) The Joint Secretary, Ministry of Environment and Forests or any officer designated by the Ministry or an agency designated by it may, after giving reasonable opportunity to the applicant of being heard, refuse to grant registration;

(7) The Joint Secretary, Ministry of Environment and Forests or any officer designated by the Ministry or an agency designated by it may cancel or suspend a registration issued under these rules, if in his / her opinion, the registered recycler has failed to

comply with any of the conditions of registration, or with any provisions of the Act or rules made there under after film an opportunity to explain and after recording the reasons therefore;

(8) It shall be the responsibility of the State Boards to monitor the compliance of conditions while according registration, and

(9) An appeal shall lie against any order of suspension or cancellation or refusal of registration passed by the Joint Secretary to the Ministry of Environment and Forests or any officer designated by the Ministry or agency designated by it. The appeal shall be in writing and shall be accompanied with a copy of the order appealed against and shall be presented within 30 days of passing of the order.

10. Responsibilities of cousumer or bulk consumer

(1) It shall be the responsibility of the consumer to ensure that used batteries are not disposed of in any manner other than depositing with the dealer, manufacturer, importer, assembler, registered recycler, reconditioner or at the designated collection centres.

(2) It shall be the responsibility of the bulk consumer to –

(i) Ensure that used batteries are not disposed of in any manner other than by depositing with the dealer/manufacturer/registered recycler/importer/reconditioner or at the designated collection centres,– and

(ii) File half-yearly return in Form VIII to the State Board.

(3) Bulk consumers to their user units may auction used batteries to registered recyclers only.

11. Responsibilities of auctioneer: The auctioneer shall –

(i) Ensure that used batteries are auctioned to the registered recyclers only-,

(ii) File half-yearly returns of their auctions to the State Boards in Form-IX; and

(iii) Maintain a record of such auctions and make these records available to the State Board for inspection.

12. Prescribed Authority: The prescribed authority for ensuring compliance of the provisions of these rules shall be the State Board. And, it shall file an annual compliance status report to the Central Pollution Control Board by 30th April of every year.

13. Duties of Central Pollution Control Board: The Central Pollution Control Board shall compile and publish the data received every year from the State Boards. It shall review the compliance of the rules periodically to improve the collection and recycling of used. lead batteries and apprise the Ministry of Environment and Forests, Government of India.

14. Computerization of Records and Returns: Ministry of Environment and Forests or an agency designated by it shall develop a system for computerized tracking of

(i) Distribution and sale of batteries;

(ii) Collection, auction, transport and re-processing of used batteries;

(iii) Sale of re-processed lead by registered recyclers; and

(iv) Sale of lead from all domestic producers or importers.

SCHEDULE

S.No.	*Year*	*Number of used batteries to be*
(i)	During first year of implementation of rules	50% of new batteries sold
(ii)	During second year of implementation of rules	75% of new batteries sold
(iii)	After second year of implementation of rules	90% of new batteries sold

Index

A
Acid rain 245
Air (Prevention and Control of Pollution) Act 254
Anti-fouling paints 95
Aquatic ecosystem 46
Arable land 21
Arsenic poisoning 133
Atmosphere 16

B
Ballast water discharges 94
Batteries (management and handling) rules, (2001) 308
Bhopal gas tragedy 127
Biodiversity 56
  value 60
  conservation 68
  threats 67
Biological reprocessing 158
Biological sequestration 292
Biological treatment technologies 180
Biomass 26

C
Carbon credits 300
Carbon cycle 15
Carbon sequestration 290
Carbon tax 305
Chemical treatment technologies 181
Chipko movement 241
Chula issue 148
Climate and health 128
Climate change 243, 275, 276
Community biodiversity registers 139
Composting 171
Consumerism 250

D
Decomposers 54
Deforestation 7, 87
  causes 8
  consequences 8
Desertification 33
  causes 34
  prevention 34
Detritovores 54
Detritus food chain 39
Diarrhoeal diseases 11
Dredging 92
Drought 13

E
Ecological pyramids 41
Ecosystem diversity 59
Ecosystem
  classification 36
  functions 38
  structure 37
EIA 104, 260
  benefits 118
  impact 113
  principles 108
  process 109
Embodied energy 233
Emission markets 304
Energy cycle 44

Energy resources 26
Energy
- from waste 162
- urban problems 232

Environmental Acts 191
Environment 1
- cancer 136

Environment (Protection) Act 253, 194
Environment education and awareness 241
Environment impact evaluation methods 114
Environment management plan 118
Environment science 2
- importance 2
- public awareness 3

Environmental auditing 117
Environmental ethics 237
Environmental health 126
Environmental information systems 263
Environmental issues 23
Environmental law
- development phases 187
- functions 185
- principles 186
- trends 187

Environmental legislation 260
Environmental management systems (EMS) 262
Environmental pollution 73
Environmental values 140
EPA 104
Estuaries 52
Eutrophication 78

F

Family welfare programme 121
Farm size 22
- credit services 22

Fertilizers 86
Floods 13
Food
- chemicals 135
- production 22
- resources 20
- web 41

Foodchain 38
- significance 40
- trophic control 51

Forest Conservation Act 258
Forest ecosystem 42, 43
Forest resources 6
- over-exploitation 7
- uses 6, 238

Fossil fuels 17

G

Gandhian way of life 242
Gene bank 70
Genetic diversity 58
Geologic sequestration 291
Global hotspots 65
Global population growth 121
Global warming 245, 272
Grassland ecosystem 45
Grazing foodchain 39
Greenhouse gases 267
Groundwater remediation techniques 179
Gully erosion 31

H

Hazardous waste treatment 173
Herbicides 86
HIV/AIDS 147
Hotspots 65

I

Import-export balance 23
Incineration 157
India
- biodiversity 63
- biogeographical classification 60
- EIA process 106
- endangered species 66
- endemic species 67
- hotspots 66
- national parks 69
- water resources 9

Infectious diseases 129
Injection wells 178
Insecticides 86

Integrated waste management 155
Intellectual property rights 139

K

Karnataka's GIS scheme 151
Kyoto's 'flexible mechanisms' 303

L

Lake habitats 48
Lake thermal structure 48
Lakes 47
Land pollution 81
Land resources 28
Land treatment units 178
Landfills 177

M

Man-made resources 5
Marine pollution 91, 92
Mineral resources 14
Mineral sequestration 292
Ministry of Environment 211
Municipal solid waste (management and handling) rules, (2000) 218

N

Narmada issue 264
Natural land pollution 91
Natural resources 5
Nitrogen cycle 14
Noise pollution 97
- control 99
- impact 102
- measurement 98
- sources 98
- standards 100

Non-renewable resources 5
Nuclear accidents 247
Nuclear holocaust 247, 248
Nutrient cycle 44
Nutrients 49
- role 49

O

Ocean sequestration 290
Oceanic circulation 18
Oceans 17
Oxygen content 77
Ozone depleting substances (regulation) rules, (2000) 211
Ozone layer depletion 246

P

Pain patency 234
Pesticides 86
Photosynthesis 16
Plant nutrients 78
Pond ecosystem 53
Population growth 21, 120
Project elephant 71
Project tiger 71
Pyramid of biomass 41
Pyramid of energy 42
Pyramid of number 41

R

Rainwater harvesting 234
Reduce, reuse, recycle principle 251
Reforesting 91
Regional climate change 283
Renewable resources 5
Resource Conservation and Recovery Act (RCRA) 165
Respiration 16
Reverse desertification 34
Rill erosion 31

S

Seed bank 70
Shantiniketan model 242
Sheet erosion 31
Ship discharges 94
Silent valley 265
Simulation modeling 117
Soil erosion 30
- prevention 32

Soil pollution
- causes 86
- control 90
- effects 89
- environmental effects 90
- types 85

Soil structure 32

Soil texture 31
Solid waste management 152
Solid waste treatment 91
Solid wastes 87
Species 66
    classification 66
Species diversity 57
Spills 94
Sterilization 123
Stream ecosystems
    threats 52
Surface impoundments 177

T

Tehri project 236
Temperature forest ecosystem 43
Terrain unit 32
Terrestrial biosphere 17
Terrestrial sequestration 291
Tropical rainforest ecosystem 43
Tuberculosis 130

U

Underground soil 88
Urban and industrial wastes 154
Urbanisation 88, 123

V

Value education 140
Vegetation cover 32

W

Waste management 160, 253
    concepts 160
    technologies 160
Waste piles 177
Waste products 250
Wasteland development 249
Wasteland reclamation 248, 249
Water (Prevention & Control of Pollution) Rules, (1975) 202
Water (Prevention and Control of Pollution) Act 256
Water conservation 233
Water cycle 44
Water erosion 30
Water pollution 73, 84
    impact of agriculture 76
    impact of industrial activities 76
    sources 74, 78
Water related diseases 11
Water resources 8, 9
Water-based diseases 12
Water-borne diseases 11
Waterfront drainage 96
Water-related diseases 11, 132
Water-related insect vectors 12
Watershed management 235
Water-washed diseases 12
Wildlife Protection Act 257
Wind erosion 31
Woman and child welfare 148
World food problems 21